# Yearbook of Sustainable Smart Mining and Energy - Technical, Economic and Legal Framework

Volume 2

The Yearbook of Sustainable Smart Mining and Energy publishes the latest articles on sustainability and digitalisation in mining and energy. These two elements ensure a modern mining and energy economy in keeping with the UN's sustainability goals: environmental protection and economic development through new technology are combined. In this way, affordable and clean energy can be achieved (Goal 7), together with responsible production to reduce environmental costs (Goal 12) and decent work and economic growth as defined in Goal 8: sustainable economic growth will require societies to create the conditions that allow people to have quality jobs that stimulate the economy without harming the environment. The articles in the yearbook show the prerequisites and possibilities for achieving these aims all over the world. The annual volumes discuss the latest developments in mining and energy in various countries, as well as problems in connection with post mining. Accordingly, the yearbook represents a valuable contribution to Springer Nature's Sustainable Development Goals (SDG) Programme. The Minister of Economy and Digitalisation of North Rhine-Westphalia, Germany Pinkwart is its patron. All articles are peer-reviewed.

Walter Frenz • Axel Preuße
Editors

# Yearbook of Sustainable Smart Mining and Energy 2022

## Technical, Economic and Legal Framework

Springer

*Editors*
Walter Frenz
LFG Berg-, Umwelt- und Europarecht
RWTH Aachen University
Aachen, Nordrhein-Westfalen, Germany

Axel Preuße
Institut für Markscheidewesen
RWTH Aachen University
Aachen, Nordrhein-Westfalen, Germany

ISSN 2661-8834 ISSN 2661-8842 (electronic)
Yearbook of Sustainable Smart Mining and Energy - Technical, Economic and Legal Framework
ISBN 978-3-031-41872-3 ISBN 978-3-031-41873-0 (eBook)
https://doi.org/10.1007/978-3-031-41873-0



This Springer imprint is published by the registered company Springer Nature Switzerland AG
The registered company address is: Gewerbestrasse 11, 6330 Cham, Switzerland

Paper in this product is recyclable.

Walter Frenz • Axel Preuße
Editors

# Yearbook of Sustainable Smart Mining and Energy 2022

## Technical, Economic and Legal Framework

Springer

*Editors*
Walter Frenz
LFG Berg-, Umwelt- und Europarecht
RWTH Aachen University
Aachen, Nordrhein-Westfalen, Germany

Axel Preuße
Institut für Markscheidewesen
RWTH Aachen University
Aachen, Nordrhein-Westfalen, Germany

ISSN 2661-8834 ISSN 2661-8842 (electronic)
Yearbook of Sustainable Smart Mining and Energy - Technical, Economic and Legal Framework
ISBN 978-3-031-41872-3 ISBN 978-3-031-41873-0 (eBook)
https://doi.org/10.1007/978-3-031-41873-0



This Springer imprint is published by the registered company Springer Nature Switzerland AG
The registered company address is: Gewerbestrasse 11, 6330 Cham, Switzerland

*In Memory of Hermann Wotruba.*

# Preface: Yearbook ofw Sustainable Smart Mining and Energy 2022

The UN Sustainable Development Goals (SDG) address a variety of global challenges regarding social, environmental and economic development issues. The mining industry is not linked to one single goal, rather, due to the mining sector's multi-layered impacts, there is a significant potential to face several of these mayor challenges. The assessment and development of sustainable processes and the implementation of new technology and methods can make significant contributions towards the Sustainable Development Goals.

This book covers several aspects of the application of Sustainable Development Goals to mining related subjects. The included works range from methods to assess and implement sustainability to discussions of legal impacts and relations as well as technological developments and outlooks.

First, the challenges and opportunities of clean energy transition in the African mining sector are described. With regard to the assessment of sustainable developments, this book includes the applications of the sieving method, the concept of thermodynamics and the United Nations Framework Classification to mining projects and case studies. The implementation of Sustainable Development Goals into academic project work and education of geo-engineers is covered as well. The legal topics contain discussions of corporate climate liabilities and extraterritorial legal responsibilities as well as an analysis of the impact of the German Federal Constitutional Court's climate decision. Two legal contributions consist of the presentations at the 11th World Congress of Constitutional Law in Johannesburg in December 2022. Important mining aspects and technological developments like proactive water management, sustainable approaches to mine closure and implementation of ventilation on demand in underground mines are described. In addition, the state and potential of the Mongolian raw materials industry is covered.

The Sustainable Smart Mining and Energy Yearbook is not only aimed at researchers and professionals, but at all who want to get an overview of current important technical and legal topics in this field.

This book is dedicated to our friend and colleague Hermann Wotruba, professor and head of the Unit of Mineral Processing (AMR) at RWTH Aachen University, who passed away in February 2023. He had gained national and international reputation as an excellent researcher and shared his expert knowledge with his students and research partners. The development of sustainable processes in mining in developing and threshold countries was an essential part of his work. His many achievements and lasting contributions will always be remembered.

Aachen, Germany
1st March 2023

Walter Frenz
Axel Preuße

# Contents

# Contents

# Thermodynamics, Conservation, and Transformation: A Conceptual Model of Sustainability in the Mining Industry

**J. Alejandro Delgado Jiménez and Oscar Jaime Restrepo Baena**

## Contents



**Abstract** A meaning of sustainability in the mining industry might be approached from the transformations that happen in territories where mining developments occur. This paper proposes a conceptual model that illustrates an alternative approach to interpret sustainability in a territory intervened by mining projects integrating the relationships and transformations of a territory capitals based on the

J. A. D. Jiménez · O. J. R. Baena (✉)
Universidad Nacional de Colombia – School of Mines, Medellín, Colombia
e-mail: jadelgad@unal.edu.co; ojrestre@unal.edu.co





W. Frenz, A. Preuße (eds.), *Yearbook of Sustainable Smart Mining and Energy 2022*, Yearbook of Sustainable Smart Mining and Energy - Technical, Economic and Legal Framework 2, https://doi.org/10.1007/978-3-031-41873-0_1

tools provided by thermodynamics. The model integrates the first and second laws of thermodynamics, the dimensions of sustainability, and a set of functional relationships through a rate process model. As a result, this chapter proposes that sustainability is a dynamic state based on a combination of capacities that guarantee the maintenance of the conditions that provide life to society.

## 1 Introduction

In this chapter it is addressed mining development as an extractive system that transforms the environments it intervenes through inflows and outflows that modify the configurations of the territory where mining occurs. Impacts generated in the extractive system allow the study of interactions between its various components.

This work provides a tool to enhance the understanding of relationships between different dimensions of the sustainability concept and the transformation of resources at the territorial scale. Based on the above, an integrating function has been formulated to interpret the evolution of sustainability in a territory intervened by mining projects which might contribute to improving planning of mineral extractive territories.

The conceptual model is based on three main concepts: sustainability, thermodynamics, and conservation. Sustainability is understood as conditions that guarantee the maintenance of life in a particular environment; thermodynamics is the science that studies the states and transformations of energy; and continuity integrates the concepts of thermodynamics into sustainability. In other words, this model is composed of three levels. The first level corresponds to the identification of the context, the second corresponds to the dynamics of the life cycle of mining projects, and the third level corresponds to the circular flow of sustainability.

Through an integrative perspective, the present model contributes to understand some restrictions in mining business, some interactions between mining development and environment that promote sustainability in the physical, social, and economic contexts where mining industry develops extractive projects.

## 2 Thermodynamics, an Approach to Interpreting Sustainability in Mining

Applying thermodynamics to extractive activities development in a territory improves the understanding of how sustainability may occur in the mining industry. The models of thermodynamics also allow recognizing valorization and transformation of capitals in the territory and interactions with mining industry. The application of thermodynamics laws to sustainability in the development of mining projects focuses on two aspects. First, to identify the types and states of energy in relation to the system under study, i.e., to know how capitals are present in a territory, and the status of these capitals. Second, to consider the energy transformations and transitions undergone by the capitals in the system extractive system, that is, to identify the transformations resulting from the development of a mining project.

To clarify the thermodynamics applications to sustainability, recall a definition of the science of thermodynamics is useful. Thermodynamics deals with the science of energy, storage, and transformation of energy,[1] and allows to describe a behavior, performance and efficiency of energy conversion systems.[2] Thermodynamics also studies conditions required for energy transformations to take place, and as a result, allows to describe which processes can occur spontaneously in nature.[3] The science of energy provides a formalism for understanding how materials respond to the forces acting in their environment. Furthermore, an equilibrium perspective defines thermodynamics as the science of systems transitions, describing energy changes and chemical composition in different equilibrium states.[4] The associated changes in the transition of equilibria are related to the extraction of "useful work," fluxes and reserves of energy.[5]

Thus, thermodynamics in this model is applied as a reference framework to conceptualize mineral reserves as energy sources, and establish interactions present in mining development activities. Thermodynamics also allows to identify through energy transitions, potential contributions of mining development to the sustainability of the intervened environment. The concepts of thermodynamic system, the first and second laws of thermodynamics are defined below.

## 2.1 *Thermodynamics System*

A thermodynamic system makes it possible to identify the components of the object to be studied.[6] A complex system can be seen as a composition of some or many smaller and simpler systems interacting with each other, which is useful in the application of thermodynamics. The proposed description can be achieved by defining system boundaries, and materials that are part of the system. Inside the system, the existing interactions can be described by balance laws, through equations that account for the changes in the system and the transfers that occur at the system boundaries.

There are two types of thermodynamic systems defined by their relationship to their environment: closed systems and open systems. In a closed system, there is a substance inside a container that does not allow mass or energy to flow across its

[1] Borgnakke and Sonntag (2013), p. 1.

[2] Dincer and Rosen (2007), p. 23.

[3] Grossman (2014), p. 16.

[4] Thess (2011), p. 18.

[5] Ayres et al. (1996), p. 2.

[6] Struchtrup (2014), p. 11.

boundaries. In an open system, the container boundaries allow the flow of mass and/or energy, which causes changes in the substance inside the system.[7]

## 2.2 First Law of Thermodynamics

The first law of thermodynamics is identified as the conservation of energy law. This principle states, that energy is neither created nor destroyed, but rather changes its form.[8] This definition provides a formal statement about energy conservation.[9] Because of this law, the energy lost by a system in a process is equal to the energy gained by its environment. By following this observation, it is possible to formulate energy balances for the studied system. This law relates the energy changes that a system undergoes by non-equilibrium processes, through two mechanisms: heat and work. The concepts of energy, heat and work are defined below.

Energy is an agent that promotes transformations in a system, and reconfigures the properties of the system components and its environment. In[10] the author defines energy as "the ability to cause change in the world, stored in matter and force fields. The conversion of energy provides the work that drives life processes and the production of goods and services."

Heat is a form of energy manifested through a transient phenomenon, which is identified over transfers across boundary between two systems at different temperatures.[11] Heat is "what is left over" in a thermodynamic process, when work is extracted from the change in internal energy.[12]

Work is a form of energy transported, associated with a force acting across distance.[13] Work is a function of the path of the process, as well as the initial and final state.[14]

## 2.3 Second Law of Thermodynamics

The second law states the difference in the quality of different forms of energy and explains why some processes can occur spontaneously while others cannot.[15] This law considers that "in any transfer or conversion of energy" within a closed system,

[7] Struchtrup (2014), p. 11.

[8] Montes et al. (2009), p. 2.

[9] Dincer and Rosen (2007), p. 8.

[10] Kümmel (2011), p. 29.

[11] Borgnakke and Sonntag (2013), p. 98.

[12] Struchtrup (2014), p. 40.

[13] Sidebotham (2015), p. viii.

[14] Borgnakke and Sonntag (2013), p. 132.

[15] Dincer and Rosen (2007), p. 9.

the probability of predicting its arrangement in a given time, decreases as the system undergoes successive transformations, i.e. its entropy increases.

Entropy is a measure of energy availability to do work in a non-equilibrium state,[16] and increases as a function of energy transfers, i.e. energy availability to do work decreases as energy undergoes a greater number of transformations. Entropy represents a depletion in the quality of energy, accumulated through successive energy transfers.

Two consequences of the second law of thermodynamics are: the continuous increase of entropy and the irreversibility of energy transformation processes. The above consequences imply that the sum of the entropy changes of a system and its environment must always be positive. Specifically, the sum of all systems and environments will be more disordered and its direction must be oriented towards thermodynamic equilibrium with some absolute maximum value of entropy.[17]

## 3 Components of Sustainability for the Mining Industry

This section describes economic, environmental and social criteria in the development of the proposed model. These criteria aim to establish a common field of concepts. The following is a conceptual approach to contribute to the discussion of sustainability in the mining industry.

### *3.1 Economic Criteria for Sustainability*

The approach to economics used for this model, refers to the application of this science to natural resources administration. This approach uses economic theory and quantitative analysis methods to address governmental policy problems regarding supply, allocation, distribution and conservation of natural resources, including the environment.[18] The particular treatment of non-renewable resources in economics focuses on defining optimal depletion periods and rates. However, an economic approach to determining the contributions of mining industry to sustainability requires taking on several challenges. These challenges involve the inclusion of uncertainty about the needs of future generations, the influence of technological change, the development of substitute materials, and market behaviors. In line with the challenges for the economics of sustainability in the mining industry,[19] provides a definition that offers some insights for approaching economic analysis:

[16] Dincer and Rosen (2007), p. 7.

[17] Dincer and Rosen (2007), p. 9.

[18] Posada and Vargas (1997), pp. 84–85.

[19] Carrizosa (1987), p. 595.

> Resources are created by people from a certain matter-energy, by looking for a certain use for them and developing the possibility of obtaining them in sufficient quantity. In this way the definition of a certain resource must necessarily be dynamic, varying with technology, availability and utility. This detail is due to the fact that in reality for any living being what is important is not the resource itself, but the service derived from its consumption.

Carrizosa in his definition, implicitly raises the renewability (or depletion) of natural resources associated with technology, market, and society. In addition, the author raises the existence of a utility function associated to natural resources that can be linked with the services in which the resource is used. This approach could be understood as a process of dynamic depletion-renewal in the long term. In the case of the mining industry, this depletion-renewal dynamic might be a combined function of new mineral deposits discoveries that replenish depleted reserves, technologies that maximize mineral extraction, development of substitute materials, optimization in the final use of materials, recycling, and market dynamics.

In contrast to this concept of dynamic depletion, there is an approach to natural resource economics, specifically the strong sustainability criterion, that considers an unalterable amount of critical natural capital, which generates a restriction on the extent of exploitation of some mineral deposits. That is, there are certain forms of natural capital whose exchange or substitution is not possible. Strong sustainability requires the maintenance of critical natural capital, due to the no substitution of natural systems necessary to maintain life or a better quality of life, and the irreducible uncertainty about the functioning of complex environmental systems.[20] Proponents of "strong sustainability" argue that technological progress can always provide technological solutions to resource depletion and degradation of the natural world in the future; therefore, exceeding a limit of natural capital should be avoided.[21]

In consideration of the above, the consolidation of sustainability in economic terms is based on the combination of the transformation of a non-renew able natural resource, accompanied by protection of strategic ecosystems, and restoration, compensation and mitigation of the impacts generated by the mining industry.

A useful approach in establishing economic criteria for the analysis of the contribution to sustainability from the mining industry can be found in neoclassical economic theory. This approach can be used to interpret sustainability by establishing a connection between intergenerational equity and economic theory, looking for a theoretic maximization of the present value of future w ell-being.[22] Intergenerational equity is constituted as an "ethic of sustainability," which defines a commitment of each generation to its successor by using resources in the present and substituting them for investment in capital and knowledge.[23]

[20] Ring et al. (1999), pp. 50–52.

[21] Hardisty (2010), p. 22.

[22] Posada and Vargas (1997), p. 108.

[23] NRC (1994), p. 9.

The economic task behind determining intergenerational equity and anticipating the future is complex with high levels of subjectivity, particularly if one combines uncertainty about the future and the multiplicity of meanings behind the concept of a future well-being. However, despite the complexity of the task of quantifying, qualifying, and maximizing future well-being, the economic analysis of the mining industry's contributions to sustainability requires the development of financial and economic models, which allow to assess a set of mining-based capitals, which could provide a 'sufficient' amount of well-being for future generations.

Faced with the challenge of guaranteeing the sustainability of future generations, the proper management of natural resources and the administration of the stock of capital that supports the capacities of the economy is a critical factor. This challenge was posed by[24] in the following terms:

> Whenever a society or economy depletes its stock of resources, it reduces its future productive capacity. Unless the stock can be renewed, or unless some substitution can be made for the depleted stock, the path is unsustainable. For a sustainable economy, the net "disinvestment" of natural capital, or any other form of wealth that contributes to economic activity, must be minimized or prevented. This does not necessarily require replacing identical stocks, as long as nations and businesses can renew or replace depreciated resources. Therefore, in the view of some experts, economic sustainability does not require the preservation or replacement of original resource stocks.

The above quotation highlights the role of the depletion of natural resources and the renewal of a stock composed of various forms of capital, to substituting the services offered by the depleted resource, mineral resources in our case. This confirms the position expressed by[25] relating the definition and use of natural resources and their dynamic process of depletion and renewal.

Faced with these economic positions that interpret the substitution of some capitals, some economists have proposed models for the management of non-renew able resources under the premise of maintain the effective resource base of the economy and the balance between depletion, technical progress, and accumulation of other forms of capital. Two theoretical models propose the following sustainability conditions:

- The first proposed by Solow in 1986 that suggested that "a society that invests the competitive rents from the extraction of exhaustible resources in reproducible capital will enjoy a steady flow of consumption over time. This is a sustainable path identified by Solow, as one "that allow s all future generations the option of being as well off as their predecessors"[26]
- The second model is proposed by Pearce and his team in 1988 which stated that "the necessary condition for sustainable development is to keep the stock of natural capital constant, more strictly the constraint requires non-negative changes in the stock of natural resources such as soil and its quality, land and surface

[24] NRC (1994), p. 7.

[25] Carrizosa (1987), p. 595.

[26] NRC (1994), p. 6.

water, biomass of land and water and the waste assimilation capacity of the receiving environments."

Some practical implications are derived from these two models. In the first model, a society needs to develop capacities to make adequate investments that guarantee competitive rents. These required capacities include application of adequate prospective methods, management and adaption of investment to market conditions to guarantee future flows and support next generations well-being. In the second model, a society needs to develop its capacities to determine the available quantity of natural capital, its adaptive capacity, its reproduction rates, and the trade-offs that can be balanced against projected extractive scenarios.

Therefore, the economic criteria considered for this model are based on the understanding of the depletion of mineral reserves as a dynamic process, which derives rents for the maintenance of future welfare flows, but also ensures the maintenance of optimal levels of natural resources that contribute to the sustainability of future generations.

## *3.2 Social Criteria for Sustainability*

A society can be understood as a set of individuals or groups that structure a relationship around common interests, practices, identities, or territories. Additionally, the social structure is not static, but responds to economic, political, technological, and environmental changes. Structural changes can alter the way in which a society organizes itself.[27]

In terms of the social impacts associated with the consolidation of sustainability, changes in the social and cultural fabric of human groups must be observed. Groups are the essence of life in society, since the human being is due to membership in a group.[28] The development of a mining project generates transformations in the surrounding human environment, caused by the arrival of alternative practices and cultures mixes, which generate changes, hard to predict in earlier stages of the project.

Some elements to contribute to sustainability from a social standpoint lies on understanding the impacts derived from mining developments and its potential implications. A theoretical framework used to explain interaction in social life is composed of three primary elements: symbolic interactionism, functional analysis and conflict theory defined by[29] as follows:

In symbolic interactionism the central idea is that symbols, things to which meaning is attached, are the key to understanding how the world is viewed and how society communicates. Through this form of interactionism society is perceived as

[27] Spitz and Trudinger (2008), p. 318.

[28] Spitz and Trudinger (2008), p. 316.

[29] Henslin (2015), p. 12.

composed of symbols, used by people to create meanings, establish their views about the world, and communicate.

In functional analysis, also known as structural functionalism, or functionalism, the central idea is that society is a whole, made up of interrelated parts that work together. In addition, this analysis considers that each part of society has a function that contributes to regulation and capacity in society. This perspective allows to understand society, based on the analysis of both the structure and its function.

Conflict theory provides a perspective that emphasizes that society is composed of groups competing for scarce resources. A simple analysis may show cooperation, but deep observation may reveal a struggle for power. This theoretical framework focuses on largescale patterns of society.

For Henslin, each perspective focuses on selected features of social life and provides a unique interpretation. The combined information from all three provides a more complete picture of social life, and mining developments derived impacts. Another important factor in relation to social sustainability focuses on determining how impacts occur. In this sense, efforts to contribute to social sustainability should seek to focus on improving social justice, distribution of benefits, and balancing burdens associated with economic activities.[30] Benefit sharing involves ongoing interaction between governments, mining companies, and local communities. The quality of these relationships, and the outcomes of the mining project, are determinants in the development of environmental and socioeconomic conflicts.

In particular, the development of a mining project must recognize the diversity and complexity of communities, rather than viewing them as a collective entity. Integrating a complex and diverse view implies to understand the direct and indirect impacts, and appreciate the importance of contextual and cultural factors that could affect community responses to mining activities.[31] Each community has unique attributes and conditions essential to effective planning for mining and community development, which deserve to avoid simplified approximations, or homogenous responses about development and recognize competing and diverse interests.[32]

Some key tools for building good and stable relationships between mining projects and communities include a culture within the mining company, which also influence the community relations. Some recommendations from:[33]

- Respect and inclusion;
- Transparency and honesty;
- Listening and empathy;
- Responsiveness and delivery on the promise;
- Goodwill, care and protection of each other's interests; and
- Clear rules and principled actions.

30 Eggert (2006), p. 188.

31 Evans and Kemp (2011), p. 1770.

32 Spitz and Trudinger (2008), p. 318.

33 Thomson and Boutilier (2011), p. 1795.

The transformations on the social dimension are also dynamic. Various social and political issues can have a significant impact on the interactions and relationships between mining operations and communities. As a result, it is relevant to emphasize the influence of the context, and to recognize that the nature of community issues will change over the life cycle of a project.[34]

## *3.3 Natural Environment Criteria for Sustainability*

The concept of natural environment is understood as the set of natural resources and natural services derived from the natural environment functioning. The qualities and quantities of the natural environment are determinants of sustainability, because they support different forms of life, provide regulation of natural services, and generate relationships with society and economy through ties of identity, culture, dependence, and resource production. Additionally, natural environment is the recipient of physical and chemical disturbances linked to mining projects development.

The natural environment contains resources in form of materials and energy, which are appropriated by society and can be considered as natural capital. This appropriation provides a valuation framework for the natural environment. Natural capital is a term used for the stock of natural resources that derives flows of resources and services.[35] Natural capital can preserve its quality and services and support biodiversity in the natural environment through management that considers appropriate harvesting, protection, investment and restoration mechanisms. Thus, the valuation of the natural environment as natural capital and its strategic management contribute to its sustainability. According to[36] there are four critical components of the sustainability of the natural environment.

- The existence of the natural environment,
- Recognition of the impact of human activities on the degradation of the natural environment,
- Understand the *trade-offs* between the costs and benefits of managing impact and benefit,
- The dependence between scientific and technological progress and effective policy actions to achieve significant results.

The sustainability of the natural environment, becomes a physical objective that considers maintaining environmental quality and the stock of natural resources. For this reason, mining development must maintain the ability of the natural environment to provide aesthetic and life support services. Therefore, emphasis should be

[34] Evans and Kemp (2011), p. 1770.

[35] DFID (1999).

[36] Miller and Spoolman (2007), p. 23.

placed on the objective of preserving the natural environment, regardless of how humans use it for their activities.[37]

The sustainability of the natural environment intervenes actors, whose interests, time horizons, and desired results must be matched as much as possible. The search for synergistic actions that contribute to the sustainability of the natural environment over the mine project time horizon and the stages of its life cycle is an opportunity for the mining industry. These opportunities can might become into an environmental strategic objective for the development of a project.

### 3.4 A Meaning of Sustainability for the Mining Industry

Three relevant aspects of sustainability to maximize the exchanges between society, economy, and natural environment are presented below:

I. The determination by society with high levels of certainty of the cost-benefit of the transformation of non-renew able natural resource reserves.
II. The reduction of the asymmetry of knowledge between a community and a mining company. This balance should contribute to identify a high proportion of mining project risks and its adequate management, through financial, technical, technological, social, and legal instruments.
III. A spatial and temporal accurate prediction of the mining impacts. This forecast should support the decision about the funds for covering mitigation, restitution and compensation of the social, economic, and natural environment.

## 4 Proposed Model

The proposed model is composed of five elements:

- A reactor model illustrating the physical interaction of a mining project with the territory;
- A circular flow of sustainability that illustrates the relationship between the economy, society, and the environment;
- The connection of the model constituted by a set of functional relationships;
- An equation of continuity that integrates fluxes and transformations; and finally,
- The interpretation of entropy in the model.

[37] Eggert (2006), p. 188.

## 4.1 Reactor Model

The reactor scheme emulates the basic model used to understand the flows present in a study system. The reactor model is used as the basis to analyze what flows exist in the context of mining developments. This model contains four components that represent the object of study:

- The mining project,
- The territory where the mining project takes place,
- Flows entering the system,
- Flows out of the system.

***The Mining Project*** For the purpose of this model, the mining project is considered a black box, and focuses on analyzing only input and output flows. This approach is useful because it focuses on the net effects that result from mining development on the territory in which the projects take place. This does not ignore the fact that the details and results of the unit operations of the mining activity are fundamental determinants of the results of the mining activity.

The reactor model should consider geographic, geological, and environmental constraints. In addition, market, business capacity, and regulatory constraints shape the scope of the mining project. In addition, natural resources and mining business constraints shape the scope and level of impacts of the mining project, define the boundaries of the reactor model, and their performance determines contributions to sustainability.

The reactor model is shown in Fig. 1. Thermodynamic qualities of materials may be interpreted in the mining project through a delta associated with the valorization of reserves. Scale is defined by the limits and constraints of the natural resource system, and the environment delineated by the mining project and its boundaries.

***The Environment*** In addition to the black box represented by the mining project, the environment in which the mining project is developed forms a system like a reactor. In this approach, the territory is a set of natural, biotic and abiotic resources; administration jurisdictions that give rise to borders; hard and soft power relations; in addition to production, regulation, support, and information dynamics. The terri-

**Fig. 1** Schematic diagram of the model—mining project. Source: Own elaboration

tory is a field, in the sense of the physical model. This field is a representation of forces distributed in space and time resulting from one or multiple particles that exert an effect on the objects under its influence. In this case the territory is a field modeled by the action of three dimensions that give shape to the field, in simile with the particles of the physical model.

- The first dimension is the anomalies of the natural environment, both renewable natural resources and non-renewable natural resources. In the case of non-renewable natural resources, the anomalies are manifestations of mineral deposits, potential originators of mining projects. In the case of renewable natural resources, anomalies refer to manifestations of high biodiversity that provide high value ecosystem services to society.
- The second dimension is anthropic anomalies, resulting from demographic concentrations. Also includes the impacts due to the territory vocations and use, such as infrastructure and modifications of the natural environment for its use.
- The third dimension is the socio-economic relations that configure power anomalies, and result from actions of the social organizations present in the territory, that seeks to take benefits from natural and anthropic anomalies.

The interaction of these three entities shapes the field that represents the territory and the baseline for a mining development site. As a result of mining development, transformations are generated on the territory, which can be interpreted from the natural, anthropic and socioeconomic dimensions of the territory. Each state of the territory at time $t$ is the result of the interaction between these three dimensions and shape the territory. In Fig. 2 is represented the black box previously presented and determines the territory in which the mining project is inserted.

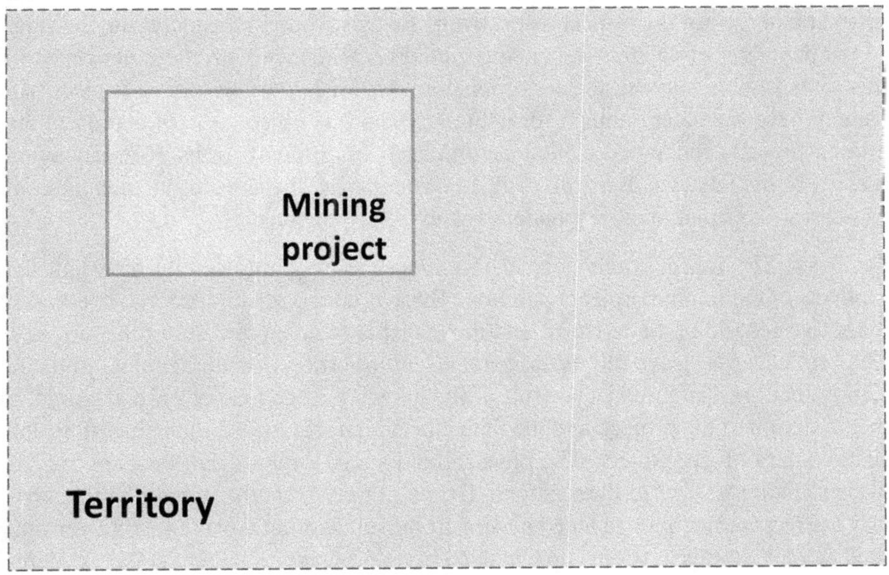


**Fig. 2** Schematic of the model—mining project and territory. Source: Own elaboration

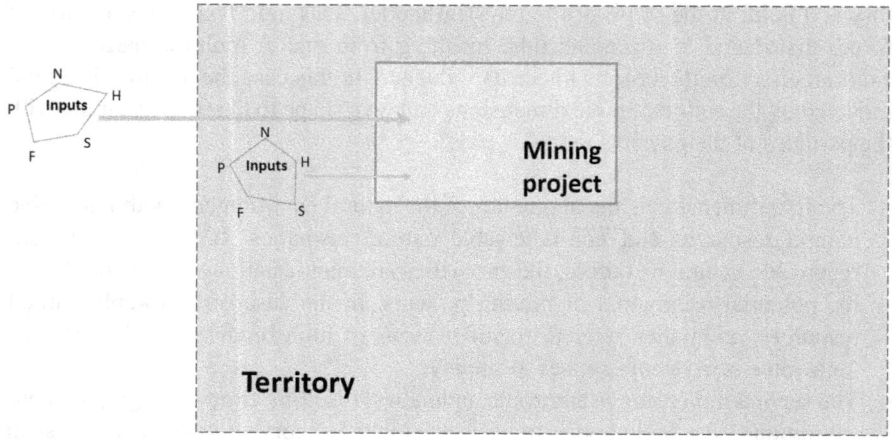


**Fig. 3** Schematic of the mining project, and inputs in a territory. Source: Own elaboration

***Input Flows*** This third component represents the transport of inputs required for the development of a mining project. Two types of flows are differentiated: those provided by the territory and those provided from outside the territory. One way to represent the flow s is through five capitals: natural, human, social, financial and physical, represented by a pentagon that reflects a relative amount for each of the capitals, as shown in Fig. 3.

In the development of a mining project in a remote area, the set of assets of external inputs will usually have a greater representation in the demand for goods and services grouped in the human, physical and financial capitals. These capitals are personnel with specialized knowledge, the installation of support infrastructure, and financing of the mining project respectively. Because of this consideration, the sides of the pentagon of capitals are unequal in correspondence with these demands. In this same remote context, the set of local assets flowing to the mining project will usually have a greater demand for natural capital, the object of exploitation in the mining project, and social capital accumulated in different forms of organization present in the territory. Because of this consideration, the sides of the pentagon of capitals are unequal in correspondence with these demands.

***Outflows*** The fourth component of this model is the outflows that represent the products of the mining project operation. These products are divided into two flows, those that remain in the territory, and those that are transported out of the territory. The products that leave the territory represent a portion of natural capital, human capital that has increased as a result of the knowledge, and experience acquired in the execution of the project, and the financial capital that is paid to the investors and shareholders of the project. The physical and social capitals outflows are low, as these remain installed in the territory. The output products that remain in the territory are represented by a reduced portion of the initial natural capital, in the form of rock wastes, compensations, and reclamations products. Human capital is repre-

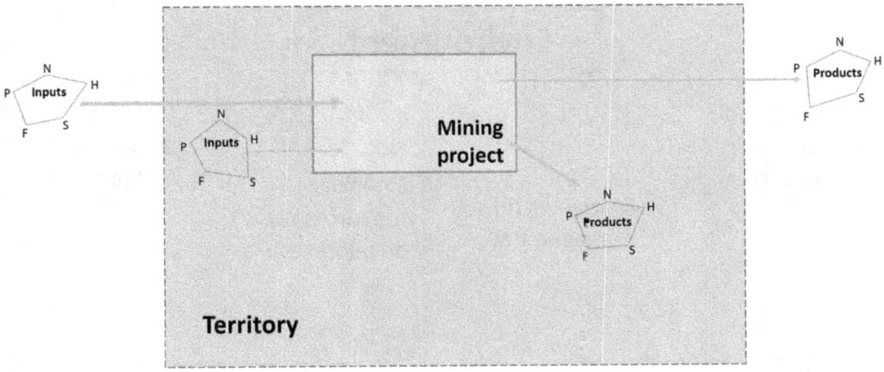


**Fig. 4** Diagram of the mining project, territory, inputs and outputs. Source: Own elaboration

sented in the training and experience acquired in the execution of the mining project. Social capital is represented by the interaction of the mining project and local society with the mining project. Financial capital is represented in the form of labor income, private investments, royalties and taxes. Finally, an important portion of physical capital remains installed in the intervened territory, the value of which will depend on the adaptations and use made.

Figure 4 depicts the reactor model as a schematic approach to illustrate the flow of resources in the form of five capitals to and from a mining project to and from local and external sources.

## 4.2 *Circular Model of Sustainability*

The relationship between the components of the concept of sustainability in the proposed model is developed by adapting the idea of the circular flow model of the economy. This idea proposes that there is a continuous flow between the three components of sustainability: the natural environment, society, and economy. In the circular flow of sustainability, three components of sustainability are integrated and their relationships generate transformations between each of the components of the model. Additionally, it is proposed that the continuous flow occurs through a set of functional relationships.

The functional relationships between the natural environment, society, and the economy generate six functional relationships (Fig. 5). Two between the natural environment, society, and the natural environment. Two between the natural environment, the economy, and the natural environment. Two between the economy, society, and the economy. Additionally, each pair of functional relationships generates a relational cycle.

The functional relationship between the natural environment and society is through flows of materials and ecosystem services, from natural capital. The

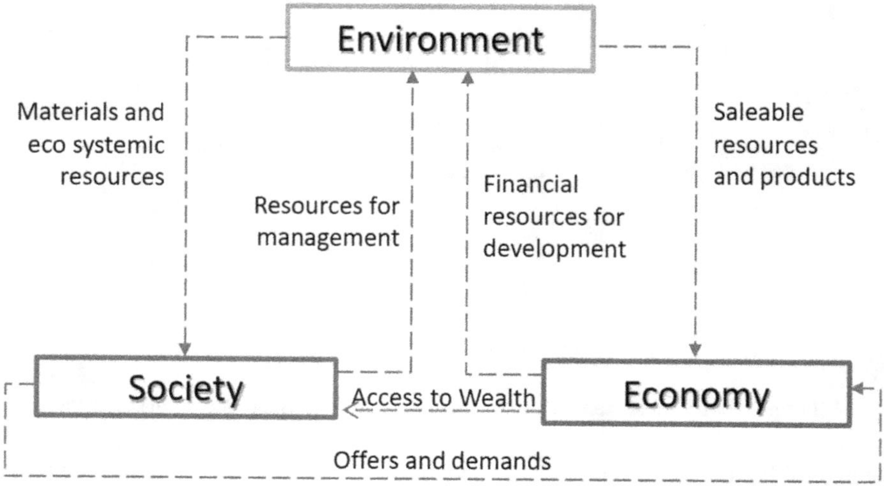


**Fig. 5** Circular flow of sustainability. Source: Own elaboration

functional relationship between society and the natural environment occurs through management actions on the natural environment through human and social capitals. This pair of functional relationships shapes the first cycle that relates the ecosystem goods and services provided by the natural environment and the knowledge and organization required to manage the use, conservation, and development of the natural environment.

The functional relationship between the natural environment and the economy occurs through the flow of natural resources traded in the market, which put some dynamism into the economy, and respond to the needs of a society. The functional relationship between the economy and the natural environment is through the generation and provision of resources to mitigate, restore, compensate, and improve the natural environment. In addition, the economy provides the resources to invest in the adaptation of the natural environment for the development of rural economic activities that provide income to communities once the mining industry has ceased its activities. This pair of functional relationships shapes the second cycle that links the provision of natural resources in the economy, and the provision of economic resources for the management of the natural environment.

Finally, the functional relationship between the economy and society is through the creation of opportunities for wealth generation through business opportunities and entrepreneurship, and the provision of goods and services to meet needs of society. The functional relationship between society and economy is through the supply of goods and services for the functioning of the economy. This pair of functional relationships shapes the third cycle that relates the effects of the economic and social dynamics generated by the development of mining projects.

### 4.3 *Functional Relationships of the Circular Flow of Sustainability*

The meanings of the functions of the relationships described in the previous section, have been defined based on the concept of ecosystem functions defined by[38] and.[39] The argument is that the functioning of society and the economy can be interpreted in the light of the functioning of ecosystems. About ecosystem functioning,[40] recommended that "the first step towards a comprehensive assessment of ecosystem goods and services involves the translation of ecological complexity (structures and processes) into a more limited number of ecosystem functions." These functions in turn, provide elemental goods and services for human well-being. The authors suggest that it is convenient to group ecosystem functions into four main categories. For[41] ecosystem functions are "the capacity of natural processes and components to provide goods and services that satisfy human needs, directly or indirectly." Following the above definitions, the conceptual model proposes to establish functional relationships between the components of the circular model of sustainability, adapting the ecosystem functions.

De Groot defined four functions: regulation, habitat, production and information. The first two groups of functions (regulation and habitat) are essential for the maintenance of natural processes and components and, therefore, are conditioned to the maintenance of the availability of the other two groups of functions (production and information). Similarly, it can be identified and defined variables of the functional relationships between society and natural environment, natural environment and economy, economy and natural environment, economy and society, and society and economy.

Following the ecosystem functions (regulation, habitat, production, and information) proposed by De Groot, and our adaptation of these functions for each of the functional relationships between the natural environment, society, and the economy, twenty-four variables can be obtained for the three cycles and shape the proposed system of relationships.

Similar to what De Groot defined for the ecosystem functions, for defining the remaining five relationships variables, the criterion will be the capacity of the processes and components of the unit of study (economy or society) to provide goods and services that satisfy directly or indirectly human needs. The result of adapting the four ecosystem functions to the relationships between society and the natural environment, the natural environment, the economy and the environment, and the economy, society and the economy is illustrated in the following Table 1.

The definitions of each one functional relationship variables are beyond the scope of this conceptual model, however, they should keep the philosophy of

[38] De Groot (1992).

[39] De Groot (2002).

[40] De Groot (2002).

[41] De Groot (1992).

**Table 1** Variables of the functional relationships of the model

| Functional R.-variables | Functional relation variables | Objective of the relational variables |
|---|---|---|
| Environment–Society | Regulation, habitat, production, information | Providing materials and eco systemic resources for sustaining society needs |
| Society–Environment | Regulation, habitat, production, information | Delivering human and social resources for managing natural environment |
| Environment–Economy | Regulation, habitat, production, information | Supplying natural resources and services to markets for monetizing natural wealth |
| Economy–Environment | Regulation, habitat, production, information | Financing environment maintenance, compensation, restauration, and recovery |
| Economy–Society | Regulation, habitat, production, information | Provide opportunities for accessing wealth and providing goods and services for society well-being |
| Society–Economy | Regulation, habitat, production, information | Wealth, knowledge, organization, workforce for the functioning of the economy |

Source: Own elaboration

interpreting the regulation, habitat, production and information functions proposed for the ecosystems.

## 4.4 Continuity Equation

To illustrate the resources transformation mechanism in the form of capitals, which are part of the proposed reactor system (Fig. 4), a continuity model defined by[42] is adopted into the conceptual model. This is a *rate process* model, which consists of a set of partial differential equations describing the conservation of mass, momentum, energy and components. The authors establish the following elements of the rate process relationship and the models of the first and second law of thermodynamics:

Two laws underlie the concept of process rates. The first law of thermodynamics states the conservation of matter and energy. The second law of thermodynamics, in the context of sustainability, states that matter and energy consumed and then rejected into the environment are of equal or worse quality than that acquired from the environment. The first law reveals that all resources, including energy, are finite and that their exploitation invokes inexorable trade-offs. The second law reveals the limitations and consequences of trade-offs. The model proposed by is represented by the Eq. 1.

**Equation 1** Rate process model

$$\frac{1}{L}\frac{dx}{dt} = \sum r_i$$

[42] Neuman and Churchill (2015).

In the model of Neuman and Churchill, $x$ represents an extensive quantity, such as mass or other measure of inputs and outputs to the process, time $t$, and $L$ a measure of the extent of the system, i.e. the boundaries of the system where analysis takes place. Additionally, $r_i$ represents various velocity mechanisms of the process, which can be positive (inputs) or negative (outputs). The model reflects the concept of conservation of the first law of thermodynamics. That is, the changes that occur in a unit of control volume for each of the components of the system under study is equal to the sum of the different accumulation mechanisms acting on the system. In his initial proposal[43] proposed five types of process rate mechanisms: they are accumulation, production, assimilation, depletion, and exhaustion. These mechanisms may be useful for future model adaptations.

In the proposed model, equality is defined between the changes of each capital and the flows generated by the circular flow model of sustainability for a specific volume. That is, the changes in the components of the system (the livelihood capitals) for unit of control volume (the territory) are equal to the sum of the flow mechanisms acting on the system (the functional relationships of the circular flow model of sustainability) as presented in Eq. 2.

**Equation 2** Balance model for the study unit of a mining project

$$\frac{1}{L}\frac{dC}{dt} = \sum r_i$$

where $L$ is the extent of the territory, $C$ represents the five livelihood capitals and $r_i$ represents the different rate mechanisms of the six functional relationships in the circular flow model of sustainability. The equation of the balance model represents the change of the capital of livelihoods as a function of time (a speed) based on the functional relationships between the components of the circular flow of the economy. In an extended way the equation is (Eq. 3):

**Equation 3** Extended equation of the balance model for a unit in the territory

$$\frac{1}{L}\frac{d}{dt}\left(C_s + C_h + C_n + C_{fc} + C_{fn}\right) = \sum R_{NE-S}, R_{S-NE}, R_{NE-E}, R_{E-NE}, R_{E-S}, R_{S-E}$$

The extended equation contains the following terms:

L: is the extent of the territory
$C_s$ = *Social capital*
$C_h$ = *Human capital*
$C_n$ = *Natural capital*
$C_p$ = *Physical capital*
$C_f$ = *Financial capital*
$R_{NE\text{-}S}$ = *Natural environment and society relationship*

[43] Churchill (1974).

$R_{S\text{-}NE}$ = *Society and natural environment relationship*
$R_{NE\text{-}E}$ = *Natural environment–economy relationship*
$R_{E\text{-}NE}$ = *Economy–natural environment relationship*
$R_{E\text{-}S}$ = *Economy–society relationship*
$R_{S\text{-}E}$ = *Society–economy relationship*

Graphically the model can be represented as:

Figure 6 can be described as the variation of livelihood capitals for each physical unit of study equals the process rate mechanisms of the circular flow of sustainability. The equality values are a schematic representation for each of the system components, the five sustainable livelihood capitals, across the six functional relationships in the circular flow of sustainability.

A positive value for the right-hand side of equation $(dx/dt)/L$, represents the rate of accumulation of quantity $x$ and a negative value its rate of depletion, in both cases by the sum of the mechanisms rates $r_i$. Process rates form an essential component of sustainability because the rate of any process must be able to be maintained over time without exceeding the innate and "natural" capacity of its environment to support it.

## 4.5 Entropy in the Sustainability Model

In the proposed model, the concept of entropy is related to the quality of energy and depletion, which results from successive transformations linked to effects of mining developments and their accumulation over time, in different dimensions of the territory. In the model, the incorporation of the concept of entropy is reflected in the decomposition for each of the variables of the functional relationships, of the energy components in terms of high and low quality, according to the results obtained.

The approach makes it possible to identify the components of the resulting transformations in the territory intervened by the mining project, and to interpret the

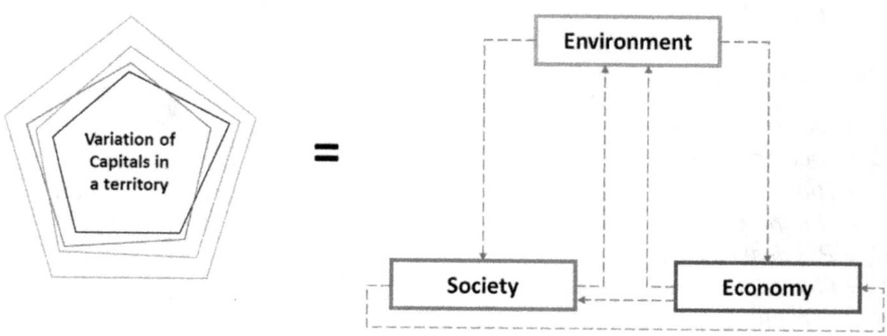


**Fig. 6** Graphical representation of Eq. 3, equation of balance for a unit of territory. Source: Own elaboration

potential changes in the quality of energy. This means that the extended equation of the balance model (Eq. 3) will have for each variable of the functional relationship, its energy components, α and β, being α energy of good quality and β energy of inferior quality. These energy components change as a result of the transformations undergone by the system.

In the proposed model, energetic changes are identified from the energetic components of the functional relationships. The entropy variable develops an additional dimension in the model. The evolution of this variable allows to know how the energy quality of the resources involved in the functional relationships changes from its main variables.

The implementation of the model requires the definition of criteria to evaluate the changes in the components and in the variables of the functional relationships. This process should be based on indicators that record the evolution of the variables and consider the results of high and low quality in the energy reflected in the intervened resources.

## 5 Conclusion

The proposed model provides a new perspective on the analysis of sustainability in mining, integrating the concepts of the first and second laws of thermodynamics, a novel circular flow of sustainability, and integrates them through an equation of continuity. This conceptual model helps to explain the consolidation of sustainability in a territory intervened by mining industry through the integrating the meaning of energy transformations in the dimensions of sustainability, operated through functional relationships and a continuity equation.

This model reinforces the idea of sustainability as a dynamic state and not a static one. In this model it has been illustrated how sustainability obeys to a combination of conditions that guarantee the maintenance of the conditions that provide life to society. In the dimension of the natural environment, the rational use of resources, continuous maintenance of natural resources, and provision of goods and services that make life possible for society's groups and biodiversity. In the dimension of the economy, the development of strong, diverse and equitable economies that fairly distribute the benefits derived from the use of natural resources and human activities. In society, understanding the characteristics that shape and identify society, the transformations promoted by the development of mining projects in its context, and the mechanisms that contribute to society's healthy functioning.

This model proposes that sustainability is the sum of capacities present in the natural environment, the economy and society to guarantee adequate living conditions in the intervened environment. Sustainability in the mining industry is an adaptive process to specific natural environment, economy, and society systems. Consequently, the development of mining projects contributes to the sustainability of the territory as far as it guarantees its existence. The development of mining projects can contribute to sustainability to the extent that new local configurations are made in a negotiated and cooperative manner with the society that inhabits it, in which the different visions, objectives and visions for the future of the territory are included.

Finally, the formulation of this conceptual model and the investigation that supports it, generates new research perspectives for the application of sustainability concepts and the development of mining projects. These results contribute to understanding the conditions that contribute to the sustainability of the territory. Future research derived from this model can be oriented to adapt the concepts of kinetics, catalysis and diffusion of chemical processes, and detailed definitions of the functional relationships' variables, which will contribute to a deeper understanding about impacts generated by extractive developments. Future research based on the present model, can also address mitigation and compensation programs, and generate effective transformations of capitals (natural, in other forms of capital) to support the maintenance of life, that is, contribute to the sustainability of the intervened territory.

## References

Ayres R, Ayres L, Martinas K (1996) Eco-thermodynamics: exergy and life cycle analysis. ISEAD

Borgnakke C, Sonntag R (2013) Fundamentals of thermodynamics. Wiley

Carrizosa J (1987) Concepto de Recurso Natural. In: Diccionario de la Naturaleza. Hombre, Ecología, Paisaje, Madrid, Espasa - Calpe, pp 595–597

Churchill S (1974) The interpretation and use of rate data: the rate concept. Hemisphere Publishing, Washington

De Groot RS (1992) Functions of nature: evaluation of nature in environmental planning. Management and Decision Making

De Groot RS, Wilson MA, Boumans RM (2002) A typology for the classification, description and valuation of ecosystem functions, goods and services. Ecol Econ:393–408

DFID (1999) Livelihoods Centre. http://www.livelihoodscentre.org/documents/20720/100145/Sustainable+livelihoods+guidance+sheets/8f35b59f-8207-43fc-8b99-df75d3000e86. Accessed 22 May 2023

Dincer I, Rosen M (2007) Exergy, energy and sustainable development. Elsevier

Eggert R (2006) Mining, sustainability and sustainable development. In: Australian mineral economics a survey of important issues. Monograph 24. The Australasian Institute of Mining and Metallurgy, p 225

Evans R, Kemp D (2011) Community issues. In: SME mining engineering handbook. Society of Mining, Metallurgy, and Exploration, p 1832

Grossman J (2014) Thermodynamics: four laws that move the universe. The Great Courses, p 27

Hardisty P (2010) Environmental and economic sustainability. CRC Press - Taylor & Francis Group

Henslin J (2015) Sociology a down-to-earth approach. Core concepts. Pearson

Kümmel R (2011) The second law of economics. Energy, entropy and the origins of wealth. Springer

Miller T, Spoolman S (2007) Essentials of ecology. Brooks/Cole, Belmont

Montes VJ, García TJ, Querol AE (2009) Termoeconomía y Optimización Energética. Fundación Gómez Pardo y Consejo Superior de Ingenieros de Minas, Madrid

Neuman M, Churchill S (2015) Measuring sustainability. Town Plan Rev:457–482

NRC, National Research Council. Commission on Geosciences (1994) Environment, and resources. Assigning economic value to natural resources. National Academy Press, Washington, p 7

Posada LLG, Vargas PE (1997) Desarrollo Económico Sostenible, Relaciones Internacionales y Recursos Minero Energéticos. Universidad Nacional de Colombia
Ring I, Klauer B, Watzold F, Mansson B (1999) Regional sustainability. Springer
Sidebotham G (2015) Heat transfer modeling. Springer
Spitz K, Trudinger J (2008) Mining and the environment from ore to metal. CRC Press Taylor & Francis Group
Struchtrup H (2014) Thermodynamics and energy conversion. Springer
Thess A (2011) The entropy principle. Thermodynamics for the unsatisfied. Springer
Thomson I, Boutilier R (2011) Social license to operate. In: SME mining engineering handbook. Society for Mining, Metallurgy, and Exploration, p 1832

# A Case Study on the Application of the United Nations Framework Classification for Resources on the Piampaludo Titanium Deposits

**Ghadi Sabra and Marilena Cardu**

## Contents



**Abstract** With the current declining trends, energy and raw materials have become focal points for sustainable development. Safeguarding mineral resources in a sustainable matter grew into a great challenge in Europe, especially for Critical Raw Materials. These strategic commodities are fundamental for the energy transition. Securing mineral supply depends on type of deposits, which determines mineral

---

G. Sabra (✉) · M. Cardu
Politecnico di Torino, Department of Environment, Land and Infrastructure Engineering, Turin, Italy
e-mail: ghadi.sabra@polito.it; marilena.cardu@polito.it


W. Frenz, A. Preuße (eds.), *Yearbook of Sustainable Smart Mining and Energy 2022*, Yearbook of Sustainable Smart Mining and Energy - Technical, Economic and Legal Framework 2, https://doi.org/10.1007/978-3-031-41873-0_2

quantity and quality. Instead, mineral project development is defined by the technical feasibility and the related environmental socio-economic viability. These factors are strongly embedded in the United Nations Framework Classification for Resources. UNFC is a global, principle-based reporting system developed in accordance with the Agenda 2030. It is based on three fundamental pillars: the environmental-socio-economic viability, the technical feasibility, and the degree of confidence in product estimates. The Titanium deposit of Piampaludo was the subject of this case study to underline the process for social and environmental considerations when reporting in UNFC. The Piampaludo Titanium deposit is one of the greatest in Europe, yet it remains bounded by serious environmental and social contingencies. The application of UNFC in this case study gives a comprehensive outlook on the project's maturity with the given circumstances, using solely publicly available data. The information provided for the Piampaludo case are classified adequately to be included in local and national mineral inventories. As a result, this classification serves as an instructive example of how UNFC can be utilized using only data accessible to the public, mostly for educational purposes. In this paper, a potential solution for the recovery of the Piampaludo Titanium is briefly explored through the implementation of biotechnology to the given resource. Finally, the United Nations Resource Management System is introduced as part of the case study to highlight its importance in sustainable resource management.

## 1 Introduction

The current increase in the global population and it's living standards indicate a movement towards undesirable wellbeing conditions for humanity. This is mainly caused by several factors that have been handled irresponsibly throughout our time on planet Earth, which inevitably lead to the obliteration of Earth's prosperity for the current generations, as well as the forthcoming. The aforementioned factors can be narrowed down to the three critical pillars responsible for human wellbeing: economy, environment, and society. If monitored and managed carefully, these pillars can constitute the foundation for the mitigation of such crisis on the near-term and the long term: thus emerging the concept of Sustainable Development. This crucial notion is built on the premise of preserving a steady economy, enriching societal wellbeing, and tending the environment. Sustainable development allows our modern society to meet its needs, without compromising the opportunities for future generations.[1]

Fortunately, in 2015, the United Nations and it's 193 member states took initiative to solidify an action plan set towards the sustainable development. The 2030 Agenda for Sustainable Development targets prosperity with regards to our planet and its inhabitants. As stated in the name, the plan aims to be accomplished by the

[1] Iberdrola (2022).

year 2030 while delivering on 17 Sustainable Development Goals (SDGs) with the intent of not leaving anyone behind.[2] These goals are intended to end poverty, to reach zero hunger, to achieve gender equality, to provide clean and affordable energy, etc. The SDGs are necessary to avoid the negative turn of our current course of direction, such as the threat of climate change, in order to allow a healthier, more protected future.

Extractive industries have the responsibility to positively impact the Agenda 2030; they are key players in delivering on the SDGs, as they are directly linked. The mining industry contributes to several SDGs on all levels of sustainable development. Mining companies active in developing countries have the ability to decrease poverty by generating employment, profits, and economic growth. In addition, mines are capable of extending beyond their lifecycle to assure a positive impact on the environment and climate through partnerships with governments and local communities.[3] However, conventional mining activities will need to adjust to a smart and sustainable approach by implementing newer greener technologies, reducing wastes, and increasing safety measures, to install good relations with local communities, and improve environmental measures. Moreover, smart mining can contribute to "Affordable and Clean Energy" by leveraging the demand for energy to supply power to lacking areas, via their communal and governmental partnerships, which enables a shared use of energy infrastructure.[4] With that being said, supply of energy, effective production, and mineral resources are in the hands of the mining industry, making it vital in achieving the 2030 Agenda for Sustainable Development.

Smart and sustainable mining shall pave the way for the delivery of the SDGs. On that note, the mining sector is left facing great challenges in meeting our mineral resources demand, as well as for the future generations. In line with the current transitional energy system, mines need to amplify their resource production in a sustainable manner. Thus, Sustainable Resource Management is critical to ensure that all mining practices are run with respect to the Agenda 2030. Sustainable resource management provides a comprehensive understanding of resource value chain, which contributes to the security of resource demand and supply. In simple terms, it is the pathway to manage resources while considering a socially, environmentally, and economically efficient development for a sustainable future.

In addition, securing resource endowments begins with a correct mapping of supply and demand achieved by sustainably managing mineral resources. Resources require a coherent and consistent definition and categorization at local, national, regional, and global scales. The sustainable, secured supply chain of raw materials also requires actual, real information as mean to fully understand the supply-chain, which in return improves decision-making. The United Nations Framework

[2] United Nations Sustainable Development (2022).

[3] Africa Renewal (2022).

[4] Africa Renewal (2022).

Classification for Resources (UNFC) is the tool to successfully report resources in line with achieving the SDGs.

## 2 UN Framework Classification for Resources

UNFC is a universally recognized and globally applicable resource classification system based on projects and principles. UNFC is a comprehensive tool for the consistent, coherent classification and the management of all types of resources. It is therefore applicable to all kinds of resources and energy projects, including mineral and anthropogenic resources, nuclear fuels, hydrocarbons, injection projects for underground storage of $CO_2$, and renewable energy. These resources are assessed on the basis of their state, as primary resources (such as minerals) or secondary resources (such as anthropogenic sources, mine tailings). Additionally, this classification system provides a framework to build local, national, regional energy and raw material policies/studies, to support governmental management laws/regulations of resources, to plan innovative industrial processes, and to assign correct capital proficiently.[5] UNFC is a threefold system, defined on sustainability (environment, social, and economic), maturity of a resource development project, and the confidence in resource availability.[6] The latter is related to the level of knowledge in the estimates of the project's product quantities present, produced and/or recoverable.

UNFC defines a project as a development or an operation where the base of sustainability, technical assessment, and decision-making are provided.[7] This tool is applicable to active projects, planned projects, even conceptual projects. However, project details are required for a better evaluation for stakeholder's decisions at a certain degree of maturity. The factors involved in the UNFC assessment provide an exhaustive outlook on conditions in the environmental, social and economic fields, including market prices, governmental settings, environmental and social consideration, maturity degrees of a project with regards to technology and industry, as well as long-term uncertainties, aligned with the requirements of the 2030 Agenda.[8]

The project-based classification of resources according to UNFC is built on three essential pillars defined by the environmental-socio-economic viability (E axis), the technical feasibility of the project (F axis), and the degree of confidence in the product estimates (G axis). Hence, UNFC is designed to maximize the requirements in applications relating to regulations founded on resource studies, financial capital allocation, resources management functions, corporate, and business processes.

[5] United Nations Economic Commission for Europe (2019a).

[6] United Nations Economic Commission for Europe (2019a).

[7] United Nations Economic Commission for Europe (2019a).

[8] United Nations Economic Commission for Europe (2019a).

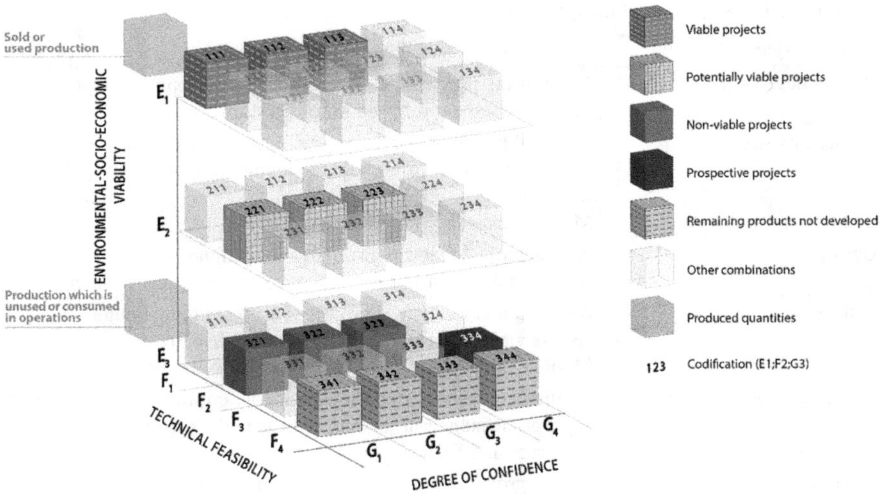


**Fig. 1** 3D representation of UNFC categories and classes (United Nations Economic Commission for Europe 2019a)

UNFC is reported on a 3-dimensional matrix, according to the three aforementioned axes, with a simple numerical coding system (Fig. 1).[9]

A combination of the axes results in the classification of the project into categories, and sub-categories. The environmental-socio-economic axis defines the favorable sustainable conditions of the project, by evaluating its viability. The F axis, the technical feasibility, describes the project's maturity degree with regard to the choice of technology and its implementation, as well as the necessary feasibility studies. This axis reflects the principles of a standard value chain management.[10] Lastly, the G axis integrates all the preliminary and on-going investigation to concur an estimated quantity of the project's product.

UNFC is designed to aptly accommodate national and regional requirements The UN has stated that several countries have taken initiatives for the application of UNFC, such as the European Union, African Union, Russia, China, India, Mexico, etc.[11] This allows the UNFC to include the complete range of raw materials and stakeholders, with the aim to meet the necessary implementation requirements for the resource industries, while complying with the Agenda 2030. Furthermore, the universality of this classification system allows it to be in harmony with other international and national classification systems, such as, for example, CRIRSCO. UNFC is therefore equipped with bridging documents that elaborate on the conversion of the classification, from one system to UNFC and vis-a-versa, with instructions that clarify the classification of estimates generated by the application of the other

[9] United Nations Economic Commission for Europe (2019a).

[10] United Nations Economic Commission for Europe (2019a).

[11] United Nations Economic Commission for Europe (2019b).

**Table 1** Stakeholders use of the UNFC[a]

| Type of organization | Useful functions of UNFC for stakeholders |
|---|---|
| Project developer | Monitor the progress of projects under development, repot to governmental entities, report to investors |
| Equity investor | Assess the viability of a project subject to investments<br>Assess project portfolio held by an initiative considering investments<br>Monitor project progress and portfolio |
| Banks | Assess project viability for loan application<br>Assess viability of portfolio for projects applying for loan |
| Utilities | Assess the viability of a project considered to be bought<br>Assess viability of portfolio for projects to be bought<br>Monitor project portfolio |
| Stock exchange, financial institutes, and ESG regulators | Confirming the reported information by entities have followed appropriate standards and credibility, including data on natural resources and ESG metrics |
| Governmental Entities and research and academia | Assess project pipeline on national and regional scales |
| Civil society and others | Assess the project's impact, importance, and development process, including alternatives |

[a] Jolović (2020)

system using the UNFC Numerical Codes.[12] Additionally, UNFC's strong suit is in the integration of all stakeholders involved in a resource project. The consideration and inclusion of all stakeholders points out the sustainable aspects of UNFC, since it aligns with to the civil society, financial institutes, government, academia, and industry (Table 1).[13]

On another note, UNFC is an important classification system for the efficient managing of national resources to ensure sufficient, affordable, and mostly clean supply of energy and raw materials for every country that is faced with the challenge of sustainable development.[14] UNFC stands out amongst the other classification systems mainly due to it overlooking the future economic, social, and environmental aspirations, which are well highlighted in the UNFC principles. This point allows UNFC to be a tangible indicator to various SDGs. Thanks to the UNFC, the "Zero Hunger" goal can be attained through the effective management of small and large scale soil nutrients, such as potassium and/or phosphorous, which are critical for the agro-food industry.[15] For the 6th SDG "Clean Water and Sanitation", UNFC can allow the monitoring, as well as management, of the impact on water systems from extraction activities, as part of a mitigation plan.[16] The UNFC also contributes to "Affordable and Clean Energy" and "Industry, Innovation, and Infrastructure" as it

[12] United Nations Economic Commission for Europe (2019a).

[13] Jolović (2020).

[14] United Nations Economic Commission for Europe (2019c).

[15] United Nations Economic Commission for Europe (2019c).

[16] United Nations Economic Commission for Europe (2019c).

directly promotes an effective development of energy resources in socio-economic terms, and the implementation of green and clean technologies in extraction processes.[17] The UNFC is a well-proven practice for sustainably reporting energy and raw materials projects. UNFC provides the necessary tools for the management of environmental issues and their mitigation challenges. In addition, UNFC is easily integrated to many sustainability reporting systems. These criteria closely link UNFC to the "Responsible Consumption and Production" and "Climate Action" SDGs.[18]

## 3 Critical Raw Materials

When it comes to mineral resources, all of them serve our living standards, one way or another. For that reason, such endowments are now more in demand than ever. Unfortunately, raw materials are exposed to supply concerns facing their fast rates of depletion. The understanding of raw material supply is critical for the sustainable development, as it is greatly influenced by environmental, social, and economical factors, which in return define market volatility and prices. On that note, the European Commission (EC) took initiative to deal with the challenge of securing an independent raw material flow with greater interest, due to shortages in regional production.[19] This initiative has been created to improve the European economy by safeguarding a sustainable flow of Critical Raw Materials (CRM) from and to national and regional sources. CRMs are natural elements with great economic importance and strategically significant for Europe, yet they express great supply risks (Fig. 2).[20]

CRMs are essential for many applications in many industries. The reason behind their criticality is in the distinctive, unique properties they possess, making them not substitutable at present times.[21] CRMs are assessed on the one hand by their Economic Importance, in terms of their added value for the manufacturing verticals and their significance in the end-of-use application, and on the other hand by their Supply Risk, which represents the factors that disrupt the supply chain. In addition, CRMs are a point of interest for the EU because they serve a crucial role in delivering on the 2030Agenda and the EU Green Deal.[22] Many of the listed materials are building blocks for green technologies and batteries, as well as mega-industries in the EU, including the transportation, defense, and medical sectors.

[17] United Nations Economic Commission for Europe (2019c).

[18] United Nations Economic Commission for Europe (2019c).

[19] Raw Materials Information System (RMIS) (2022).

[20] European Commission (2017).

[21] CRM Alliance (2021).

[22] CRM Alliance (2021).

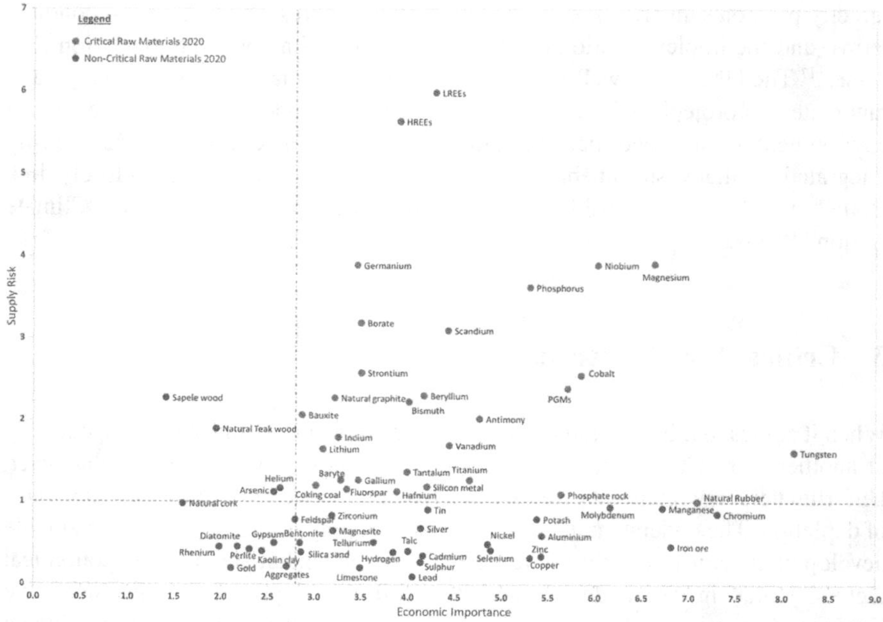


**Fig. 2** List of CRMs 2020 (Raw Materials Information System (RMIS) 2022)

Titanium exudes notable physical and chemical properties that are useful in today's society. This durable metal is perceived with great interest, increasing its demand in various sectors, such as industry and medical. In 2020, Titanium made the list of CRMs by the EC, due to its growing economic interest. Furthermore, this material has been increasing a great deal of supply risk over the years.[23] The properties possessed by Titanium eliminate the option of substitutes to the extent possible. In fact, various materials have been tested as Titanium substitutes, such as lithopone, but resulted in lower quality products with greater environmental impact.[24]

As per the sustainable development, Titanium plays a significant role, in the form of Titanium Dioxide, in delivering on the SDGs. It has been proven that Titanium Dioxide is essential for renewables and for energy efficiency.[25] As stated before, the policies and standards set by the European Green Deal are built on the pillars of the Agenda 2030 and the Paris Agreement. To elaborate more, achieving the climate goals of the European Green Deal requires a shift towards an energy sector that relies more on renewables, which in return require improved technologies, innovations, and infrastructures.[26] Titanium Dioxide is a major actor in the production of renewable energy and in enhancing energy efficiency.

---

[23] European Commission (2017).

[24] King (2022).

[25] Titanium Dioxide Manufacturers Association (2020).

[26] European Commission (2020a).

In Europe, buildings are ranked the greatest single energy, with consumptions up to 40% in energy, and accounts for 36% of Europe's Green-House-Gases.[27] Applying a Titanium Dioxide coating on the exterior of a building significantly lessens interior temperature creeps by directly reflecting solar infrared rays, which contributes to a remarkable decrease in the "Urban Heat Island Effect", as well as eliminating the need for interior cooling, to the extent possible.[28] Thus, the application of a Titanium-bearing pigment improves energy efficiency and reduces energy consumption in buildings.

Titanium is also at the foundation of innovative and improved renewable energy technologies. In the context of Europe, renewables are estimated to account for 32% of global energy consumption by 2030.[29] Renewable energy targets could be catalyzed with the photocatalytic property of Titanium Dioxide. Nowadays, this Titanium property is strongly adapted in solar energy technology, which will inevitably play an integral part in their future improvement. Studies have proved that Titanium Dioxide is the most efficient photocatalyst, especially when paired with Hematene, which is prone to improve solar-fuel generation.[30] Titanium Dioxide is also being currently used for clean air and purification of water.

An additional field of benefit in the use of Titanium Dioxide comprises battery innovation. Within the context of the European Green Deal, an important regulation set by the EC consists of a sustainable circular value chain for batteries, which leads to the reduction in emissions. Titanium Dioxide is a fundamental compound for the enhancement of battery efficiency. It extends battery storage, as well as the life expectancy. On that note, studies have showed that Titanium Dioxide batteries are capable of recharging up to 70% of the total capacity in 2 min, while maintaining a lifespan of 20 years.[31] These Titanium-bearing batteries are valuable for reversible reactions, enabling more durability, which in return reduces battery recycling feed.[32]

# 4 Case Study: Piampaludo Titanium Deposits

## *4.1 Geology of the Area*

The area under study comprising the Titanium deposits is located in Piampaludo, in the province of Savona, in Northern Italy. The Piampaludo site hosts one of the greatest Titanium Dioxide deposits in Europe, yet it remains bounded by social and environmental contingencies (Fig. 3).

---

[27] European Commission (2020b).

[28] European Commission (2020b).

[29] European Commission (2020a).

[30] Overton (2018).

[31] Liu et al. (2013).

[32] Titanium Dioxide Manufacturers Association (2020).

**Fig. 3** Pimpaludo deposits location, marked by the red pin (44°28′36.42″N, 8°35′42.45″E)

The Piampaludo site belongs to the extended mountain chain of Monte Tariné, which is part of the Voltri Massif within the Southeastern terminal unit of the Ligurian Alps. This geologic unit, referred to as the Voltri unit, belongs to the Western Alps, and is located on the Piedmont-Liguria border. The Titanium ore takes form in rutile concentrations, found in eclogitic rock masses. The Voltri unit is geologically complex due to the variations in the chain of events. In particular, the unit metamorphosed under peak conditions with pressures ranging from 18 to 22 kbar and temperatures of about 500–600 °C, typical of the eclogite facies.[33] These settings have formed the meta-gabbro rock mass, rich in rutile, garnet, Na-amphiboles, and Fe-rich minerals, which are found in the Voltri unit. The

[33] Federico et al. (2014).

**Table 2** Stratigraphy of the Voltri unit

| Formation | Location | Geology | Structure |
|---|---|---|---|
| Monte Tobbio Peridotites | Northeastern part of Voltri | Serpentinized peridotite + lherzolite + harzburgitic bodies + dunite lenses + bands of pyroxenite | Remnant mantle texture, mainly tectonic foliations |
| Bric del Dente Serpentinites | Dominant lithology of the Voltri Unit | Serpentinite + serpentine-schist + antigorite, magnetite + chlorite + olivine + diopside + tremolite + ankerite + Ti-clinohumite | Relic texture of the original peridotite. Multiple folds with shear bands |
| Colma Metagabbros | Southern part of the Voltri unit | Leucocratic metagabbro + garnet – Cr-mica + omphacite + glaucophane + albite + tremolite + Mg-chlorite + epidote + traces of white mica + titanite + talc + oxydes | Relic eclogitic paragenesis |
| Rossiglione Metabasites | Southeastern part of the Voltri unit | Ca-amphibole + chlorite + albite – epidote + minor Fe-Ti oxides + titanite + talc + biotite + calcite + white mica. Local Na-amphibole + rutile + garnet | Layered or foliated textures of melanocratic and leucocratic facies. Original textures erased by metamorphism and deformation |
| Turchino Calcschists | Western part of the Voltri unit | Micaschist + quartz-micaschist + carbonate schist with quartz + white mica + calcite + chlorite + biotite + pyrite. Local garnet + chloritoid + mica | Re-equilibrated green-schist facies conditions. Different deformational events across all outcrops |

formation of rutile allowed the rock body to become a Titanium ore. The different geologic formations of the metamorphic Voltri unit, along with their mineralogy and structure, are highlighted in Table 2. The Voltri unit dates to the Jurassic/Cretaceous period, and the Piampaludo Titanium deposits are aged around the time periods of Cretaceous through Oligocene.[34]

Deeper into the mineralogy, the Piampaludo eclogites are primarily composed of rutile, Na-, Al-, Ca-, and Fe-silicates, as well as concentrations of pyroxenes, garnet, and glaucophane.[35] The mineralogy of the eclogites has resulted from the deformation of the oceanic crust in the presence of meta-sediments and ophiolites, covered by Oligocene sediments originating from the Piedmont Tertiary Basin.[36] The eclogites of the area display very hard and compact properties.

[34] Federico et al. (2004).

[35] Mancini et al. (1979).

[36] Capponi (2002).

The Piampaludo rutiles are found at millimetric dimensions and, in some cases, as aggregates in the order of centimeters.[37] The crystals are observed within the range of 5–10 µm, with a clear metallic luster and a variation in color in between brown and yellow. The Titanium Dioxide concentrations within the rutile and the eclogites show an ore grade of 6%, with enrichments up to the double. In addition, the Piampaludo rutiles account for 9 Mt of orebody.[38] On the same note, other estimates based on field tests show an increase in the ore weight to approximately 400 Mt at greater depths.[39] According to these estimates, the Piampaludo site deposits certainly belong to the largest deposits of Titanium in Europe. The rutile aggregates are in the order of 1–2 mm thick at around 65% concentrations, and are mostly directed along the orientation of the foliation.[40] The orebody is characterized as massive and highly fractured, with a low schistosity. Moreover, geophysical data determined that the orebody dips 20°South, with a morphology of a large concordant lens.[41] On a different note, about 30 to 40% of the orebody pertain to concentrations of garnet (Fig. 4).[42] The mining of the available garnet, as a secondary mineral to rutile, can improve the economic interest of the Piampaludo site, since garnet is very attractive to the jewelry industry (Table 3).

## 4.2 C.E.T Srl Mining Plan

In the context of technical advancement and geological investigation, the "*Compagnia Europea per il Titanio*" (C.E.T Srl) earned mining concession for the Piampaludo field in 1985. The permit was transferred to C.E.T Srl from "*Mineraria Italiana Srl*", who earned the concession to mine the Piampaludo Titanium in 1976, by the Ministry of Industry of Italy. C.E.T Srl managed to carry preliminary feasibility studies during the permitted period, which lead to the development of a comprehensive mining plan. The feasibility study was supported by surface developments, such as adits and trenches. Additionally, the technical and geologic investigation included sufficient rounds of coring and drilling. The geology of the Piampaludo field was mapped throughout the investigation phase. These tests have also acquired substantial information for the geophysical and chemical analyses. The initial developments conducted by C.E.T included a pilot shaft, to strengthen the test results linked to exploration assessment. It is safe to say that C.E.T has gathered sufficient information on the ore, in which it has been supported by previous studies done at Piampaludo. The data increased knowledge for the estimation of the ore grade and

[37] Liou et al. (1998).

[38] Liou et al. (1998).

[39] Force (1991).

[40] The Diggings (2022).

[41] The Diggings (2022).

[42] Gorga (2004).

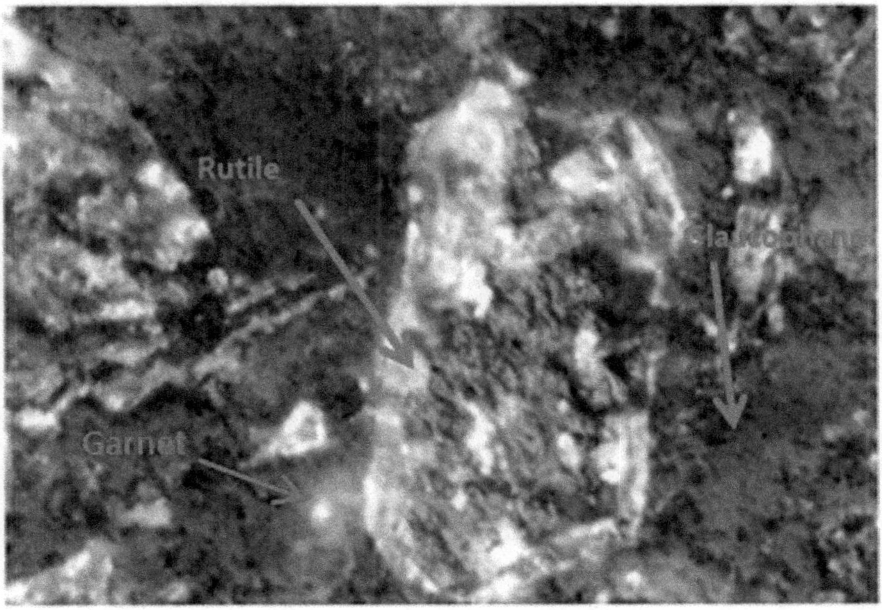


**Fig. 4** Microscopic view of Piampaludo eclogites (ca. ×20) (Gorga 2004). Lorem

**Table 3** Piampaludo titanium-ore properties

| Location | Piampaludo, Sassello, Savona, Liguria, Italy |
|---|---|
| Unit | Voltri Unit |
| Formation | Rossiglione Metabasites? |
| Age | Jurassic to Cretaceous |
| Overburden | Oligocene clastic sediments |
| Geologic setting | Metamorphosed Oceanic Crust, Ophiolites and Metasediments |
| Rock type | Metamorphic Rocks |
| Host rocks | Eclogites |
| Formation conditions | T = 500–600 °C<br>P = 18–22 kbar |
| Physical features | Massive, very compact, hard rocks |
| Ore mineral | Rutile |
| Ore mineral chemistry | $TiO_2$ |
| Ore occurrence | Aggregates about 1–2 mm wide |
| Ore weight | 9 Mt |
| Ore grade | 6% (enrichments up to 12%) |
| Ore morphology | Large concordant lens |
| Ore orientation | 20° dip to the south |
| Secondary minerals | Garnet, Glaucophane, Titanite, Ilmenite, Serpentine, Talc, Magnetite, Actinolite |

**Table 4** C.E.T Srl proposed mining plan for Piampaludo

| C.E.T Srl proposed mining plan as part of the Piampaludo prospect project | |
|---|---|
| Type of mining | Surface |
| Mining method | Open Pit |
| Mining technique | Drill & Blast |
| Surface area | 90 Hectares |
| Maximum pit slope | 60° |
| Operating days per year | 265 |
| Operating shifts per day | 2 shifts of 8 h each |
| Production | 10,000 t/d |
| Production unit cost | 1.37 $/t ore[a] |
| Waste rock | 48.4% |
| Ore mining features | |
| Ore hardness | Hard Rocks (Eclogites) |
| Length | ~1800 m |
| Width | 500 m |
| Thickness | 300 m |
| Wall-rock alteration | None |
| Ore control | Fracturing |
| Latest ore record | 1991 |

[a] Subject to conversion into today's market price

quantity. As a result, an exhaustive mining plan was developed by C.E.T Srl for the extraction of rutile, with garnet as a secondary product. The said plan is comprised of detailed information on production, mine lifecycle, transportation, market, and waste management (Table 4). According to C.E.T's mining plan, mining activities can be maintained for 90 years with a production rate of 10,000 t/d of ore, which equates to 163,240 t/y of rutile produced, at depths up to 500 m. Mining wastes comprising the garnet were foreseen to be conveyed 3 km by gravity flow tunnels towards the coast of Liguria, with a maximum flowrate of 9400 t/d.

# 5 UNFC Classification of Piampaludo Project

## *5.1 G3 Classification on the G Axis, the Degree of Confidence*

Sufficient studies and tests have been carried out to strengthen the level of knowledge on the geology of the Piampaludo field, as well as the Titanium ore mineralization. Mapping, developments, and field and laboratory analyses done over the years have resulted in confident estimations on the ore grade, weight, and quantity. Substantial rounds of drilling were performed to better understand the subsurface geology of the Piampaludo field. It is noteworthy to mention that this site was a subject of prospecting projects, which are generally tied with detailed exploration evaluations. Geological and technical data on the Piampaludo Titanium ore are accessible for the public.

According to UNFC-2019, product quantities of prospecting projects are classified as G4. However, the Piampaludo Titanium ore quantity estimates are assessed with solid results from field tests and developments, which allowed the estimations to be at high to moderate levels of confidence. With that said, the Titanium ore has high percentages (90%) to equate the estimated quantities. On the other hand, the garnet quantities are estimated with less certainty because of the fewer available information, rendering the confidence in the estimates at moderate degrees (about 50%). It is critical to mention, for the accuracy and reliability of the assessment, that these estimates are based on data lastly updated in 1991 For the purpose of assessment quality, a conservative approach was followed to classify the Piampaludo project as G3 on the degree of confidence, the G axis.

The classification of the product quantities is prone to higher levels of confidence with newer data acquisition and testing with the implementation of more modern, accurate tools and technologies for ore detection and exploration methods.

### 5.2 *Piampaludo's Social and Environmental Constraints*

It has been established that the economical aspect of the Piampaludo Titanium deposits pertains to high levels of financial attractiveness, yet the ore has not been recovered since its discovery. This Titanium ore is located in a natural preserved region, which caused the project to be tied with social and environmental challenges. The issues have emerged from the conventional social ideologies regarding mining and its harm to landscapes and ecology. All mining operations involved with the Piampaludo project, including exploration, have been ceased right when the region entered the realms of the *Beigua* Natural Park, in 1995. Located in Liguria, the *Beigua* regional natural park is the largest protected park in Italy.[43] Lots of factors have assured its protection but mainly its outstanding geology, from outcrop features to the preserved fossils. The *Beigua* GeoPark intends to conserve its geological history to safeguard the natural heritage covered by this area; this aims at protecting the environmental resources by managing social and economic development, according to the Italian law on protected areas (L. 394/91). It is noteworthy to mention that the *Beigua* Geopark belongs to the Europe's Natura 2000 network, mainly because it hosts bird species that are on the edge of rarity.[44]

In addition to its geodiversity, the *Beigua* Natural Park amasses a wide range of fauna and flora, with 26 different types of ecosystems. There are considerable

[43] Cellerino (2021).

[44] Cellerino (2021).

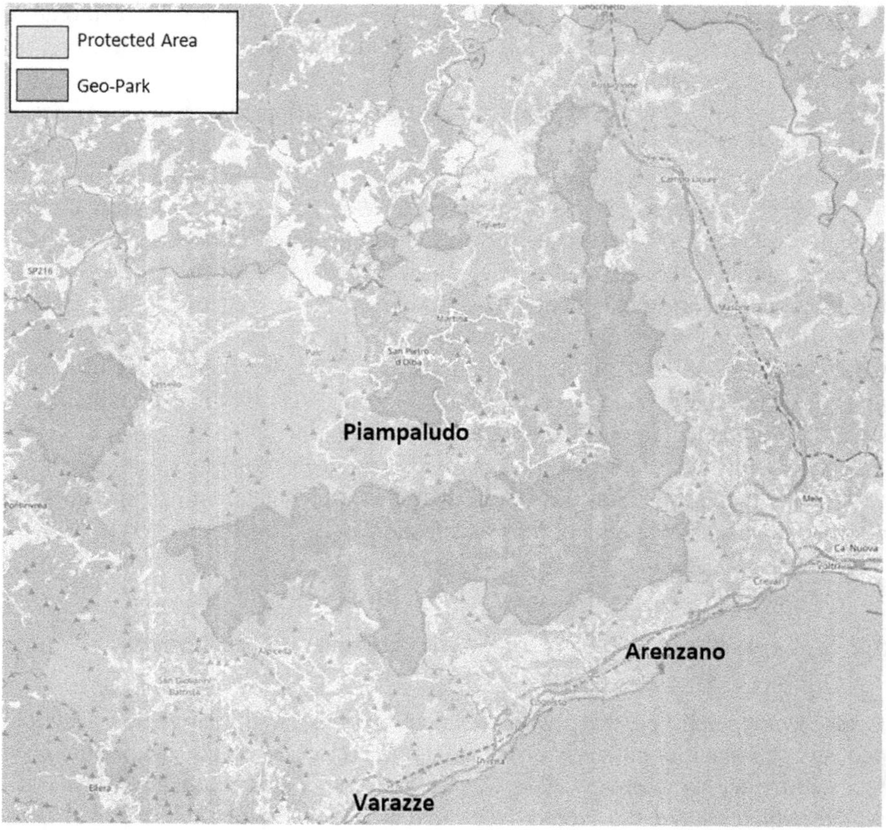


**Fig. 5** *Beigua* Geo-Park (Parcobeigua.it 2022)

wetlands at the park that exhibit significant environmental mediums for many organisms of natural value.

Both, geo- and bio- diversity observed at the *Beigua* Natural Park are focal points for social and environmental concerns to preserve these natural settings. These concerns are also supported by the touristic and historic values of the Park, with geo-tourism museums and roman artifacts. To elaborate more, the social and environmental aspects are well supported from the Ligurian region, neighboring municipalities, and environmental activists and associations opposing any mining operations within the realms of the *Beigua* Park. Another environmental concern factoring in is the fact that the Piampaludo Titanium are partially intercepting the Natura 2000 borders of the region (Fig. 5). As a consequence, all the social and environmental factors involved in protecting the area have forced excavation in Piampaludo to cease for the moment. For that purpose, several legal actions were

**Table 5** Permitting timeline of the Piampaludo project with respect to E axis

| Period | Event | E axis category |
|---|---|---|
| *1976* | Mineraria Italiana Srl granted mining concession | E1.3 |
| *1985* | Mining concession transferred to C.E.T srl | E1.3 |
| *1991* | C.E.T srl request 20 years concession renewal from the Ministry of Industry, no answer to date | E2 |
| *1995* | Area included in the Beigua Park any mining or related activity was interrupted | E3.3 |
| *2014* | Court objects to C.E.T's request to obtain new mining research permit | E3.3 |
| *2015* | Court objects to C.E.T's appeal for the annulment of the inadmissibility of the application relating to the mining research activity | E3.3 |
| *2015* | Court rejects C.E.T's request to activate the procedure for the EIA for geological surveys | E3.3 |
| *2020* | Court rejects C.E.T's plea to challenge court's decision, entrusting the recourse to Violation and/or False Application of Law | E3.3 |
| *2021* | Regional Council of Liguria confers in favor of C.E.T, granting the research permit on the mainland of solid minerals (Titanium, garnet and associated minerals) for 3 years | E3.2 |

taken by the opposing stakeholders against mining companies interested in this deposit. [45]

On a different note, it is interesting to mention that, according to the European Commission (EC), mineral exploration is yet possible in Natura 2000 sites, given that the resiliency of ecosystems and the integrity of the entire site are not severely disturbed.[46] The conditions to mine in a Natura 2000 site are often satisfied with near- and long- term strategic planning of the exploitation operations, with the inclusion of all stakeholders to avoid future disagreements and remain in line with sustainable development principles and policies.[47]

## 5.3 *E3.2 Classification on the E Axis, the Environmental Socio-Economic Viability*

The Piampaludo Project is currently in hold of exploration permits, which was granted to C.E.T srl by the regional council of Liguria in 2021. Table 5 represents the chronological events related to the Piampaludo field exploration, with respect to the socio-environmental-economic viability of the project. The Piampaludo project began when 20 years of mining concession was approved for Mineraria Italiana Srl to recover Titanium in 1976. The concession was transferred to the current owners

[45] T.A.R. Liguria (2020).

[46] European Commission (2019).

[47] European Commission (2019).

C.E.T srl in 1985. Since 1991, C.E.T srl has been requesting for additional 20 years of mining licenses from the Ministry of Industry in Italy, yet the social and environmental constraints have delayed the procedure.[48] The Piampaludo area was a subject of various regulations for its protection, e.g. Establishment of the *Beigua* Regional Natural Park by regional law no.16, 9th of April 1985. In parallel, the Ligurian regional law "n.12/95" structured and supported its protective functions over the land included in the *Beigua* Park, approximately 50% of the Piampaludo field, which was originally the targeted mining basin. For over 35 years, the land has become an environmentally protected area for its nature and ecosystem.[49]

With respect to the series of events, legal dispute resurfaced in the fall of 2014 when C.E.T srl applied for a new exploration permit for the Piampaludo titanium, garnet and associated minerals. [50] Currently, the project is approved for exploration rights only on the half (around 229 ha in surface area) that is not included within the *Beigua* Park domain, under act n. 1211 of 02/26/2021. [51] C.E.T. srl has complete exclusivity for exploration. Additionally, this permit allows C.E.T srl to develop preliminary studies with the intention of evaluating the areal and surface distribution, and to determine the rutile concentrations present in the indicated half. The exploration permit is granted till 2024. [52]

For the environmental standpoint, C.E.T's planned investigation activities and studies do not alter the natural state of the field under study. The exploration strategy is planned with minimal negative environmental impacts and landscape disturbance; C.E.T srl will not disturb any natural processes of the site, and will not include damaging technics, e.g. drilling and coring. Sampling shall be carried out on foot, as well as the geological and structural surveys and mapping, using the existing trails.[53] C.E.T is likely to undertake EIA and SEA to identify and mitigate environmental concerns, and solidify their case for future recovery.

From the economic side, Piampaludo's Titanium deposits certainly exhibit great economical interest, with the measured ore-grade and weight. The Piampaludo project is estimated to generate approximately 500 M€ in royalties to the region of Liguria, with the deposits valued between 400 and 600 B€. With the current value of rutile per ton and the rutile grades at Piampaludo, the deposit are estimated to be valued at minimum 120 B€.[54]

At present, the development of the project is not assured to be environmentally-socially-economically viable due to insufficient information to impede the project. On the other hand, due to the ongoing resistance towards the project and the recent permit granted to C.E.T, it is premature to ascertain the project's

[48] Cellerino (2021).

[49] Cellerino (2021).

[50] T.A.R. Liguria (2020).

[51] T.A.R. Liguria (2020).

[52] T.A.R. Liguria (2020).

[53] Gorga (2004).

[54] Cellerino (2021).

environmental-socio-economic viability. Accordingly, the project is classified as E3.2 on the E axis.

### 5.4 F2.2 Classification on the F Axis, the Technical Feasibility

The necessary feasibility studies and scoping for the Piampaludo Titanium project are well highlighted in the preliminary plan proposed by C.E.T srl (Table 4). In effect, these studies are a point of reference for the upcoming exploration scope by C.E.T srl with their newly claimed exploration right. Granted that the Piampaludo remains an exploration project, the prefeasibility planning provides sufficient information on the probability for progression in developments while upholding a clear resource definition. This allows the project to be categorized with more advanced classes for the technical feasibility. Nonetheless, the project's technical feasibility needs to be supported with more studies and tests.

At present, activities for the advancement in development at Piampaludo are yet pending for the justification by the sustainable viability of the project, which may result in significant delays for development.

On the other hand, C.E.T's preliminary developments include site advancements in early stages such as shaft and adits. These developments have provided substantial data on the Titanium ore, which is considered a promising prospect. The stakeholder's willingness to invest increases for that matter, rendering the Piampaludo project more outstanding.

In addition, C.E.T demonstrates relatively high production rates in their mining plan (approximately 163,000 t/y), with a mine life of 90 years. It is clear that the most viable excavation method apt for the Piampaludo project is through drill and blast in open-pit mining. However, with Piampaludo's current circumstances, this mining method does not comply with the site laws.

Thus, the Piampaludo Titanium project has been given a conservative classification for the technical feasibility, F2.2 on the F axis.

## 6 Exploration with an Alternative Solution

The classification of the Piampaludo Project according to UNFC informs the stakeholders on the status of the project with regards to its environmental-socio-economic viability, technical feasibility, and degree of confidence in the resource estimates. However, the question remains on the potential of Titanium recovery from the Piampaludo project, while respecting the sustainable development. The application of conventional mining techniques does not comply with the project's current sustainable scope, given the social and environmental oppositions.

Nowadays, the mining industry has invested in the research of new, green mining methods and technics to respect the sustainable development. Biomining has emerged in recent years as a potentially green mining method with economic, environmental, and social benefits.

In brief, biomining is built on the premise of extracting the ore through the natural interaction between metals and microorganisms e.g. bacteria. In the appropriate setting, certain bacteria have the ability to leach on metals as they dissolve them. This method is highly encouraged for critical metals recovery, especially in environmentally-sensitive areas. The entire mining operation includes metal processing simultaneously; the bio-dissolved metals are extracted as a pregnant solution from the medium through a filtering process.[55]

Environmentally speaking, in-situ applications of biomining do not consist of the physical excavation of the ore, meaning that disruption of the site's geodiversity is minor. In the context of the Piampaludo Titanium deposits, biomining can potentially preserve the essence of its Geo-Park, where the lithology remains untouched. In fact, this method is seen to be more environmentally convenient in comparison to conventional methods. Biomining only produces substances from bacterial activities, such as metabolites. Furthermore, in-situ biomining covers less volumes, which marks a decrease in pollution risks for the surface, air, and waters. As a result, biomining can take place in sensitive areas with a range of climate conditions. However, an environmental concern associated to this method involves the risk of diffusing by lose the injected solutions into the surrounding settings, e.g. rivers or lakes that may get contaminated, but this matter can be carefully mitigated with proper studies and planning. It is noteworthy to mention that biomining is primarily used as a remediation technique for environmental purposes, which means that its purpose is to clean the medium from contaminants such as oils and toxins. With that in mind, biomining serves to protect the existing biodiversity within the targeted medium. Another environmental advantage linked to biomining is that it does not emit sulfur dioxide.

The application of biomining at Piampaludo may be more socially accepted in comparison with conventional techniques, since the main opposing point is the harm that mining causes to the biodiversity and geodiversity.

From the economic perspective, biomining envelops two operations under the same process, extraction and processing, which leads to significant reductions in development and transportation costs, and in return, energy and infrastructure cuts. In comparison to conventional processing methods, in-situ biomining is more cost-efficient. The implementation of biomining can initiate a project with low capital investments with relatively high production rates. In the context of production, this method allows control over the capacity in production, meaning that the operation is flexible to stock prices.

The recovery of the Piampaludo Titanium deposits can be potentially activated with biomining in the foreseeable future. Given the current factors, biomining could place the Piampaludo project at better environmental-socio-economic viability classes in the UNFC.

---

[55] Angloamerican.com (2021).

On the other hand, in-situ biomining must be assessed taking into account the precautionary principle. This method is yet not well established in the mining world and the introduction of new substances to an existing ecosystem requires further testing in order to avoid hazards.

Alternatively, using the existing microorganisms in the medium could be possible by injecting the catalyzing substance to improve microbial activities: hence, biomining. However, the project must be scaled down to a E3F3G2 classification according to UNFC, because it remains as a concept plan where no developments occurred.

## 6.1 Biomining: A Nature-Based Solution

With the implementation of biomining as resource recovery method, the Piampaludo project can be perceived as a solution to promote the in-place biodiversity. The introduction of the proper injects in the medium, e.g. with the appropriate pH levels, the microbial life can develop in better, enriched conditions, which enhances and accelerates the Titanium recovery present in the medium. This approach reduces the risk of the precautionary principle by only enriching the existing microbial biome. At the same time, the surrounding organisms of the ecosystem suffer no damage, and use the increase of microbes for a symbiotic interaction. The amelioration of microbial living conditions is associated with positive impacts to the biodiversity and the wildlife. Microorganisms are essential for a strong biome and a healthy immune system.

By improving the biodiversity in Piampaludo and recovering the Titanium, the Piampaludo project becomes a nature-based solution. Other than supporting the biodiversity, this nature-based can benefit mankind. The recovery of Titanium from Piampaludo serves a great deal for the health sector, since it is essential for the production of prosthetics, heart-valves, and medicines. Eventually, extracting the Piampaludo Titanium saves humans and animals. Additionally, the nature-based solution improves the sustainable prosperity in the context of carbon neutrality and economy. Approaching the Piampaludo project as a nature-based solution with application of biomining becomes a service to humankind, animals, biodiversity, and the environment.

Orienting the Piampaludo project in the direction of a nature-based solution to recover the Titanium as a service must be managed by a comprehensive resource management system to guarantee a responsible governance over the project life cycle. The United Nations Resource Management System (UNRMS) is capable of delivering the project while staying in accord with the Agenda 2030. Applying UNRMS principles to the Piampaludo project is the sustainable way forward, in order to demonstrate alternatives means of proceeding with the project with respect to the environmental-socio-economic viability. UNRMS principles provide reasonable requirements for the sustainable management of a project. UNRMS is an extension to the UNFC. UNFC provides information on the project's maturity,

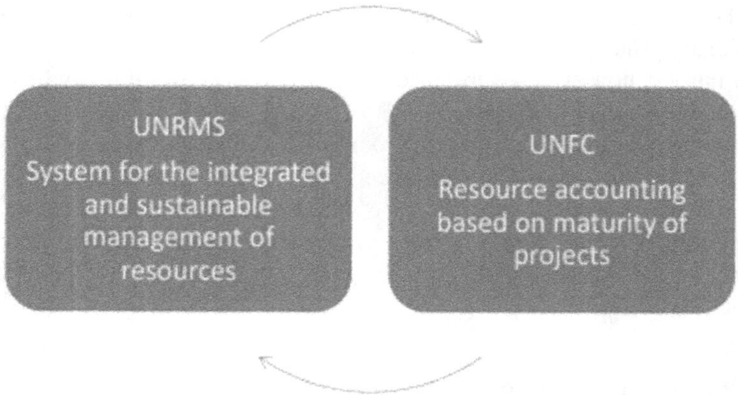


**Fig. 6** UNRMS, UNFC and project-level application (United Nations 2020)

whereas UNRMS works towards improving the project's maturity. UNRMS is a principle-based framework with specifications and guidelines that consider the dynamic timeline and space of a resource project while responding to the challenges in terms of social and environmental impacts. UNRMS is developed to guide stakeholders the sustainable pursuit of their goals.

In the context of the Piampaludo project, UNRMS offers an exhaustive methodology to social and environmental challenges facing the project by transforming the Titanium recovery as a public good. With the application of UNRMS, social acceptance is prone to increase, which results in the sustainable development of the project with the necessary investment. The classification of the Piampaludo project according to UNFC provided the information related to Piampaludo's challenges and opportunities. In parallel, UNRMS assists as a toolkit in the development of the project over its life time while complying with the UNFC data (Fig. 6).

## 7 Conclusion

Piampaludo's Titanium deposits located in Liguria, Italy, are one of the greatest Titanium deposits in Europe. The project offers substantial production potentials. From an economic point of view, the estimated ore grades and weights of the Piampaludo Titanium deposits have the ability to remarkably reduce Titanium supply chain disruptions in Italy and across Europe. Moreover, the development of the Piampaludo project can seriously contribute to the Ligurian economy, and to the possible extent, to the Italian GDP.

On another note, the Piampaludo Titanium deposits are partially located in the *Beigua* Regional Natural Park, that is acknowledged by UNESCO as a Geo-Park and pertains to the Natura 2000 network. The natural characteristics of the *Beigua* Park have raised various challenges with regards to mining activities within the park realms, including the available Titanium deposits. Social and environmental constraints have forced the Piampaludo project to refrain from moving forward. From a social perspective, the opposing associations, municipality, and neighboring communities stand against mining in Piampaludo, due to the elevated risks associated with disturbing the natural landscapes and ecosystems of the park.

The Piampaludo project was classified according to UNFC to highlight the contingencies linked to it. The classification takes into account all environmental-socio-economic aspects, technical maturity of the project, and the estimated quantities of the ore. UNFC addressed all issues related to the Piampaludo Titanium deposits, from a sustainable standpoint. UNFC is a resource reporting tool that gives stakeholders information on the project's maturity. The Piampaludo project was given an E3.2F2.2G3 classification according to UNFC. The classification UNFC provided an evaluation of Titanium quantities and the project maturity status. In addition, the UNFC assessed the sustainable aspects tied to the Piampaludo project, with respect to permitting granted by the Ligurian municipality to C.E.T srl. Thus, all the information associated with the data highlighted in the UNFC covered the full understanding of Piampaludo's constraints and opportunities.

In addition to the classification, a solution for the Titanium recovery at Piampaludo while complying with the circumstances and the sustainable development, was proposed. Introducing biomining as a potential alternative may be the way forward for the development of the Piampaludo project. While causing no harm to the natural landscape and lithology, this method can recover Titanium through the use of microorganisms. It is certain that biomining is more environmental in comparison with conventional mining methods. The application of this method could improve microbial life in the ecosystem, which in return enhances the biodiversity of the area through natural symbiotic relations.

The Piampaludo project could be perceived and established as a nature-based solution, using the Titanium recovery as a service. This approach falls under the principles of UNRMS. UNRMS system provides guidelines and rules to support resource extraction in line with the sustainable development. With the use of existing microbial life present in the medium, their activities can be enhanced and accelerated for the recovery of the resource. UNRMS is important for Piampaludo project because it builds reasonable means of dealing with the social and environmental challenges facing the project, while recovering the Titanium from the site as a public good. With this sustainable approach, all stakeholders can convert to accept this plan to develop the project.

Lastly, in addition to UNRMS guidelines and practices, more technical tests need to be held, on site and in laboratories for the proper application of Biomining on the Piampaludo Titanium. Biomining remains in preliminary phases worldwide. Sampling the eclogites hosting the Titanium ore is a crucial step to evaluate various factors implicated in the biomining process. The necessary tests include the evaluation of the rock hardness, permeability, and the bacterial life available on field. The latter serves to select the proper solution and its composition for optimum results of metal recovery. As mentioned, proper method planning and investigation lead to the sustainable recovery of the Piampaludo Titanium deposits.

## References


Africa Renewal (2022) How can mining contribute to the Sustainable Development Goals?. https://www.un.org/africarenewal/news/how-can-mining-contribute-sustainabledevelopmentgoals#:~:text=How%20can%20mining%20contribute%20to%20the%20Sustainable%20Development%20Goals%3F,By%3A&text=Mining%20companies%20have%20the%20potential,growth%20in%20low%2Dincome%20countries. Accessed 22 May 2023

Angloamerican.com (2021) Bioleaching Definition & Process. https://www.angloamerican.com/futuresmart/stories/our-industry/mining-explained/mining-termsexplained-a-to-z/bioleaching-definition-andprocess#:~:text=Bioleaching%20(or%20biomining)%20is%20a,such%20as%20bacteria%20or%20archaea. Accessed 22 May 2023

Capponi G (2002) Structural and metamorphic signature of alpine tectonics in the Voltri Massif (Ligurian Alps, northwestern Italy). Ecl Geol Helv

Cellerino F (2021) Titanio, il giacimento nel parco ligure del Monte Beigua che fa gola da 45 anni - Il Fatto Quotidiano. Il Fatto Quotidiano. https://www.ilfattoquotidiano.it/inedicola/articoli/2021/04/13/titanio-il-giacimento-nel-parco-ligure-del-monte-beigua-che-fa-gola-da-45-anni/6161673/. Accessed 22 May 2023

CRM Alliance (2021) Critical Raw Materials. https://www.crmalliance.eu/critical-raw-materials. Accessed 22 May 2023

European Commission (2017) Directorate-General for Internal Market, Industry, Entrepreneurship and SMEs, "Methodology for establishing the EU list of critical raw materials: Guidelines". Publications Office, Luxembourg

European Commission (2019) Directorate-General for Environment, "Guidance document on non-energy mineral extraction and Natura 2000: a summary", Publications Office. https://data.europa.eu/doi/10.2779/985239. Accessed 22 May 2023

European Commission (2020a) Commission Staff, "Stepping up Europe's 2030 climate ambition Investing in a climate-neutral future for the benefit of our people". https://eurlex.europa.eu/legal-content/EN/TXT/?uri=CELEX:52020SC0176. Accessed 22 May 2023

European Commission (2020b) Energy efficiency in buildings – consultation on 'renovation wave' initiative. Publications Office. https://ec.europa.eu/info/law/better-regulation/haveyour-say/initiatives/12376-Commission-Communication-Renovation-wave-initiative-for-the-building-sector. Accessed 22 May 2023

Federico L et al (2004) Exhumation of alpine high pressure rocks: insights from petrology of eclogite clasts in the Tertiary Piedmontese basin (Ligurian Alps, Italy). Lithos

Federico L et al (2014) Geology of the Pontinvrea area (Ligurian Alps, Italy): structural setting of the contact between Montenotte and Voltri units. J Maps 2014. https://doi.org/10.1080/17445647.2014.945749. Accessed 22 May 2023

Force E (1991) Geology of titanium-mineral deposits. Geol. Soc. Spec Paper 259

Gorga R (2004) Pseudomorfosi di rutilo da cloromelanite nelle rocce eclogitiche di Monte Tarme presso S. Pietro d'Olba (Appennino settentrionale). Springer. https://link.springer.com/article/10.1007/BF02904459?noAccess=true. Accessed 22 May 2023

Iberdrola (2022) What is the 2030 Agenda for Sustainable Development?. https://www.iberdrola.com/sustainability/committed-sustainable-development-goals/what-isagenda-2030. Accessed 22 May 2023

Jolović B (2020) Case Study on the Application of UNFC to energy and groundwater resources in the Republic of Srpska, Bosnia and Herzegovina. Unece.org. https://unece.org/fileadmin/DAM/energy/se/pdfs/UNFC/proj/integrated_water_RM/Case_studies/UNFC_Case_Study_Republic_of_Srpksa_Bosnia_and_Herzegovina__2020.pdf. Accessed 22 May 2023

King H (2022) Uses of titanium metal and titanium dioxide. Geology.com. https://geology.com/articles/titanium/#:~:text=Titanium%20is%20a%20familiar%20metal,weights%20about%20half%20as%20much. Accessed 22 May 2023

Liou J et al (1998) Mineral parageneses in the Piampaludo eclogitic body, Gruppo di Voltri, Western Ligurian Alps. Schweiz Mineral. Petrography Mitt

Liu S et al (2013) A flexible TiO2(B)-based battery electrode with superior power rate and ultralong cycle life. SCBE Journal Articles. https://hdl.handle.net/10356/102844. Accessed 22 May 2023

Mancini A et al (1979) Valorization of new titanium resource; titaniferous eclogites. Mineral Economics, World Mining Congress

Overton G (2018) 2D iron-ore material hematene improves solar-fuel generation potential. Laser Focus World. https://www.laserfocusworld.com/detectors-imaging/article/16571785/2dironore-material-hematene-improves-solarfuel-generation-potential. Accessed 22 May 2023

Parcobeigua.it (2022) Parco Naturale Regionale del Beigua. http://www.parcobeigua.it/. Accessed 22 May 2023

Raw Materials Information System (RMIS) (2022) CRM list 2020. Rmis.jrc.ec.europa.eu. https://rmis.jrc.ec.europa.eu/?page=crm-list-2020-e294f6. Accessed 22 May 2023

T.A.R. Liguria (2020) "European Company for Titanium - C.E.T. S.r.l. c. Liguria Region and a. Environment Mining research for titanium, garnet and associated minerals within an area Protected", vol 200

The Diggings (2022) Piampaludo Titanium Prospect. https://thediggings.com/mines/usgs10255368. Accessed 22 May 2023

Titanium Dioxide Manufacturers Association (2020) The role of titanium dioxide in enabling renewables and energy efficiency. tdma.info. https://tdma.info/the-role-of-titanium-dioxide-inenabling-renewables-and-energy-efficiency/. Accessed 22 May 2023

United Nations (2020) United Nations Economic Commission for Europe (UNECE), "United Nations Resource Management System: An overview of concepts, objectives and requirements"

United Nations Economic Commission for Europe (UNECE) (2019a) United Nations Framework Classification for Resources Update 2019

United Nations Economic Commission for Europe (UNECE) (2019b) Application of the United Nations Framework Classification for Resources Case studies. Unece.org. https://unece.org/sites/default/files/2020-12/E_ECE_ENERGY_109_WEB.pdf. Accessed 22 May 2023

United Nations Economic Commission for Europe (UNECE) (2019c) United Nations Framework Classification for Resources and Sustainable Development Goals. Unece.org. https://unece.org/sustainable-energyunfc-and-sustainable-resource-management/unfc-and-sustainable-development-goals. Accessed 22 May 2023

United Nations Sustainable Development (2022) Climate Change - United Nations Sustainable Development. https://www.un.org/sustainabledevelopment/climate-change/. Accessed 22 May 2023

# A Sustainability Approach to Optimize the Closure of Continuous Surface Lignite Mines

**Christos Roumpos, Francis Pavloudakis, and Zach Agioutantis**

## Contents



C. Roumpos
Public Power Corporation of Greece, Department of Mining Engineering and Closure Planning, Athens, Greece
e-mail: c.roumpos@dei.gr

F. Pavloudakis
University of Western Macedonia, Department of Mineral Resources Engineering, Kozani, Greece
e-mail: fpavloudakis@uowm.gr

Z. Agioutantis (✉)
University of Kentucky, Department of Mining Engineering, Lexington, KY, USA
e-mail: zach.agioutantis@uky.edu


W. Frenz, A. Preuße (eds.), *Yearbook of Sustainable Smart Mining and Energy 2022*, Yearbook of Sustainable Smart Mining and Energy - Technical, Economic and Legal Framework 2, https://doi.org/10.1007/978-3-031-41873-0_3

**Abstract** Mine closure is a crucial procedure of every mining operation and should be planned from the early beginning of the mine life cycle to ensure sustainable mining and post-mining operations. In this framework, the sustainability goals require an integrated mine planning of all mining stages, from the exploration to the environmental reclamation of mining works and the post-mining land uses. Furthermore, considering the last stages of a mining project, knowledge obtained from the mine life cycle and its management is essential for providing information on all aspects of closure planning and execution. Therefore, the closure objectives should be defined early, during the initial stages of a well-organized closure plan, and incorporated into the relative mine planning stage.

In the present study, a sustainable closure mine approach is investigated for the continuous surface lignite mines by considering the optimization of the lignite production in combination with the environmental reclamation of mining areas and the spatial planning of post-mining land uses and by evaluating socio-economic factors related to the mining operations. In the optimization context, emphasis was placed on the viable development of infrastructure to ensure sustainability. The approach was based on a statistical analysis regarding the correlation of social and economic indicators with lignite production and a dynamic optimization model for the remaining surface mining operations. More specifically, population, employment, gross domestic product, and gross added value are some of the social factors that were investigated for the broader area of the Ptolemais mines in Western Macedonia, Greece. The closure objectives include the identification of the potential risks and opportunities, the designing of closure activities, the estimation of closure costs, the evaluation of the successful closure activities against closure objectives, as well as the identification of the necessary supporting actions for the mine closure and post-mining land uses.

## 1 Introduction

Sustainable development is the development that meets the needs of the present without compromising the ability of future generations to meet their own needs. [1] In this context, sustainability deals with the intrinsic uncertainty that characterizes the future. [2] A complex, but very important dimension of sustainable development is the set of principles that applies to large-scale projects and activities, like the exploitation of a large lignite field for electricity production. The anticipated impacts of such an operation should be considered throughout the mine life cycle to achieve sustainable management. It is a critical goal for large-scale surface mining projects with a long life span, such as coal extraction by continuous surface mining. The sustainability assessment of surface coal mines constitutes a multiparametric

[1] United Nations (1987).

[2] Hardisty (2010).

process and depends on multiple factors. They may refer to optimal exploitation of the mines under operation and the optimal mineral extraction in the closure phase, efficient use of lignite resources, efficient transportation of lignite to power plants and waste to dumping sites (optimization of the transportation network), environmental management, and, ultimately, sustainable socio-economic development. Additionally, the optimal utilization of the exhausted mines, the optimal post-mining land uses, the smart water management, and the optimal utilization of mining equipment are some of the sustainability principles that should be applied during the mine closure procedures.

In recent years, the sustainability aspect in the mining industry has been continuously growing, considering the environmental principles that should accompany such processes. Environmental, economic, and social issues are considered sustainable development aspects. [3] The study of Burchart-Korol et al. (2014) [4] focused on the development of an algorithm that combines environmental and socio-economic parameters to quantify sustainability, and a respective ranking of sustainability in coal mines was developed.

The sustainability aspects are expressed through specific indicators which constitute the monitoring tools of sustainability that governs the mining processes. Sustainable development indicators in mining represent the tools used by mining companies to assess the sustainability performance of mining procedures. [5] The mining sectors where the sustainability principles should be generally applied include the exploitation of mineral and natural resources, mining operations, environmental management, and socio-economic development. More specifically, the sustainable development indicators of surface mining projects could focus on seven fundamental pillars: (a) optimal utilization of mineral resources, (b) waste management, (c) water management, (d) utilization and management of by-products, (e) land use management, (f) equipment management, (g) management of mine infrastructure. In addition, the temporal evolution of socio-economic parameters, such as regional domestic product (GDP), employment, and investment in exploration related to coal mining operations, is significant since they can be used as indicators of the degree to which the extractive industry contributes to the achievement of some of the 17 goals of sustainable development, such as no poverty, zero hunger and decent work. [6] Furthermore, the mining parameters could be controlled by respective sustainable indicators from the environmental point of view, such as the environmental reclamation of the mining areas, the thermal energy provided to the cities in close vicinity to the mines and power plants, the economic sources for archaeological excavation in the mining areas, or the use of pumping water for irrigation of the areas near mining operations.

---

[3] Burchart-Korol (2011), pp. 107–110; Que et al. (2021), 11800.

[4] Burchart-Korol et al. (2014), pp. 5–11.

[5] Azapagic (2004), pp. 639–662.

[6] United Nations A/RES/70/1 (2015).

This chapter aims to investigate a way (methodology) to assess the sustainability of continuous surface lignite mining operations in their closure phase and to demonstrate a corresponding optimization model considering the lignite production combined with the environmental reclamation and the post-mining land uses. Section 2 presents the main aspects of sustainability in continuous surface lignite mining projects. The third section describes the materials and methods of the analysis. The fourth section demonstrates and discusses the analysis results regarding the case study of West Macedonia lignite mines. Finally, Sect. 5 provides the conclusions of the chapter.

## 2 Sustainability during Closure of Continuous Surface Lignite Mines

Continuous surface lignite mining operations are long-term projects that face many challenges related to sustainable development and require integrated mine planning and management. The main challenges are associated with environmental, economic, technical, and social conditions and their interrelations.

Figure 1 shows a multidimensional approach to the continuous surface lignite mining projects, describing many crucial issues related to the closure phase of mine operations, especially in the cases of an early mine closure before the depletion of the lignite reserves. In this context, mine planning is considered the core activity for the optimization of mineral extraction, taking into account all mining phases, from the exploration of the remaining deposits and the development of optimization models to the design and scheduling of the main mining operations (excavation, transportation and dumping) through continuous (bucket wheel excavators, belt conveyors and spreaders), non-continuous (mainly backhoe excavators and trucks) or a combination of continuous and non-continuous mining equipment.

Mine planning also includes mine closure and post-mining reclamation, waste management, planning for new land uses, investigating new water uses, optimizing scheduling and utilization of mining equipment, planning the installation of renewable energy projects in the reclaimed mining areas or optimizing the utilization of exhausted mine pits in areas outside pit lakes. In an integrated approach, environmental considerations, social assessment, risk analysis, economic interests, legal applications, circular economy and sustainability issues, geotechnical investigations, knowledge management and mining and post-mining hazards should also be incorporated into the mine planning model.

The sustainability issues of surface coal mines include multiple factors. They may refer to the sustainable exploitation of mineral and natural resources, mining operations, environmental management, and socio-economic development. The environmental impact assessment analysis may combine with the production planning strategy aiming at the sustainable planning and development of the mining

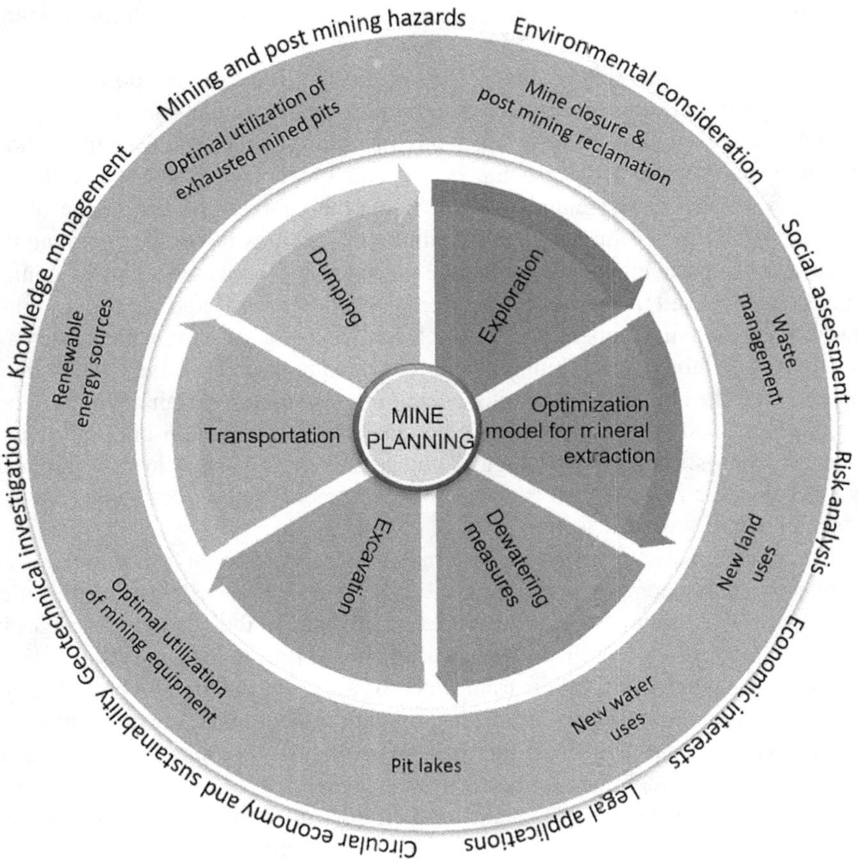


**Fig. 1** The basic elements of an integrated approach to the closure of continuous surface lignite mining projects

project throughout the entire mine life cycle, which ends with the mine closure phase. [7]

Mine closure, including decommissioning of retired facilities, has been rated among the top five operating risks in mining. [8] Mine closure could be attributed to resource depletion, the economic unprofitability of the resource extraction or a much lower market price, or the high safety risk of mining-related activities. [9] In this context, the sustainability of mine closure is achieved by considering depletion components. [10] Tools that are used for monitoring and managing performance in

[7] Roumpos and Papacosta (2013), pp. 645–651.

[8] World Risk Report (2018).

[9] Liu et al. (2021), p. 13473.

[10] Central Mine Planning and Design Institute (CMPDI) - Ministry of Coal/Sustainable Development Cell (2021).

mining operations are the key performance areas (KPAs) and key performance indicators (KPIs). [11]

Although the term "mine closure" implies the end of the mining operations, the preparation of mine closure constitutes a complex procedure that should be planned from the early beginning of the mining activities and should be followed throughout the mine life cycle to achieve complete sustainability. [12] The early integration of mine closure into mine planning would enable the continuous supervision of the procedures and the evaluation of its feasibility, as well as it would encourage the community to participate. [13] It could take place during the whole series of mining studies and related projects: exploration concept/scoping study, pre-feasibility study, feasibility study, operation phase, closure, and post-closure, with an increasing level of certainty as the mining operations, tend to end. [14] If mine closure starts too late after the beginning of mining operations, the mining sustainability will not be environmentally and socioeconomically guaranteed. [15] Mine closure is attributed to several causes: economic, geological, and geotechnical reasons due to equipment or mechanical failure, regulatory pressure, government policy, community opposition, and due to several reasons. [16]

The lignite mines closure period also includes the preparation for the anticipated changes in the economy at a mining area. In particular, the workforce of the cities in the vicinity of mining activities needs to be informed of the upcoming changes in the local economy and supplied with new skills in due time to meet the requirements of new working activities. In this framework, as a transformation of a broader mining area into a non-mining economy, the population should be adequately qualified for new types of employment. In this way, economic transformation will be gradually obtained. Although the mining industry increases the economic growth of the area, it also shapes human personalities, their way of thinking, and their culture. Therefore, during the phase-out of the lignite mining projects, the inhabitants who are employed in the mining sector should be trained to change their traditional thinking about a long-life working in the mining sector and open their minds toward new sustainable perspectives. The latter presupposes holistic support from the local authorities and stakeholders. For example, interactive educational seminars could be organized related to renewable sources, agricultural development, forest development, or other future activities.

In this framework, attention must be drawn to the spatial management of areas where coal mines in the closure phase are located, and unexploited coal deposits will remain. In these cases, the conceptual framework of circular economy is ideal

[11] International Council on Mining and Metals (ICMM) (2020).

[12] Yaylacı and Düzgün (2017), pp. 837–849.

[13] International Council on Mining and Metals (ICMM) (2019).

[14] Finucane (2008), pp. 171–182.

[15] Finucane (2008), pp. 171–182; International Institute for Environment and Development (IIED) (2002).

[16] Laurence (2006), pp. 285–298.

for planning post-coal management of facilities and infrastructure. Pactwa et al. (2021) [17] indicate directions for sustainability-oriented actions on reusing the remnants of closed mines and reshaping the mining areas, which will be transformed under the EU energy transition strategy. This issue involves numerous legal implications related to the process of closing a mine. In addition, it is related to three major post-mining problems: (i) mine pits must be safely closed, (ii) as mine facilities and buildings may be reused they must be included in all post-mining development plants, and (iii) mining waste is still a probable threat for the environment and requires more attention due to the legal implications that may arise.

Pactwa et al. (2021) [18] also present an approach that defines post-industrial facilities as a resource. This is consistent with the principles of sustainable development and extends the meaning of a circular economy. Furthermore, they noticed that, due to the high emissions allowances cost, coal-depended energy enterprises should be expected to undergo liquidation. This development may increase the number of abandoned mines and industrial facilities. Therefore, the early preparation of solutions to the related social, environmental, and economic challenges may contribute to implementing good practices at the mine closure stage. This is particularly true in cases that have not been managed comprehensively and systematically so far.

On the same subject, Garvey et al. (2022) [19] claim that spatial management of coal mining areas in the closure phase involves the fair distribution between regions of both benefits and burdens associated with energy transition. Just transition strategy must go beyond the mitigation of employment impacts. Nevertheless, the contents of this strategy are determined by the perceptions of policymakers and key stakeholders about justice. Thus, there is concern that the energy transition may not address or reproduce existing patterns of injustice. This is particularly true in regions where the land planning policy was historically spatially inequitable and has led to lasting socio-economic impacts. To make energy transition more spatially fair is necessary to stop administrative fragmentation across spatial scales and lack of coordination. At the same time, policymaking interventions must be spatially targeted, always considering the needs of the entire area under investigation.

As the most crucial part of spatial management of mining areas in the closure phase, the reclamation of mine surfaces is decisive for the post-mining era. Land suitability evaluation regarding the selection of the most appropriate land uses could be based on suitable algorithms and a multi-objective optimization to deal with the complexity of spatial constraints. [20] Also, ranking systems have been developed for sustainable mine reclamation, which combines the environmental with the socio-economic parameters. [21] A significant step for planning post-mining land uses is the investigation of their spatiotemporal changes in the mining area and the correlation

[17] Pactwa et al. (2021), 167.

[18] Pactwa et al. (2021), 167.

[19] Garvey et al. (2022), 102630.

[20] Palogos et al. (2017), pp. 491–498.

[21] Pavloudakis et al. (2020), 3995; Spanidis et al. (2021); Spanidis et al. (2022a), 713.

with the technical or other mining parameters. [22] Regarding the new land uses of mining areas, renewable energy sources installed in reclaimed mining land could be considered sustainable. [23] Although many studies have been conducted in the context of sustainable land uses, the repurposing phase of mines has been less investigated related to other phases regarding the development of sustainability indicators. [24] During the closure phase of the mining operations, emphasis should also be given to the available mineral deposits regarding their management and planning.

Water management in operating and exhausted mines is also a crucial issue that defines sustainability during mining operations and the post-mining phase. The water pumping of mines may have environmental implications, an issue that concerns the mining company and the planning and licensing authorities. [25] Thus, water management is followed by European legislation, which the European Parliament and Council adopt in the form of directives (water framework directive, groundwater directive, and environmental quality standards directive). The management of the final pit lakes is also a crucial phase of mining water management. Related research was conducted regarding the mine pit lakes development in the Amyntaion and Kardia exhausted lignite mines in Western Macedonia, Greece, to define the land use repurposing decisions. [26] In the corresponding studies, the authors concluded that the broader areas around the lakes are suitable for several new post-mining projects, including renewable energy sources. In addition, suitability surveys have been employed for the development of a hybrid pumped hydropower storage (HPHS) for the excess energy of the electric power grid [27] in the Kardia lignite mine, as it will contribute to the EU Green Deal. In practice, the open-pit mines can be used as lower or upper reservoirs for the implementation of the HPHS technology.

Decommissioning in mining areas is a necessary procedure, but it has several socio-economic effects. [28] In particular, the local unemployment in the area increases, and, usually, the key economic sector collapses. However, the optimal utilization of mining equipment would prevent these consequences. For instance, the utilization of mining equipment could be optimized by considering moving the equipment to other mining areas to increase work efficiency or by beneficial alternative uses and recycling. In addition, a valuable utilization of part of the equipment in the post-mining period could include an industrial museum for visitors or educational purposes. Also, the decommissioning procedure includes the removal of the mining area of any equipment that is considered a hazard to the total environment. [29]

[22] Paraskevis et al. (2021), pp. 72–89.

[23] Antoniadis et al. (2022).

[24] Perdeli Demirkan et al. (2022), 66.

[25] Stemke and Wieber (2022), pp. 280–291.

[26] Louloudis et al. (2022a), pp. 458–474; Louloudis et al. (2022b), 3558.

[27] Krassakis et al. (2023), p. 593.

[28] Fleming and Measham (2015), pp. 203–216; Parkins and Angell (2011), pp. 39–55.

[29] International Atomic Energy Agency (IAEA) (1994).

Furthermore, the knowledge of by-products extracted from the mining processes is critical and has become part of the general mining strategy. [30] Therefore, it is crucial to ensure economic feasibility by exploiting these by-products during the mining operations phase or after mine closure if they have been suitably dumped. Two main by-products derived from the mining operations in the Greek surface lignite mines are gypsum and ash, which are used for construction materials. Apart from this, the potential of fly ash for developing rare earth elements (REEs) has been investigated. From the mineralogical and chemical analyses, it was derived that the fly ash produced from a power plant has a high concentration of REEs. [31] In addition, the produced gypsum could be used in mineral beneficiation. [32]

Another crucial issue that should be incorporated into the mine closure procedures is minimizing risks. Although this issue should be applied to all mining operations stages, more attention should be paid to the mine closure phase. More specifically, a detailed geotechnical investigation is required to prevent stability issues or other geoenvironmental problems and to ensure successful future land uses (e.g. pit lakes). [33]

Finally, the depletion time of surface coal mines and the long-term prosperity of the local and regional economies is a crucial issue that concerns both the local community and the society at large and constitutes the crucial combination that should be considered in the context of the just transition. Therefore, a transition to an electricity sector that will be based on renewable energy sources presupposes early and well-prepared mine strategic planning. All the scientific aspects need to be concentrated into an integrated system, which constitutes a knowledge management system adaptable to the requirements of sustainability. [34]

## 3 Materials and Methods

### *3.1 Study Area and Investigated Social and Economic Parameters*

The case study refers to the region of Western Macedonia, Greece (Fig. 2), where lignite exploitation began in 1957. Three surface lignite mining areas were developed in this region, in the corresponding lignite deposits of Ptolemais, Amyntaion and Kleidi. The total lignite production until the end of 2022 was 1757 million tonnes, with total excavations of 8783 million $m^3$ and an overall stripping ratio (waste to lignite) of 4.17 $m^3/t$. Mine closures started with the Kleidi mine (2012),

[30] Espí (1999), pp. 37–41.

[31] Tsachouridis et al. (2021).

[32] Vitti and Arnold (2022), pp. 49–54.

[33] Kavvadas et al. (2022), pp. 278–296.

[34] Spanidis et al. (2022b).

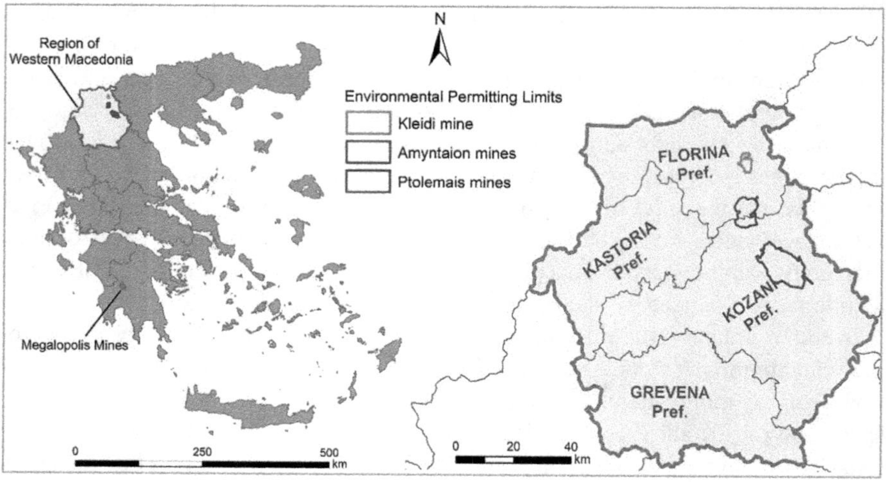


**Fig. 2** Western Macedonia region and surface lignite mines, according to the environmental permitting limits

followed by the Amyntaion mine (2021). Following the recent decisions regarding the lignite power plants phase-out in Greece until 2028, the Western Macedonia region was included in the first Just Transition Fund (JTF) Program adopted by the European Commission. [35]

The regional sustainable development of the Western Macedonia region in Greece, was investigated concerning long-term surface lignite mining operations for electricity generation by examining the evolution and conducting regression analysis of the following production, social and economic parameters: (a) the lignite production from the mines of the region, (b) the regional GDP of the region, (c) the gross value added (GVA) of the region, (d) the population of the region, (e) the employment in manufacturing (including mining), (f) the total employment in the region, (g) the water pumping from the mines of the region and (h) the rate of change of overall consumer price index (OCPI) for Greece. The economic and social data for the regression analysis referred to the period from 2000 to 2019. They were all obtained from the Hellenic Statistical Authority (ELSTAT) for the region of Western Macedonia except the rate of change of the OCPI, which concerns all of Greece, while the mining data were obtained from the Public Power Corporation (PPC) of Greece. The main research question refers to the impacts of mining activities and their implications for sustainable development goals, considering the related parameters.

[35] E.C. (2022).

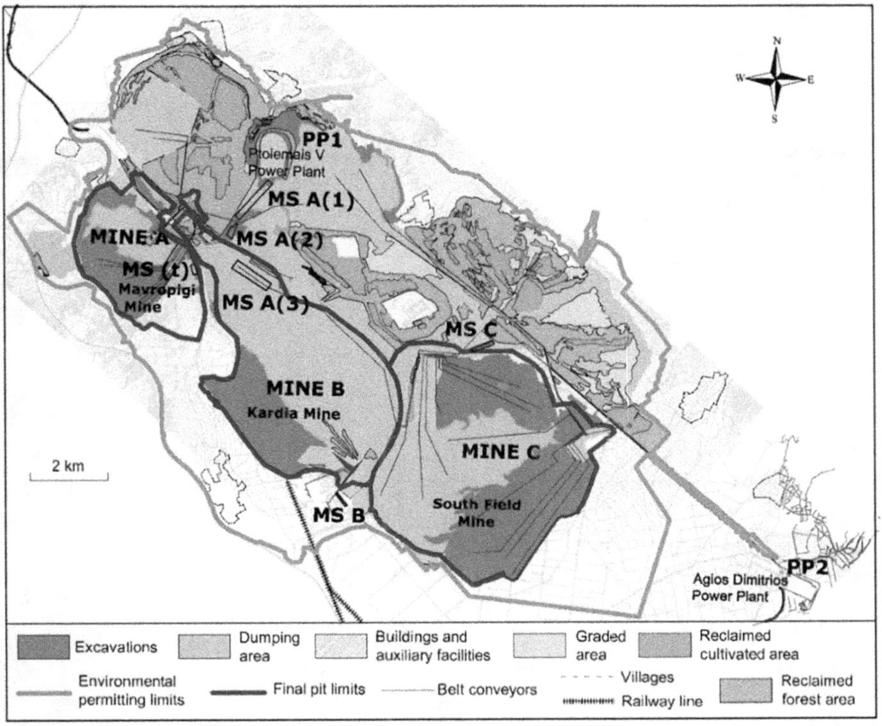


**Fig. 3** Map of the lignite surface mining complex of the Ptolemais basin, Northern Greece, with current land uses (MS: Mine Stockyard)

## 3.2 *Optimization Model of the Ptolemais Continuous Surface Mines*

This analysis below focuses on the optimization model of the complex of Ptolemais mines (Fig. 3), with three lignite mines (Mavropigi-Mine A, Kardia-Mine B, and South Field-Mine C) in operation until recently, based on the data of the earlier phase-out of lignite mines. The applied analysis as an integrated approach to the optimization of mining operations is presented in Fig. 4, where the proposed model is divided into sub-models. The optimization of lignite production in relation to mine reclamation and the planning of new land uses are one of the most critical stages of the mine closure phase, provided that they define the future perspectives of the mining area. In the context of mining area repurposing, land planning is an indispensable part of mine planning, based on transforming the exhausted and reclaimed mine area into new land uses.

In this framework, one of the strategic steps of the applied approach is the optimal combination of the lignite production of the mines in operation and the consumption scheduling of the power plants. In this direction, a probabilistic approach

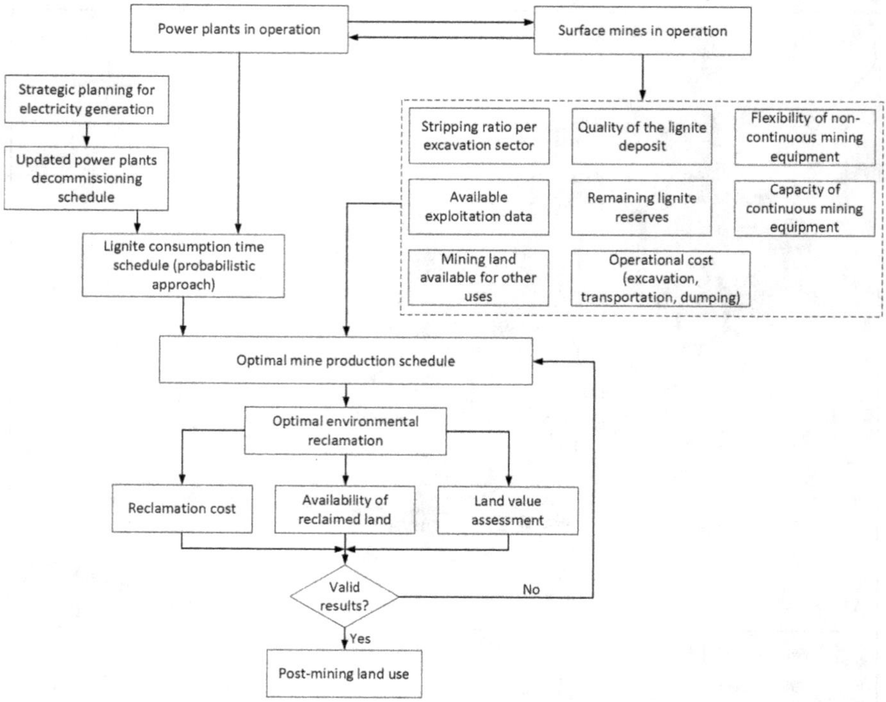


**Fig. 4** Flow chart of the optimization model considering lignite production, environmental reclamation, and post-mining land uses

was adopted by making scenarios of lignite consumption and production considering the mining parameters, such as the remaining lignite reserves in each mine and the corresponding stripping ratio per excavation sector, the operational costs, the capacity of continuous mining equipment, the required quality of the run-of-mine lignite, the available exploitation area, the mining land available for other purposes, and the flexibility regarding the supplementary operation of non-continuous mining equipment. The available exploitation area is a crucial issue in combination with the areas planned for other uses.

# 4 Results and Discussion

## *4.1 Statistical Regression and Correlation Analysis*

The initial statistical regression analysis of the selected parameters was based on the linear model of the ordinary least square (OLS) method. The results of the linear regression can be seen in Fig. 5. According to these results, the variables that are significantly related to lignite production in Western Macedonia are the population

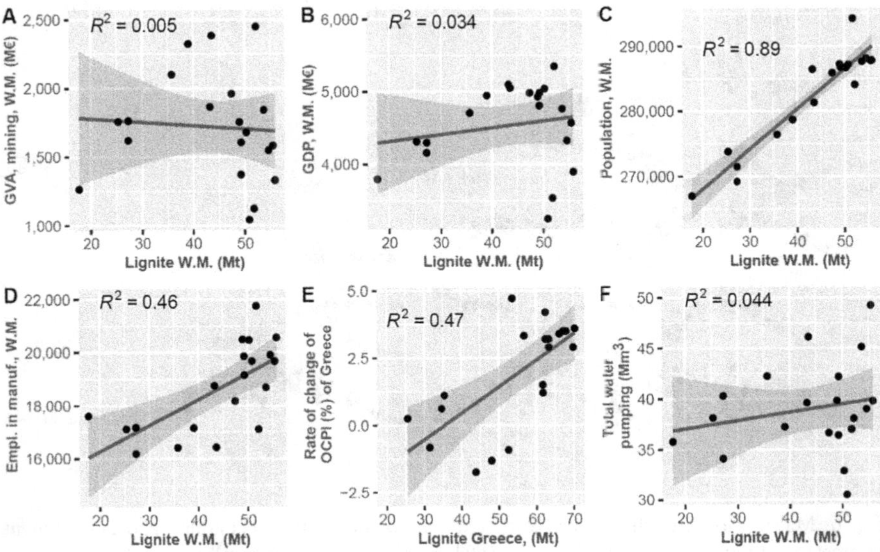


**Fig. 5** Graphical representation of the linear regression model. (**a**) GVA in mining—Lignite production in Western Macedonia, (**b**) Population-Lignite production in Western Macedonia, (**c**) GDP—Lignite production in Western Macedonia, (**d**) Employment in manufacturing (including mining)—Lignite production in Western Macedonia, (**e**)—Rate of change of the OCPI (%)—Lignite production in Greece, (**f**) Total water pumping (Mm$^3$)—Lignite production in Western Macedonia (W.M.: Western Macedonia)

of Western Macedonia (p-value 0.000), the employment in manufacturing (including mining) in Western Macedonia (p-value 0.001), and the rate of change of OCPI (%) of Greece, which is correlated with the total lignite production of Greece, which, in a percentage that ranges between 75–85%, is shaped by the lignite production in the Western Macedonia region (p-value 0.000).

The next step of linear regression was to determine if the model explained the variance of the variables. According to the data of Fig. 5, the parameter of lignite production that is statistically related to employment in manufacturing and the rate of change of OCPI cannot explain the variation of these parameters as the R-squared value is very low. In contrast, the regression between the parameter of population and that of lignite production in Western Macedonia has a high R-squared, which signifies that the model describes very well the population variance.

The results of further correlation analysis based on Spearman's statistical analysis are presented in the matrix plot of Fig. 6, which displays the pairwise relationship of the main production, employment and economic parameters of Western Macedonia. The corresponding Spearman correlation coefficients between the parameters of Fig. 6 are presented in Table 1. From the results of the correlation analysis of Fig. 4 and Table 1, it is concluded that employment in the manufacturing

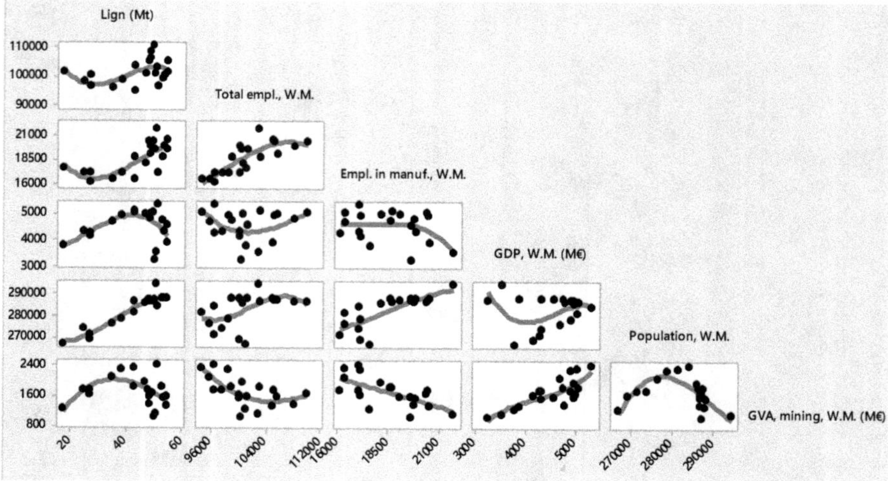


**Fig. 6** Matrix plot with the indicative correlations of production, economic and employment parameters (Lign.: Lignite production, Total empl.: Total employment, Empl. in manuf.: Employment in manufacturing including mining, W.M.: Western Macedonia)

**Table 1** Spearman correlation coefficients between the parameters of Fig. 6 (each cell presents Spearman's rho and the respective p-value)

| | Lign (Mt) | Total empl., W.M. | Empl. in manuf., W.M. | GDP, W.M. (M€) | Population, W.M. |
|---|---|---|---|---|---|
| Total empl., W.M. | 0.343 | | | | |
| | 0.139 | | | | |
| Empl. in manuf., W.M. | 0.633 | 0.830 | | | |
| | **0.003** | 0.000 | | | |
| GDP, W.M. (M€) | 0.023 | 0.021 | −0.220 | | |
| | 0.925 | 0.930 | 0.352 | | |
| Population, W.M. | 0.890 | 0.504 | 0.821 | −0.096 | |
| | **0.000** | **0.024** | **0.000** | 0.686 | |
| GVA, mining, W.M. | −0.287 | −0.579 | −0.693 | 0.744 | −0.444 |
| | 0.220 | **0.007** | **0.001** | 0.000 | **0.050** |

sector and Western Macedonia's population strongly correlate with lignite production in Western Macedonia.

It is also interesting to examine the temporal evolution of some parameters included in the correlation analysis during the 20-year period 2000–2019. Figure 7 presents the evolution of lignite production and employment in manufacturing (including mining) in the region of Western Macedonia for the period from 2000 to 2019. While the decline in lignite production has constantly been accelerating, jobs in this sector have been declining at an almost steady rate as a result of various

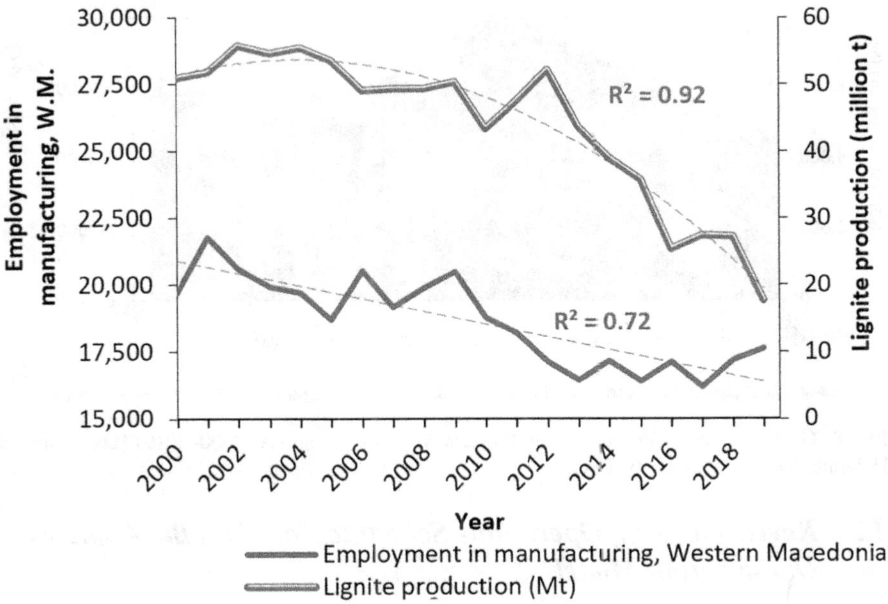


**Fig. 7** Evolution of lignite production (R-squared value from polynomial regression) in relation to employment in manufacturing, including mining (R-squared value from linear regression), in the region of Western Macedonia between 2000 and 2019 [Data source: Hellenic Statistical Authority, PPC]

strategies applied for the lignite mines and power plants to reduce production costs and permanent staff and to increase seasonal staff, outsourcing, etc.

Figure 8 shows the GDP change in Greece and the Western Macedonia regions from 2000 to 2019. It is concluded that in the crucial period 2008–2013, the country's economic crisis did not affect Western Macedonia. In 2010 the country's GDP began to decrease, while the GDP of Western Macedonia continued to increase until 2012. Furthermore, the rapid decrease in GDP of Western Macedonia and mainly of the Kozani Prefecture began with the decrease in lignite production. In fact, the investigated period is divided into two sub-periods: The first lasts until 2005, and it is characterized by successive lignite production records. This sub-period is the end of a long period of increased electricity demand that started in the 1980s rather than the beginning of the developments that followed. The second period is characterized by dramatic changes in national energy policy, which in turn affected the economy of the Western Macedonia region. The fiscal crisis of 2009 resulted in a significant reduction in electricity demand. At the same time, the entry of natural gas into the domestic energy market, the rapid development of RES, the support of private investments in the energy sector, and the increase in the cost of lignite-based power generation due to the EU emissions trading system weakened the role of lignite for the local economy and highlighted alternative production activities, such as photovoltaic parks.

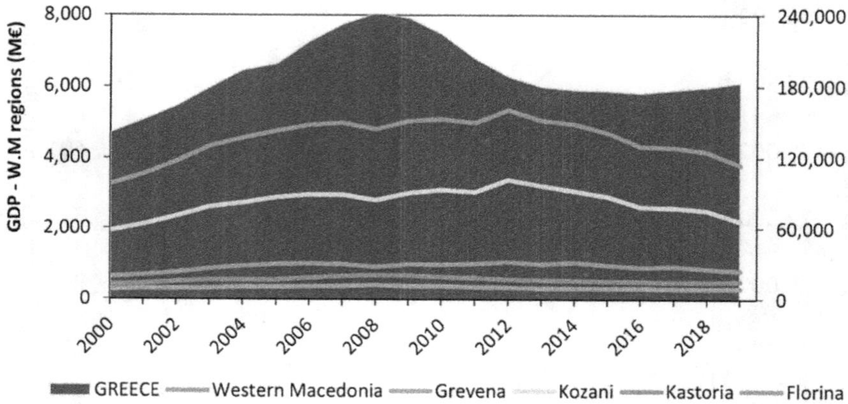


**Fig. 8** GDP change in Greece and the Western Macedonia regions (2000–2019) [Data source: Hellenic Statistical Authority, PPC]

### *4.2 Revised Mining Operations Schedule Based on the Applied Optimization Model*

Based on the results of the statistical analysis and the social and economic impacts of the lignite mines closure, the optimization model described in Sect. 3 for the Ptolemais mines operation until the phase-out of the two power plants (Agios Dimitrios and new Ptolemais V), according to a National Energy and Climate Plan, concluded to a revised mining operations schedule with only two mines in operation, Mavropigi (Mine A) and South Field (Mine C).

The revised mine planning included:

(1) A detailed annual production schedule of each lignite mine bench from the two mines
(2) The forecasting of run-of-mine lignite quality
(3) The areas of developing new mine excavation areas in case of an increase in lignite production
(4) The schedule of mining equipment movements and utilization
(5) The revised project management of the technical infrastructure (railway lines, roads, rivers, etc.)
(6) The geotechnical and hydrogeological investigation of mine closure and post-mining reclaimed areas
(7) The schedule of the environmental reclamation of post-mining areas and the estimation of the corresponding cost
(8) The annual availability of reclaimed land
(9) The spatiotemporal planning of new land uses

Figures 9 and 10 show the changes in the network of lignite mines and power plants connections. The main targets of the revised network planning were (a) to increase the flexibility and the availability of the transportation system, (b) to

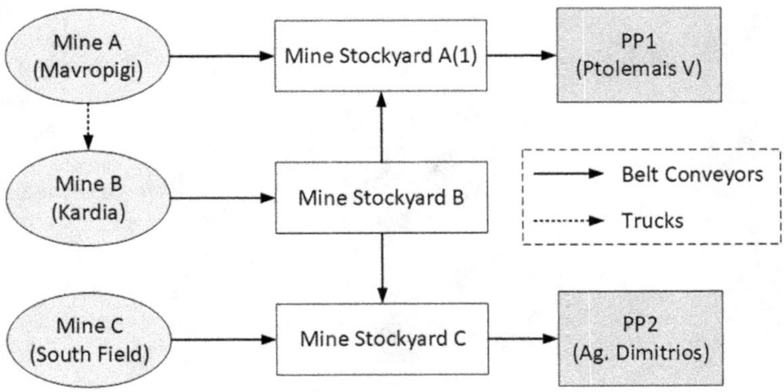


**Fig. 9** Network of lignite mines and power plants in the Ptolemais mining area based on the exhaustion of the lignite deposits (the location of the lignite mines and stockyards are shown in Fig. 3)

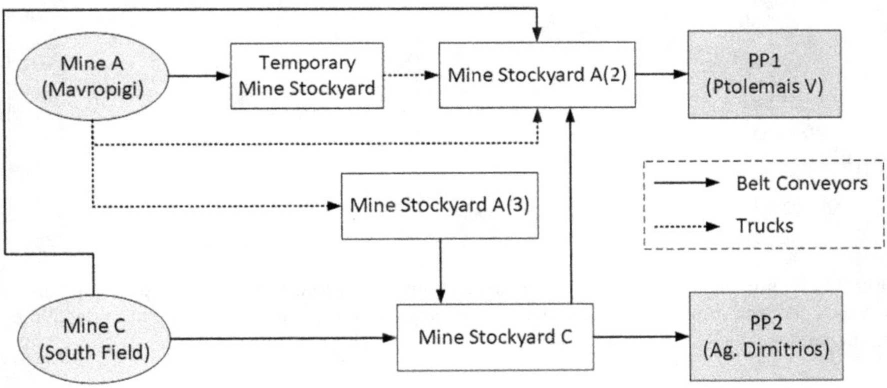


**Fig. 10** Network of lignite mines and power plants in the Ptolemais mining area in the revised power plants phase-out schedule (the location of the lignite mines and stockyards are shown in Fig. 3)

improve the homogenization of the lignite in the mine stockyards and, therefore, to improve the quality of the lignite feeding to power plants, and (c) to reduce the transportation costs.

The corresponding availability of the reclaimed land to public entities until the middle of 2025, based on the applied production and reclamation model, is shown in Fig. 11. In addition, a conceptual model for the balance between the transferred land and the corresponding reclamation cost was applied to investigate the land availability schedule.

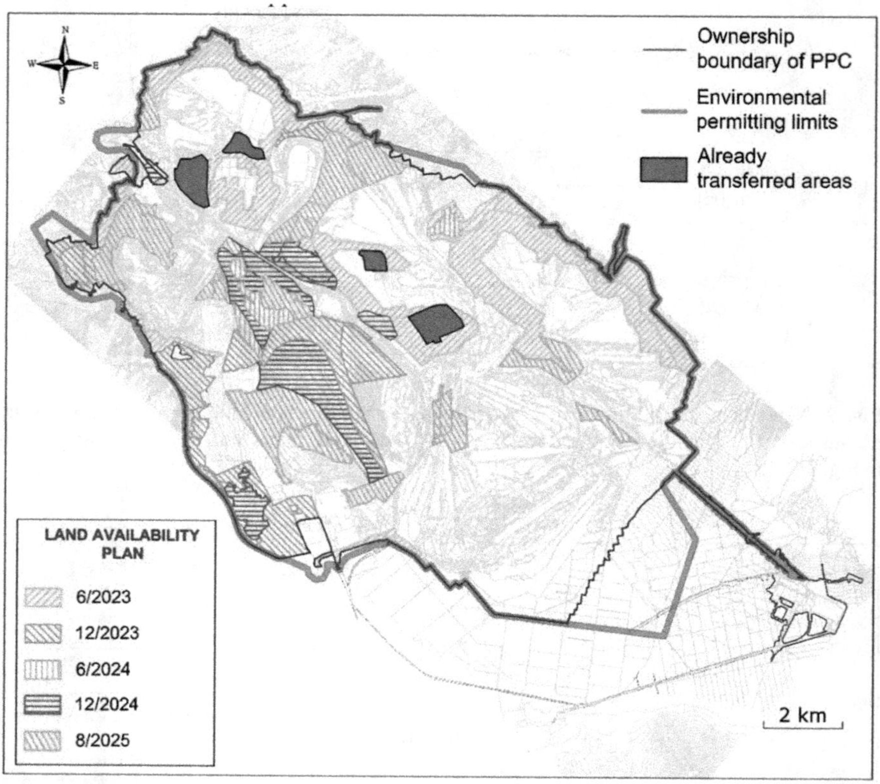


**Fig. 11** Future mine land availability to the just transition according to the revised mine planning (areas without coloring will be used for the remaining mining operations and other future activities in the mining area)

## 4.3 Public Acceptance and Land Management

Regarding the public acceptance of the mining land transformation at a regional level and the energy transition process in Western Macedonia, the main stakeholders are presented in Fig. 12. So far, local stakeholder interventions have shown a widespread reluctance to accept an accelerated process toward lignite phase-out. In addition, almost all local stakeholders agree on the following: [36]

- No alternative economic activity could unequivocally replace the jobs and wealth the local lignite industry offered for decades. This practically means that the new production model of Western Macedonia must be based on a strongly differentiated production level in terms of complementarity.

[36] Ziouzios et al. (2021), 115.

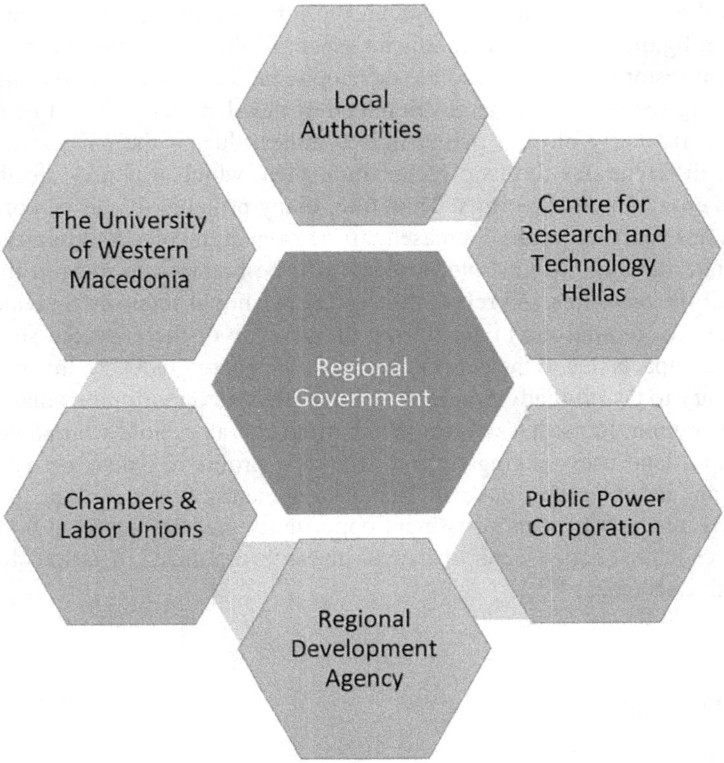


**Fig. 12** The main stakeholders related to the energy transition process in the Western Macedonia region

- From the primary sector and manufacturing to renewables and hydrogen technologies, new productive activities will not be labor-intensive but knowledge-intensive, which means that investing in research and innovation is a critical success factor.
- There must be a balance between external direct investment and the development of an intra-regional production capacity to avoid future mono-dependencies.
- The rehabilitation of depleted lignite mines, spatial planning, the promotion of licensing simplification and the launch of infrastructure projects are recorded as important prerequisites by the stakeholders.

It is becoming increasingly clear to all the involved parties that the energy transition is not just about choosing the best technical solutions, but relies heavily on human habits and behaviors and should ensure that every region and workers is included. This applies especially to Western Macedonia, where the demand for a generalized productive reconstruction accompanies the energy transition. Consequently, participatory decision-making, political commitment, transparency, and trust are crucial to a successful and fair energy transition.

Regarding mining land management, the rehabilitation of lands affected by long-term lignite mining is a significant aspect of the just transition in terms of ecological restoration and economic development. The discussion on mine land repurposing scenarios and the development of new land uses should be emphasized that the early closure of the lignite mines, due to the energy transition strategy, differentiates the land reclamation plan, which was prescribed in the previous environmental permits. Therefore, many projects should be completed within a few years, requiring increased efforts from all involved parties as well as appropriate funding. To this end, projects will be financed to reclaim the mine lands. All the measures and relevant investments should focus on a greater geographical area around the mines. Given the severity of the expected social and economic impacts due to the enormous size of the energy industry, this provision is necessary to take full advantage of the strengths and opportunities that characterize the region. Regardless of the above, mine operators hold a broad portfolio of potential land uses ranging from maximizing profits to balancing ecological restoration and financial targets. The development of photovoltaic parks in reclaimed mine areas is an investment opportunity that has attracted both giant enterprises in the energy sector and small investors organized in various forms of cooperative schemes. [37]

## 5 Conclusion

The depletion of coal surface mines due to the energy transition policies adopted by many countries worldwide makes it difficult to achieve sustainable development goals at a regional level, at least in short to medium term, without appropriate measures associated with the sustainability of the related mining projects.

From the statistical analysis considering the lignite production data and social and economic parameters of the Western Macedonia mining region, it was concluded that there is a strong correlation between the annual extracted lignite and employment in the manufacturing sector as well as the population. Moreover, the results of further correlation analysis indicated that delays in implementing certain site-specific measures to support the emergence of new productive activities and maintain social cohesion lead local economies to a lower balance level, which is restored through the migration of scientists, reducing wages, or other social and economic impacts. In this framework, the effects of the lignite mine closure should be faced as a challenge and an opportunity for proper planning of mine closure, following the principles of sustainable development and circular economy as well as incorporating material, energy and other indicators flow analysis, which is expected to absorb most of the adverse socio-economic impacts.

---

[37] Pavloudakis et al. (2020), 3995.

The presented proper mine closure planning was based on the optimization of the mining operations, considering that advanced revised planning of environmental reclamation is needed in combination with optimal planning of new land uses in the mining area. In this context, the main targets of the revised mine planning to obtain optimal remaining operations and closure of continuous surface lignite mines in the Ptolemais basin, Northern Greece, were (a) to increase the flexibility and the availability of the transportation system, (b) to improve the homogenization of the lignite in the mine stockyards and, therefore, to improve the quality of the lignite feeding to power plants, and (c) to reduce the transportation costs. In addition, from the applied model, a suitable spatial mine land availability schedule through time to the just transition was derived.

Finally, from the analysis of the main stakeholders regarding the public acceptance of the mining land transformation at a regional level and the energy transition process in Western Macedonia, Greece, it was concluded that participatory decision-making, political commitment, transparency, and trust are crucial to a successful and fair energy transition.

## References


Antoniadis A, Roumpos C, Anagnostopoulos P, Paraskevis N (2022) Planning RES projects in exhausted surface lignite mines – challenges and solutions. In: Proceedings, International Conference on Raw Materials and Circular Economy (RawMat 2021), 5–9 September. https://doi.org/10.3390/materproc2021005093. Accessed 22 May 2023

Azapagic A (2004) Developing a framework for sustainable development indicators for the mining and minerals industry. J Clean Prod 12:639–662. https://doi.org/10.1016/S0959-6526(03)00075-1. Accessed 22 May 2023

Burchart-Korol D (2011) Application of life cycle sustainability assessment and socio-eco-efficiency analysis in comprehensive evaluation of sustainable development. J Ecol Health 15:107–110

Burchart-Korol D, Krawczyk P, Czaplicka-Kolarz K, Turek M, Borkowski W (2014) Development of sustainability assessment method of coal mines. J Sustain Min 13:5–11. https://doi.org/10.7424/jsm140402. Accessed 22 May 2023

Central Mine Planning and Design Institute (CMPDI) - Ministry of Coal/Sustainable Development Cell (2021) Status of Environmental Sustainability in Coal Mines 2019-2020

E.C. (2022) E.U. Cohesion Policy: €1.63 billion for a just climate and energy transition in Greece. https://ec.europa.eu/commission/presscorner/detail/en/ip_22_3711. Accessed 22 May 2023

Espí J (1999) Minerals as a natural resource Tecno Ambiente (Spain). Year IX 89:37–41. (in Spanish)

Finucane SJ (2008) Thinking about the end before you start - integrating mine closure planning into feasibility studies and environmental and social impact assessments. In: Fourie AB, Tibbett M, Weiersbye IM, Dye PJ (eds) Proceedings Third International Conference on Mine Closure, Australian Centre for Geomechanics, Johannesburg, pp 171–182

Fleming DA, Measham TG (2015) Income inequality across Australian regions during the mining boom: 2001–11. Aust Geogr 46:203–216. https://doi.org/10.1080/00049182.2015.1020596. Accessed 22 May 2023

Garvey A, Norman JB, Büchs M, Barrett J (2022) A "spatially just" transition? A critical review of regional equity in decarbonization pathways. Energy Res Soc Sci 88:102630. https://doi.org/10.1016/j.erss.2022.102630. Accessed 22 May 2023. ISSN 2214-6296

Hardisty PE (2010) Environmental and economic sustainability. CRC Press. https://doi.org/10.1201/EBK1420059489. Accessed 22 May 2023

International Atomic Energy Agency (IAEA) (1994) Decommissioning of Facilities for Mining and Milling of Radioactive Ores and Closeout of Residues, Technical No. STI/DOC/010/362, 362, Vienna, Austria

International Council on Mining and Metals (ICMM) (2019) Integrated mine closure, good practice guide, 2nd edn. London, UK

International Council on Mining and Metals (ICMM) (2020) Annual Report of International Council on Mining and Metals, 2019

International Institute for Environment and Development (IIED) (2002) Minerals and sustainable development mining for the future. Geneva, Switzerland

Kavvadas M, Roumpos C, Servou A, Paraskevis N (2022) Geotechnical issues in decommissioning surface lignite mines – the case of Amyntaion mine in Greece. Mining 2:278–296. https://doi.org/10.3390/mining2020015. Accessed 22 May 2023

Krassakis P, Karavias A, Zygouri E, Roumpos C, Louloudis G, Pyrgaki K, Koukouzas N, Kempka T, Karapanos D (2023) GIS-based assessment of hybrid pumped hydro storage as a potential solution for the clean energy transition: the case of the Kardia Lignite Mine, Western Greece. Sensors 23:593. https://doi.org/10.3390/s23020593. Accessed 22 May 2023

Laurence D (2006) Optimisation of the mine closure process. J Clean Prod 14:285–298. https://doi.org/10.1016/j.jclepro.2004.04.011. Accessed 22 May 2023

Liu H, Wu Q, Chen J, Wang M, Zhao D, Duan C (2021) Environmental impacts related to closed mines in inner Mongolia. Sustainability 13:13473. https://doi.org/10.3390/su132313473. Accessed 22 May 2023

Louloudis G, Louloudis E, Roumpos C, Mertiri E, Kasfikis G, Chatzopoulos K (2022a) Forecasting development of mine pit lake water surface levels based on time series analysis and neural networks. Mine Water Environ 41:458–474. https://doi.org/10.1007/s10230-021-00844-5. Accessed 22 May 2023

Louloudis G, Roumpos C, Louloudis E, Mertiri E, Kasfikis G (2022b) Repurposing of a closed surface coal mine with respect to pit lake development. Water 14:3558. https://doi.org/10.3390/w14213558. Accessed 22 May 2023

Pactwa K, Woźniak J, Dudek M (2021) Sustainable social and environmental evaluation of post-industrial facilities in a closed loop perspective in coal-mining areas in Poland. Sustainability 13:167. https://doi.org/10.3390/su13010167. Accessed 22 May 2023

Palogos I, Galetakis M, Roumpos C, Pavloudakis F (2017) Selection of optimal land uses for the reclamation of surface mines by using evolutionary algorithms. Int J Min Sci Technol 27:491–498. https://doi.org/10.1016/j.ijmst.2017.03.008. Accessed 22 May 2023

Paraskevis N, Servou A, Roumpos C, Pavloudakis F (2021) Spatiotemporal interactions between surface coal mining and land cover and use changes. J Sustain Min 20:72–89. https://doi.org/10.46873/2300-3960.1053. Accessed 22 May 2023

Parkins JR, Angell AC (2011) Linking social structure, fragmentation, and substance abuse in a resource-based community. Community Work Fam 14:39–55. https://doi.org/10.1080/13668803.2010.506030. Accessed 22 May 2023

Pavloudakis F, Roumpos C, Karlopoulos E, Koukouzas N (2020) Sustainable rehabilitation of surface coal mining areas: the case of Greek lignite mines. Energies 13:3995. https://doi.org/10.3390/en13153995. Accessed 22 May 2023

Perdeli Demirkan C, Smith NM, Duzgun S (2022) A quantitative sustainability assessment for mine closure and repurposing alternatives in Colorado, USA. Resources 11:66. https://doi.org/10.3390/resources11070066. Accessed 22 May 2023

Que CT, Nevskaya M, Marinina O (2021) Coal mines in Vietnam: geological conditions and their influence on production sustainability indicators. Sustainability 13:11800. https://doi.org/10.3390/su132111800. Accessed 22 May 2023

Roumpos C, Papacosta E (2013) Strategic mine planning of surface mining projects incorporating sustainability concepts. In: Proceedings, 6th International Conference on Sustainable Development in the Minerals Industry (SDIMI 2013), 30 June–3 July 2013, Milos Island, Greece, pp 645–651

Spanidis PM, Pavloudakis F, Roumpos C (2021) Introducing the IDEF0 methodology in the strategic planning of projects for reclamation and repurposing of surface mines. In: Proceedings, International Conference on Raw Materials and Circular Economy (RawMat 2021), 5–9 September. https://doi.org/10.3390/materproc2021005026. Accessed 22 May 2023

Spanidis PM, Roumpos C, Pavloudakis F (2022a) A methodology combining IDEF0 and weighted risk factor analysis for the strategic planning of mine reclamation. Minerals 12:713. https://doi.org/10.3390/min12060713. Accessed 22 May 2023

Spanidis PM, Roumpos C, Pavloudakis F, Servou A, Paraskevis N (2022b) Developing knowledge management in mine reclamation projects. In: Proceedings, 16th International Congress of the Geological Society of Greece, Sp. Publ. 10. Patras, Greece

Stemke M, Wieber G (2022) Closure of German hard coal mines: effects and legal aspects of mine flooding. Mine Water Environ 41:280–291. https://doi.org/10.1007/s10230-021-00842-7. Accessed 22 May 2023

Tsachouridis A, Pavloudakis F, Kiratzis N (2021) Development of rare earth elements separation processes from coal fly ash. In: Proceedings, International Conference on Raw Materials and Circular Economy (RawMat 2021), 5–9 September. https://doi.org/10.3390/materproc2021005069. Accessed 22 May 2023

United Nations (1987) Our Common Future (Report of the World Commission on Environment and Development United Nations. No. A/42/427)

United Nations A/RES/70/1 General Assembly Distr. (2015) Transforming our world: the 2030 Agenda for Sustainable Development. Resolution adopted by the General Assembly on 25 September 2015

Vitti C, Arnold BJ (2022) The reprocessing and revalorization of critical minerals in mine tailings. Min Metall Explor 39:49–54. https://doi.org/10.1007/s42461-021-00524-6. Accessed 22 May 2023

World Risk Report (2018) Edition, n.d., Mining Journal

Yaylacı ED, Düzgün HŞ (2017) Evaluating the mine plan alternatives with respect to bottom-up and top-down sustainability criteria. J Clean Prod 167:837–849. https://doi.org/10.1016/j.jclepro.2017.07.248. Accessed 22 May 2023

Ziouzios D, Karlopoulos E, Fragkos P, Vrontisi Z (2021) Challenges and opportunities of coal phase-out in western Macedonia. Climate 9:115. https://doi.org/10.3390/cli9070115. Accessed 22 May 2023

# Lifecycle of Mine Water

**Mareike Bothe-Fiekert, Angela Binder, Sandra Pamela Nowosad, Florence Apollo, and Oliver Langefeld**

## Contents



**Abstract** Water is an unequal distributed resource and the increase in demand in business and industry will lead more frequently to regional and temporary shortages, which will be exacerbated by global warming. As a major water user, the mining sector is expected to promote and support the implementation of the United Nations Sustainable Development Goals (SDGs). Sustainable and circular water

---

M. Bothe-Fiekert (✉) · A. Binder · S. P. Nowosad · F. Apollo · O. Langefeld
Technische Universität Clausthal, Institut für Bergbau, Clausthal, Germany
e-mail: mareike.bothe-fiekert@tu-clausthal.de; angela.binder@tu-clausthal.de; sandra.nowosad@tu-clausthal.de; florence.apollo@tu-clausthal.de; oliver.langefeld@tu-clausthal.de





W. Frenz, A. Preuße (eds.), *Yearbook of Sustainable Smart Mining and Energy 2022*, Yearbook of Sustainable Smart Mining and Energy - Technical, Economic and Legal Framework 2, https://doi.org/10.1007/978-3-031-41873-0_4

management is critical for the mining industry to achieve a 'social licence to operate' and long-term profitability (Kunz and Moran, J Clean Prod 84:469–474, 2014). Following the guiding principle of the circular economy, the blue mining concept was developed at the Clausthal University of Technology, which stands for mining operations that are able to work in an energy-efficient and ergonomic way, aim for full automation and support sustainability by promoting the continued use of mine sites during and after extraction (Kellner and Langefeld, "Blue Mining" - the future of mining, 2013). In this chapter, dynamic water management in the sense of this concept is focused. Based on the life cycle of a mine current challenges, trends and research projects are presented.

## 1 Introduction

Water is a vital resource that is limited, but available in a natural cycle on earth. The amount of freshwater renewed annually exceeds global demand many times over. Nevertheless, the unequal distribution of usable resources and the increase in demand in business and industry will lead more frequently to regional and temporary shortages, which will be exacerbated by climate change. Rapid growth in industrial demand for water in mining basins and high purchasing power for water licenses are creating conflicts over water allocation across sectors, which is likely to escalate in the future.[1] As a major water user, the mining sector is expected to promote and support the implementation of the United Nations Sustainable Development Goals (SDGs). Sustainable and circular water management is critical for the mining industry to achieve a 'social license to operate' and long-term profitability.[2] Priority areas for the mining sector in achieving a sustainable future include ensuring sufficient water for mining operations (SDG 6 Clean Water and Sanitation), improving mine resilience to climate-related hazards (SDG 13 Climate Change Action), and mitigating or avoiding significant adverse impacts of mining activities on land and water resources and the surrounding environment (SDG 3 Health and Wellbeing and SDG 15 Life on Land). Water use and management were ranked as the most important sustainability issue in all mining regions. Robust management strategies that are consistent with the SDGs and advance sustainable mining practices are needed that can address the challenges of climate-influenced water scarcity and unregulated discharge.[3] In high developed operation for example in the Australian and New Zealand gold mining industry, the use of water is an integral part of the entire mining process, from the initial extraction of the ore to its final processing. The way in which water is managed on-site effects the running costs, extraction rates, water usage and environmental impacts. Therefore, it is crucial to implement a Water

[1] Gao et al. (2017), pp. 1009–1020.

[2] Kunz and Moran (2014), pp. 469–474.

[3] Gao et al. (2017), pp. 1009–1020.

Efficiency Management Plan which integrates all components of mine planning, operation and closure to optimize site water management and reduce environmental impacts. Neglecting environmental responsibility can also lead to significant reputational damage with financial implications.

Retroactive socio-economic and biophysical rehabilitation of abandoned mines is far costlier than preventing these impacts through intelligent mine design and technologies and ongoing monitoring and maintenance. This approach is followed by the Blue Mining concept initially presented 2013 at the sixth international conference on Sustainable Development in Mineral Industry (SDIMI). In 2013, the Blue Mining concept was presented for the first time at the sixth international conference on sustainable development in the minerals industry in Greece. Blue mining stands here for mining operations that are able to work in an energy-efficient and ergonomic way, aim for full automation and support sustainability by promoting the continued use of mine sites during and after extraction.[4] Blue Mining is the holistic approach to defining responsible mining by reaching beyond the extraction of raw materials and utilizing the amenities of the mines during and after production based on an early stage planning approach. Thus, a sound, safe and ergonomic mine practice is defined as that which maximizes the positive effects of mining while minimizing negative impacts by the early but constant development of multiuse. Incorporating those principles, Blue Mining stands for a completely new step to stage of mine planning, because these aspects have to be planned starting from a very early stage of the mine life. This planning includes the life cycle of mine water with the aim of aligning the available water resources with the supply and demand at each phase of the mine's operation.

The basis for improving water management in the mining and resource processing sector is to develop an understanding of how much water is used in the various processes, how much wastewater is generated and of what quality. During the presented stages, the role of water changes and different aspects need to be considered. Unique, finite deposits represent the base for each mining operation. In comparison to other operations, this limits the life of an operation. On the other hand, the deposits as a base for the operation must be found and explored before the start of the operation. This leads to a characteristic stage of a lifecycle. Those are presented with the qualitative development of costs and production in Fig. 1. Beyond any doubt, the sequence of the five main stages with several substages represents an ideal case. In practice, most operations don't follow this linear sequence and changes back and forward through the stages can be seen. For example, the rise in commodity price can lead to a change from declining to increasing production or even from post-closure to production. To give an overview of this role, the aspects are elaborated based on the presented stages, conclusions are made and research needs are identified.

[4] Kellner and Langefeld (2013).

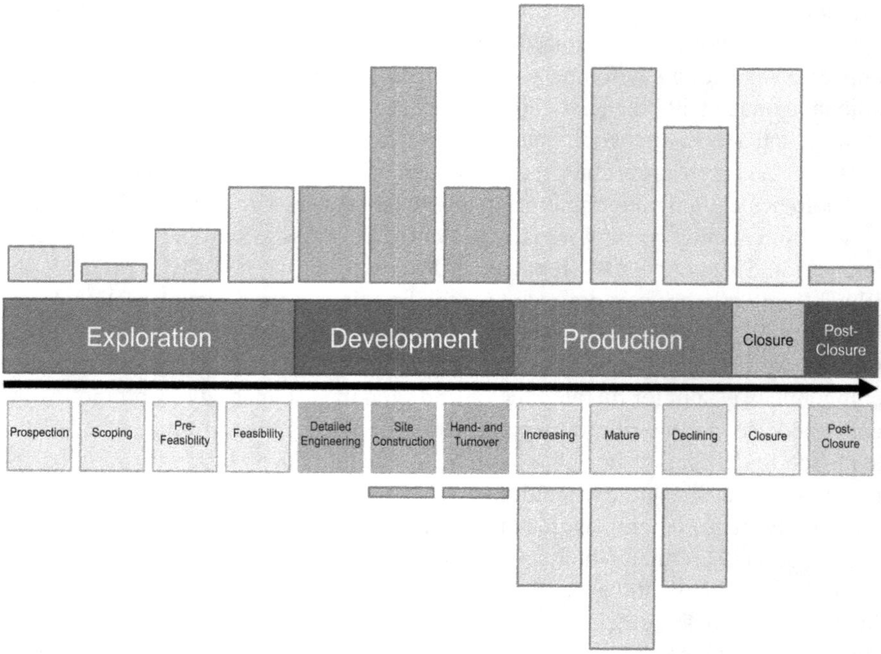


**Fig. 1** Stages of a mine lifecycle with qualitative costs (top) and production (bottom)

## 2 Exploration and Mine Planning According to the Blue Mining Principle

An extended planning process that includes the allocation of relevant information is an important aspect that distinguishes mining from other industries. During the exploration and planning phase, water is both a subject of investigation and planning, as well as influencing the exploration itself. While the focus is primarily on the former, responsible water management practices must be considered in order to protect the existing water system. The following basic objectives generally apply to water management in mining: Mine planning, design, operation and closure must be conducted in a manner that minimizes the uncontrolled intrusion of water into the mine, the water consumption, the degradation of water quality and the impacts on the region's water resources, while maximizing the reuse of water and all potential uses.[5] The interactions between mining activities and water resources are site-specific and exhibit complex system behavior characterized by feedback effects and changes in management strategy that can lead to unexpected system consequences.[6] Hence, water management in mining operations must be constantly adapted aligned

[5] Department of Water Affairs and Forestry (2008).

[6] Northey et al. (2016), pp. 1098–1116.

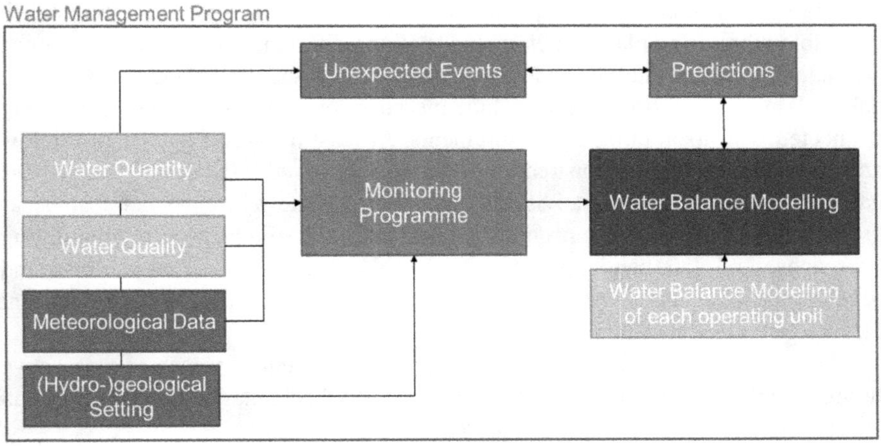


**Fig. 2** Components of the water management program

with the life cycle of the mine. Each phase requires different approaches to water management. Revising and renewing the water management program and related permits is an ongoing process.[7] Water Management is defined as a practical management structure with a strategy and goal that is often environmentally driven.[8] Optimal water management includes probabilistic water balance models, sufficient monitoring of the surrounding hydrological conditions and the water quality and quantities of the mine relevant to each phase (Fig. 2). According to More et al. (2020) the traditional spreadsheet-based water management approach is gradually being replaced by a dynamic system approach. This modelling approach offers a flexible way to react to or demonstrate transient and probabilistic events. Not only predictions can be made about the trends in water quality and quantity, but also complex system correlations and feed-back effects can be evaluated. Therefore, this is a powerful tool to assess the overall performance of the water management system. Due to sensitivity and risk analyses, sensitive units in the water management system, units with large fluctuations and units where a system failure causes the greatest damage can be identified. Changes in the mining process can be flexibly included in the considerations and their effects on the mine's water balance can be visualized. Moreover, the system behavior under extreme conditions (e.g. extreme weather events) can be investigated and safety constructions can be planned.[9] Common software tools used in mine water balance calculations are GoldSim, MATLAB, Simulink, Geochemists Workbench, and PHREQC.[10] Some of the different programs are presented and described by Punkkinen et al. (2016a, b). Since the

[7] Punkkinen et al. (2016a, b).

[8] Mueller (2015).

[9] Punkkinen et al. (2016a, b).

[10] More et al. (2020).

underlying data determines the quality of a model or plan, it is essential to establish a suitable monitoring plan, which includes all the different water types at the mine site, namely, natural (surface and groundwater), waste and used waters (tailings and dams). Water monitoring should include measurements of physicochemical quality connected with local climate measurements. At least the water quality of outflow and discharge at sites downstream should be monitored.[11] To better analyze the gathered information, it is advisable to investigate the hydrogeochemical background. An overview of current developments and trends in new monitoring approaches and data handling is provided in the Sect. 3.4. Careful planning and development of accurate water management with a holistic approach is an important factor for successful mining operations.[12]

The aim is to move from a reactive water management approach to a proactive future-oriented water management. There are several advantages of starting to plan water management measures at an early stage:

- Reduced risks and environmental impacts through early preparation for extreme situations and changes in the water balance.
- Reduction of fresh water consumption (optimization of water recycling)
- Reduction of wastewaters
- Reduction of costs
- Increase of social acceptance

## 2.1 Exploration and Baseline Studies

As a mine site is part of a catchment area in a meteoric water circulation system with its surface and groundwater subsystems[13] the site specific-factors affecting the water balance differ. Environmental baseline studies are the initial step to provide unaffected data for the proposed side.[14] Good collaboration with the relevant authorities and determination of the submitting legal entity for an integrated water use license prepare the work as well. The purpose of the prospecting, exploration, conceptualization and planning stage is the first high-level assessment of the water management requirements for the mine.[15] At this stage any background information that is available is collected to understand the hydrogeological setting of the mine site as well as possible, to identify possible risks and develop an initial water balance. Therefore, meteorological data, geological data, environmental information (e.g. strategic area environmental assessment) and all water resources monitoring and modeling reports (both quantity and quality) and possible water quality for

---

[11] Gao et al. (2017), pp. 1009–1020.

[12] State of Victoria, Department of Primary Industries (2014).

[13] Gao et al. (2017), pp. 1009–1020.

[14] Punkkinen et al. (2016a, b).

[15] Department of Water Affairs and Forestry (2008).

watersheds and specific mining areas target acquisition.[16] These surveys are not only crucial to obtain permission for a mining project they are also necessary to assess the environmental impact of the mining operation and the water management objectives. The objectives of water management depend on the current climate conditions. In arid regions, the challenge is ensuring optimal water supply during mining, so mine water management aims to reuse and recycle as much water as possible and reduce waste water. On the other hand, in humid areas, it is important to have adequately sized reservoirs, diversion systems to separate fresh and process water, facilities to prevent uncontrolled discharge, landslides or liquefaction due to tailings releases during flood periods.[17] In order to minimize environmental impacts and contamination, there are a number of important issues related to groundwater systems that need to be considered and addressed already in the prospecting phase like the drilling of exploration boreholes.[18] If not planned and performed correctly, these could lead to significant long-term impacts on the water balance and the migration of contaminants. These issues contain the impact of exploration boreholes on ground and surface water resources, geochemical compositions of waste horizons generated and tailings management.[19] Geochemical investigations should also be used to predict the development of the water qualities. Predictions during exploration are generally generic and avoid assumptions about future engineering or mine design. But, initial assumptions should include the quality and quantity of the water. Qualitative predictions include assessing whether acidic conditions will develop with the attendant release of metals and acidity to mine drainage. In fact, predictions of whether acidic conditions will develop at high sulfur (often acid-producing) and low sulfur (often nonacid producing) are often straightforward. When qualitative predictions indicate that Acid Rock Drainage (ARD) is likely to develop without mitigation, options for prevention have to be considered. At the end of this phase, a low-level water management plan is developed, which should include details of borehole sealing and closure of trenches, bulk sample pits and adits and measurements to minimize their impact on the surface and groundwater regimes.[20] At the end it should be assessed whether anthropogenically created water-rock contacts and altered redox conditions due to the mining process will lead to water qualities that cannot be treated economically. If there is a risk of environmental impact, the mining project should not be pursued even though the prospects for the mine are promising. Of course, after-care measures should be developed for the exploration area. In the event of a positive decision to continue the mining project, consideration should already be given during exploration to whether the drilled exploration boreholes could be reused as groundwater monitoring stations.

[16] Department of Water Affairs and Forestry (2008).

[17] Punkkinen et al. (2016a, b) and CSIRO (2013).

[18] Department of Water Affairs and Forestry (2008).

[19] WRC (2015).

[20] Department of Water Affairs and Forestry (2008).

## 2.2 *Pre-feasibility*

The most important intermediate step in mining project evaluation is the preliminary feasibility study determining whether a mineral deposit is economically viable for mining. Based on the exploration data, the reserves are estimated, preliminary mining plans and technical assessments are compiled, sufficiently precise enough to allow an evaluation of different project configurations (e.g. different mining methods, optimizing overall water balance, water reuse and recycling plans, synergies) including the first version of the water balance model and monitoring for different future lifecycle phases.[21] The mine's water requirements and potential water supply sources are determined and other water use requirements are identified in consultation with the relevant regulatory authorities. The pre-feasibility study phase is typically an iterative process that is progressively updated and modified as exploration and engineering design progresses. During each iteration of the study, additional work should allow pre-development activities to progress to a stage where a full feasibility study is possible. Pre-feasibility studies should identify the critical issues and risks that need to be resolved in the final feasibility study.[22] In terms of water management, in this phase an initial first-order water management model with monthly time steps is developed, which includes basic meteorological and water quantity and quality data, including a regional borehole census for the mine.[23] In addition to the initial planning of the water systems, a monitoring concept is developed.

The first-order water management model should be implemented into the pre-feasibility mine plan and should be aligned with the preliminary groundwater plan, the waste management plan and the mine closure objectives. During the study the first-order water management model will be used to

- provide preliminary sizing and costing of the water management infrastructure,
- provide first order assessment of water supply and water demand for the mine,
- identify synergies and conflicts of water and energy management,
- serve input to the risk and impact assessment.[24]

The pre-feasibility initially unites the water and mining experts, their data and plan. In common planning, all possible end users of the treated water can be identified and evaluated. A synergy-focusing approach considers the additional supply for regional public drinking water supply, water for agricultural usage such as irrigation. Despite the high treatment costs are still high, this approach can realize some additional income, and supports the regional development.

---

[21] Punkkinen et al. (2016a, b).

[22] New Zealand Petroleum & Minerals (2017).

[23] Department of Water Affairs and Forestry (2008).

[24] Department of Water Affairs and Forestry (2008).

### 2.3 Feasibility

The feasibility study is focused on a project configuration from the pre-feasibility study to determine the technical and economic feasibility within the accuracy limits of the available data.[25] During this stage, a detailed water management model should be developed and integrated into the feasibility level mine plan, the detailed groundwater model, geochemical modeling and the detailed waste management plan.[26] The combined examination is intended to prevent important interrelationships of the various investigation tasks at a mining site from not being considered, which have a significant influence on the feasibility of the project. Otherwise, the project's overall feasibility appears to be achievable until the details are considered together. Further development of water management model will provide detailed information on water supply and the quantity and quality of discharge of the mine. During the feasibility study, simple deterministic monthly simulations are performed, involving water sufficiency and deficiency fluctuations during the seasons to calculate the water supply and water circuits of the water treatment plants and all operational units required for the operation of the mine. All related matters will be more accurately defined (Quality and quantity of the water sources (groundwater, surface water, rainfall, evapotranspiration), treatment demands, discharges). The model should be used to refine the water monitoring plan and to size the water infrastructure to encompass the requirements of the mine plant and water treatment facilities aligned with the regulatory requirements.[27] Climate change requires the consideration of unpredicted weather events like droughts or heavy rainfall events which require the calculation of minimum and maximum water volume needed in basins, the minimum and maximum discharge as well as changes in chemical conditions to identify possible needs to store water during surplus moths or to design an additional storage basin. The size of the water infrastructure will be checked after the submission of the Integrated Water Use License Application (IWULA). Information from the model can also be used as input for the Environmental Impact Assessment (EIA) process, the Environmental Management Plan (EMP) and the Socio Economic Impact Assessment (SIA).[28] The overall mine plan should consider wide-ranging water issues from initial design and construction, through operation (including slope depressurization, blasting impacts, dewatering volumes and chemical composition), to water balance for the mill and tailings facility and water issues at final closure.

In terms of synergetic uses of the water and energy system, measures formulated in the pre-feasibility phase are specified and enhanced based on an updated assessment considering its performance in terms of economy, environment, social aspects, resource efficiency, and safety.

[25] New Zealand Petroleum & Minerals (2017).

[26] Department of Water Affairs and Forestry (2008).

[27] Punkkinen et al. (2016a, b).

[28] Department of Water Affairs and Forestry (2008).

# 3 Development & Production

If exploration and planning phases have led to a conclusion that the project is viable and will be implemented, the planned measures are first implemented in development and then in production. The focus thus shifts from the planning to the operational. Monitoring and constant updating of models and planning are key for integrated operations. Hence, those areas are additionally addressed after this section.

## *3.1 Detailed Engineering*

As in this stage, the detailed design of all operational areas is conducted, also, the detailed design of the water infrastructure (dams, pumps, pipelines, etc.) is conducted and should also include obtaining approval for the construction of the water management infrastructure in particular any dams with a safety risk.[29] Depending on the size of the mine site and the extent of processing activities, tailings ponds can cover several hectares and be several tons deep. During the planning of tailings dams the recommendations of the International Commission on Large Dams (ICOLD), should be followed. The water management and monitoring program will progress to become a detailed description of the water system and the mine site and will provide more detailed information on estimated flows from the various units in the plant and mine area. The monitoring program should include the yearly amount of water intake, the amount and quality of water to be led to the tailing ponds, seepage and water flows in treatment facilities.[30] The detailed planning should already include concrete for the after-use of the water infrastructure after the closure of the mine for instance, the use as dynamic water reservoirs, for geothermal energy production, or energy storage systems. These after-use opportunities must be identified as early as possible to take the appropriate engineering measures from the outset in order to avoid expensive reconstruction measures during the closure process. All of this should be incorporated into a detailed mine plan, including details on underground mine service water systems, mine cooling systems, plans for handling water inflows and the civil design with the mechanical, electrical and instrumentation design work. The IWULA document and the license requirements (e.g. water use license) create the baseline for the Integrated Mine Water Management Plan (IWMP) for the mine during this stage and provides details on the water management and monitoring plan.[31] If necessary the water management model should be adjusted as well as the mine closure planning and closure costing details.

[29] Department of Water Affairs and Forestry (2008).

[30] Punkkinen et al. (2016a, b).

[31] Department of Water Affairs and Forestry (2008).

## 3.2 Site Construction and Commissioning

Beginning with the construction of the mine site, the water infrastructure is constructed as well and the first environmental impacts emerge (changes in topography, surface runoff, hydraulics of aquifers). The first monitoring data on quality and quantities are collected and reported to the authorities The minimum amount of data collected is set by the licenses and permits of the mines, but environmentally responsible mining companies have to go further.[32] The established monitoring plan should be reviewed and adjusted with increased knowledge of the water system and measured environmental impacts. At this stage, regular monitoring and audit reports are provided to mine management and construction staff establishing a solid cooperation for the production stage. Water management models should also be updated as needed.[33]

## 3.3 Production

Mines will operate for decades and will need to regularly and continuously collect critical information such as water quality and quality during operation and incorporate it into current water balance models and water management plans. The collected data will be used to recalibrate and update the mine water management model according to the operational requirements of the mine, to prepare monitoring and audit reports, to report to the regulatory authorities against the requirements of the IWMP and as feedback to stakeholders in the catchment.[34] The basis for optimal water management is not only a precise understanding of how much water is used in the different phases, but in the different operation units, and how much wastewater is generated and of what quality. During the extraction process, water management encompasses all of the waters and water cycles on a mine site as well as affected waters in combination.[35] This includes in general three categories of mine site (Fig. 3):

1. Process water flowing in the plant, process effluent waters
2. Natural waters/fresh water (groundwater, surface waters, precipitation, evaporation)
3. Waste waters (tailings, dams, ponds, seepage, used waters, treatment facilities)

The waters of the first category have their own closed cycle and require their own adapted monitoring system. Accurate monitoring enables the calculation of the fresh water and waste water quantity of the treatment plant. The waters of the

[32] Punkkinen et al. (2016a, b).

[33] Department of Water Affairs and Forestry (2008).

[34] Department of Water Affairs and Forestry (2008).

[35] Punkkinen et al. (2016a, b).

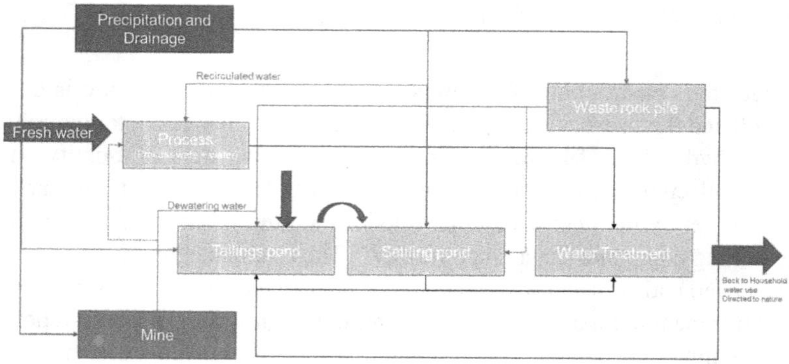


**Fig. 3** Overview of different mine water sources and streams (modified after Kauppila et al. 2011 and Punkkinen et al. 2016a, b)

second category can be used as a resource for operation or must be separated. Either way, the quality and quantity of the natural waters must be constantly monitored. The regular monitoring of surface and groundwaters in combination with climate measurements increases the reliability of the water management of the mine site and supports the forecasting of hydrogeological conditions and the preparation for unexpected situations. The most challenging are the waters of the third category as these pose the greatest risk of environmental hazards.

During the operation, the water management model should be used not only to minimize water consumption and wastewater generation by optimizing cycles, but also to review and design the water management measures required for mine closure in order to make financial provision for closure costs.[36] If in the course of changes in mine operations certain operation units are closed it is economically reasonable to rehabilitate them immediately. This has the advantage that they could be made available for another operational use and the required personnel and equipment are already in place and revenues are generated from the ongoing operation, which can be directly invested in the rehabilitation. Nevertheless, financial reserves for closure and renaturation should be built up during operation. Where unacceptable contaminant loads are predicted, the water management and treatment alternatives required to bring those to an acceptable level must be evaluated and it must be verified that the financial provision for such water management/treatment is included in the financial provision for mine closure. With each rehabilitation and remediation, the water monitoring and closure plans are revised and the water management plans for the closure phase are updated. Although mines are required to prepare preliminary closure plans only five years prior to closure, the operational phase should be used proactively to ensure that all relevant data are collected for the development and calibration of predictive models so that a reasonably reliable assessment of the required closure measures can be generated. This requires that careful data

[36] Punkkinen et al. (2016a, b).

management is undertaken from the outset, and that all existing mine residue facilities and mine workings are thoroughly characterized in terms of their physical and geochemical heterogeneity and their probable post-closure water balances.[37]

## 3.4 *Monitoring Challenges, Data Handling and Modern Approaches for Solving Them*

As mentioned, the role of monitoring and modeling increases highly after planning and exploration. Hydrological monitoring is an essential guide to evidence-based decision-making necessary for sustainable water resource management and governance.[38] Water monitoring can be classified into, quantity and quality which can be done both offline and online. Quantity measurements relevant to mine-water monitoring, precipitation and evaporation, can be obtained from a weather station. The discharges and groundwater levels are often measured manually. However, as mine water management requires more detailed information, online monitoring of mine water discharges and pressure sensors in groundwater monitoring pipes and weirs is becoming commonplace.[39] Continuous monitoring of the flow and water level is recommended. Nevertheless, this does not preclude periodic visual inspection to confirm the operation of the sensors and the monitoring program.

Realtime data on mine water results into a transparent and dynamic assessment of the water management and enable a faster response to quantitative and qualitative system changes. In order to have real-time data, digital technologies need to be incorporated into the data collection and analysis. However, it is not always possible or required to include life data, especially as communication in underground mines is a challenging task. The DECOMDA communication system developed at the Technical University Bergakademie Freiberg, for example, offers a solution for free underground communication without a complex and expensive network infrastructure. Also, the dynamic modelling approach needs a central database that allows the collected data to be analyzed in combination. According to More et al. (2020) wireless sensor networks (WSNs) can add significant value to mine water monitoring, as they require relatively low start-up and maintenance costs. The networks can be used for automated measurement systems, data collection of physicochemical parameters for management and treatment of mine water. A WSN network is composed of nodes and a base station. The nodes provide measurement, processing and communication functions while the base station is responsible for collecting and accessing all measurement data from the nodes and can sometimes provide gateway services to manage the data remotely. The data is collected, processed, stored and then transmitted to the monitoring server for analysis. Furthermore, data security

[37] WRC (2015).

[38] Weeser et al. (2018), pp. 1590–1599.

[39] Punkkinen et al. (2016a, b).

poses a major challenge. Sensitive content must be adequately protected and data preservation must be guaranteed in the event of system failures. Only a few mines are utilizing digital technologies and their experiences show that they are profiting substantially which is a sign that there is need to shift to digitization.[40] Digitization which might at the beginning be an expensive investment can do hand in hand with citizen science to minimize initial investment cost. Afterwards, both can operate simultaneously, with citizen science data as a control once full digitization is achieved. Some of the advantages of citizen science are its capability of increasing data collection and analysis at minimal costs, citizen science can argument project scope across temporal and spatial scales,[41] hence facilitating the observation of otherwise difficult to quantify phenomena. Other advantages of citizen science include education through improving scientific literacy and engaging with the public. As more information becomes available on water sources, use, rates, etc., monitoring measures will need to be further developed. With increasing potential of modeling and computing, digital twins and intelligent water management system become the new state of the art.

## 4 Mine Closure and After-Care Measures

After mining operations ceased, the landscape has changed drastically and with it the hydrogeological setting. The aim of aftercare and rehabilitation is to leave the site in a condition that will ensure a sustainable legacy for the surrounding communities and their future generations after mining.[42] Therefore, the mine closure plan should be an integral part of the mine life cycle starting from the first day of the initiation of a mining project followed by its inherent perpetual task. It is an iterative process that has to be reviewed periodically to assess the suitability of proposed measures to the development of the mine, the environment and the society. Mine closure has traditionally mainly focused on the rehabilitation and reclamation of mining regions by controlling and monitoring disturbed water, air and soil levels to a level acceptable to all stakeholders. Nevertheless, this perspective needs to change from focusing only on minimizing negative impacts, into a holistic approach that involves a closure and Post-Mining strategy able to offer extended opportunities of economic and social development. To implement this holistic approach does not only bring benefits to the society and environment but also to the mining company since new strategies, as progressive closure, could be implemented resulting in a reduction of closure costs. Furthermore, according to the Integrated mine closure model of ICMM, mine closure planning should be developed and reviewed with all stakeholders throughout the life of mine and consists of eight interactive elements.

---

40 More et al. (2020).

41 McKinley et al. (2017), pp. 15–28.

42 Punkkinen et al. (2016a, b).

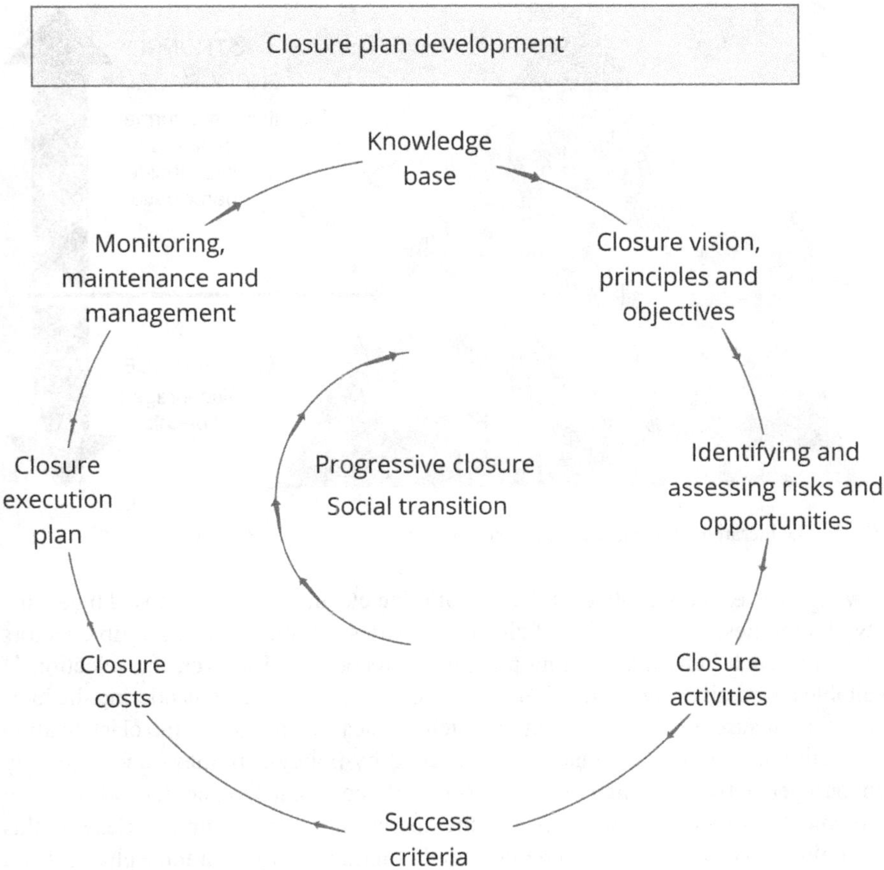


**Fig. 4** Elements of closure planning (simplified after ICMM 2019)

The elements that compose a successful mine closure plan as shown in Fig. 4 start the cycle by gaining deep knowledge in the related ecosystem, environment and society and defining a closure vision, principles and objectives in cooperation with the stakeholders. For each defined objective, a risk and opportunities assessment will be conducted and this will result in the definition of the concrete closure activities and related success criteria for evaluation. Later, the calculation of closure costs and definition of the closure execution plan takes place and can be conducted. After closure execution, the monitoring, maintenance and management of results of closure activities is carried out providing new information nurturing the cycle and allowing continuous success evaluation of the measurement. Conducting a closure execution plan as integrative part in the life cycle of a mine allows the implementation of progressive closure and social transition so risks, opportunities and reactions of the ecosystem and society to the measurements can be assessed and managed on time, if necessary.

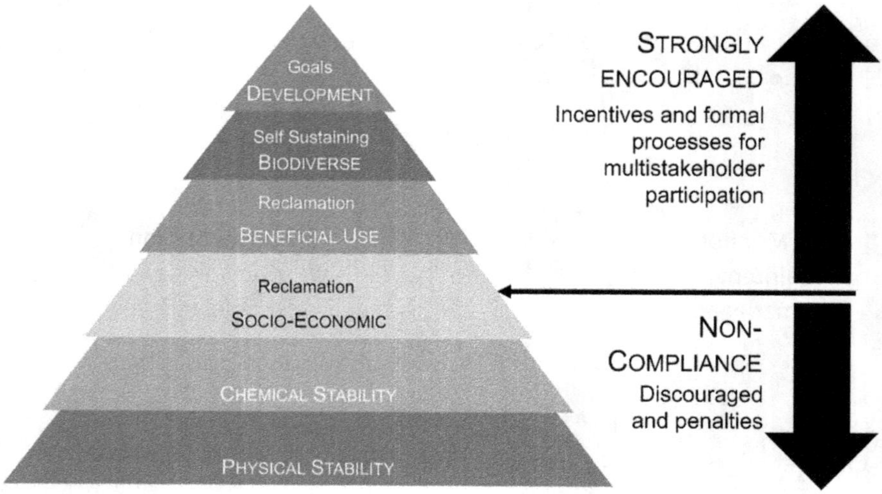


**Fig. 5** The hierarchy of closure needs (APEC 2018)

A closure execution plan is the heart of mine closure and is composed by a variety of activities. The definition of closure activities is influenced by multiple factors typical to each individual mining project and resource. Moreover, the selection of suitable alternatives for each individual project should succeed considering the hierarchy of closure needs,[43] as in Fig. 5 which defines two majors in the classification of activities: non-compliance or required-by-policy-activities and strongly encouraged activities which extension is usually not regulated and depends only on the objectives and internal policies of the operating mining company. Base on this pyramid, every closure execution plan should fundamentally guarantee physical and chemical stability, and promote a socio-economic smooth transition. In general, the transition to Post-Closure should generate new opportunities for the region.

The closure stage of a mine does not end after all closure activities have been implemented, it develops into a post-mining phase and integrates all after-care measurements necessary to guarantee a sustainable social, environmental and economic development of former mining regions. These regions are from this point onwards named regions in transition or post-mining regions. The first term refers to regions currently transitioning from a mining-based economy to other industries, at this stage the extraction has completely been finalized and closure activities have been implemented. The second term refers to regions that have already faced closure, are currently dealing with post-mining challenges and opportunities and long-term control and monitoring activities are been conducted.

In general, the transition to Post-Mining should generate new opportunities for the region which can only be reached under a holistic approach and represent a

[43] APEC (2018), p. 104.

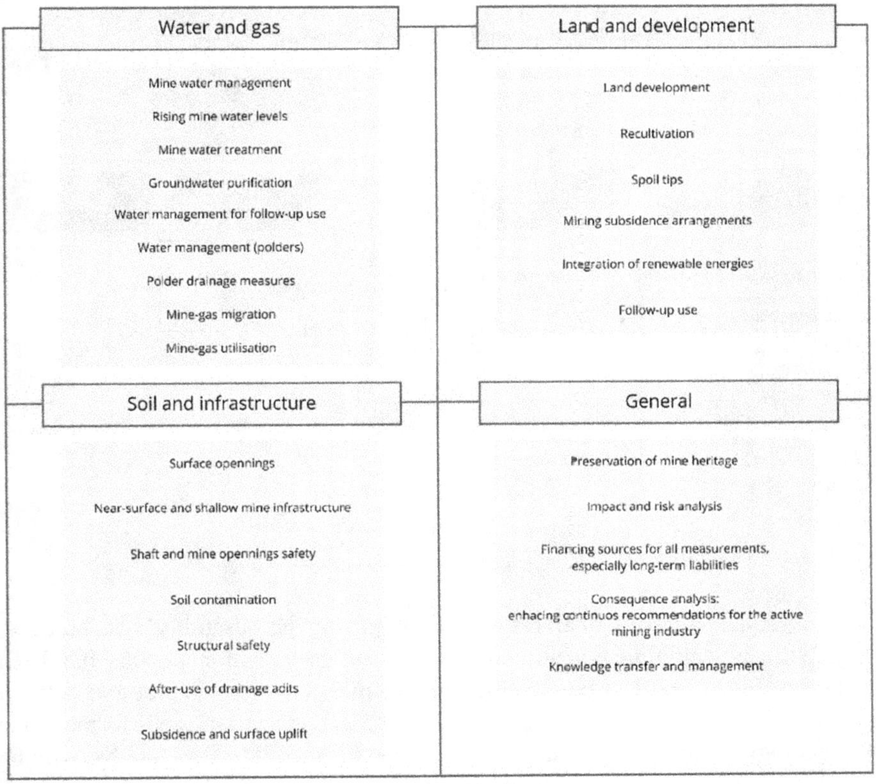


**Fig. 6** Post-mining themes (modified after Melchers et al. 2020)

perpetual task involving all stakeholders.[44] In a modern society where, sustainable processes are becoming a standard to which emerging generations lay high value at, the correctness and success of the implemented post-mining measurements can directly influence the willingness of other societies to support or not upcoming mining projects. To achieve successful strategies, planning of post-mining measurements has to be integrated since early stages of a mine, provide sustainable and long-lasting solutions to the four post-mining themes and built up the closure vision and objectives. The four post-mining themes,[45] as in Fig. 6, classify the general challenges post-mining regions can face after closure into four major categories: water and gas, land and development, soils and shafts, and general challenges.

Germany counts with a vast tradition in mining and in the last thirty years has faced mine closure in different stages and developed post-mining measurements in former mine sites and in research. Furthermore, based on the Blue Mining concept, the closure of a mine site does not represent anymore the abandonment of the site

[44] Kretschmann (2020), pp. 1401–1409.

[45] Melchers et al. (2020), p. 346.

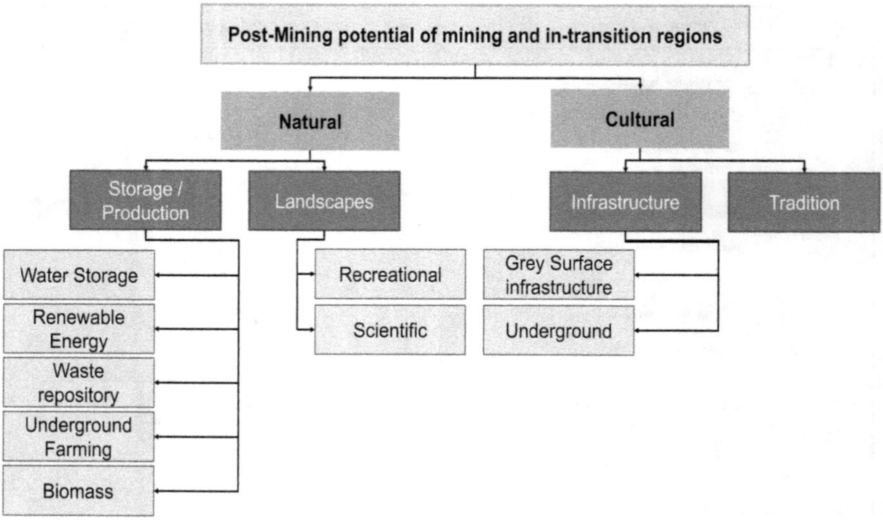


**Fig. 7** Post-mining potential of mining and in-transition regions

but a subsequent after-use of all resources maximizing their usability is encouraged. Hence, the post-mining potential of mining and in-transition regions has been defined in Fig. 7 and can be classified in two major groups: natural and cultural potential. The natural potential enhances the natural characteristics of a former mine site (geology, location, environment, etc.) and associated infrastructure by defining its suitability to host development-relevant solutions. Hereby, the two sub-categories are the after-use as storage facilities, and the remediation and reclamation of mining landscapes for recreational, scientific and educational purposes. Through the years, different initiatives have been evaluated for underground and surface mines, some examples are the after-use of open pits and drainage adits as water reservoirs, the after-use of surface and underground cavities for energy storage from renewable sources under the concept Power-to-X, and even the use of former underground salt mines as final waste repositories for radioactive materials. Meanwhile, the cultural potential strengthens mining culture and tradition through the subsequent use of infrastructure and by the support to maintain mining traditions in the regions.

The selection of the best-suitable closure and post-mining strategy is influenced by the factors that influence closure activities as in Fig. 8 and is a task that must be carefully evaluated since it defines and influences the future of the affected region. Moreover, water as a vital resource represents a key player for regional development of post-mining regions and can only be achieved through the implementation of dynamic and integrative water management systems.

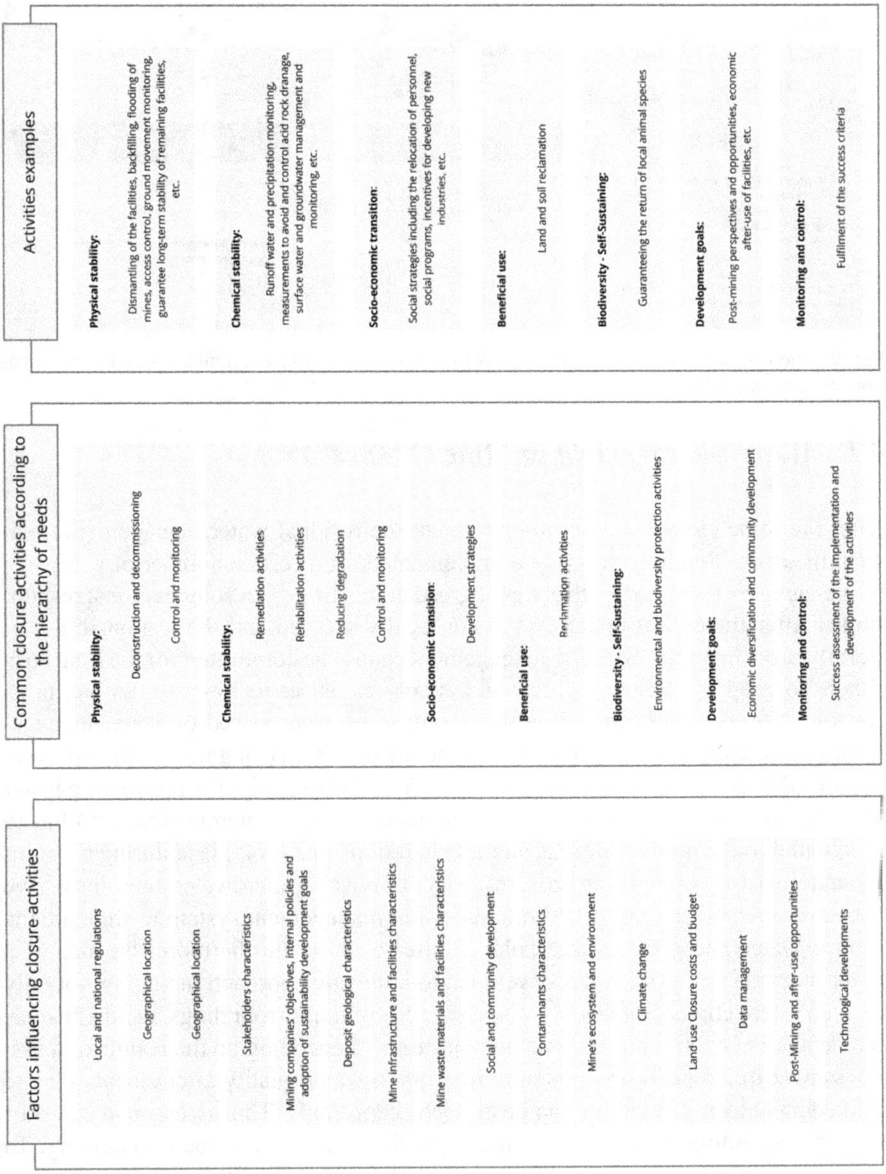


**Fig. 8** List of major factors influencing closure activities, common list of activities classified according to the hierarchy of needs and activities examples

**Fig. 9** The resource-pathway-receptor model to properly identify contaminants and contamination sources and define action areas

## *4.1 Water Management in Mine Closure*

After the mine closure has been authorized, the residual water management/treatment measures that may need to be implemented for a certain number of years will be managed by third parties that were agreed at this time.[46] In some regions, responsibility after the end of operations is not legally secured and the responsibility of eternity tasks in related to water management cannot be committed for the long term either. To keep maintenance, costs and energy expenses as low as possible, post-closure water management should be gravity driven and based on water management structures resilient to physical and chemical erosion (long life span) to prevent possible defects of equipment (e.g. pumps). As a starting point the pre-mining water baseline (climatic conditions, yearly precipitation rates, watercourses, etc.) has to be updated to the new conditions and all alterations that took place during the mining activity (diversion of watercourses, water removal, groundwater pumping), have to be documented and integrated into the water management system to simulate the surface water and groundwater balance after closure. Furthermore, the use of a water management system has developed to a holistic approach including not only the ecosystem characteristics, but also the needs of the surrounding catchment areas and dependent communities and other end-users. Depending on the results of a first assessment the impacts and potential impacts in water quality after closure can be defined and closure activities can be defined accordingly. This assessment can start with the beginning of the mine by using predictive models for better planning. To prevent future water contamination and guarantee water quality, a source-pathway-receptor model can be used to define closure activities.[47]

Figure 9, describes the three stages of this approach and considering the results on identification of receptors (human and environmental) urgency and specific activities can be determined. In the presence of contaminants, activities such as

[46] WRC (2015).

[47] ICMM (2019), p. 132.

isolation of the contaminants, surface water diversion, contact prevention passive treatment or active treatment of water can be defined. Depending on the waste material, water balance and potential impact on receptors, long-term water treatment is required. However, the isolation of waste material and contaminants from water sources tend to be more rentable since it is hard to forecast the required time water treatment will be required. In the case this cannot be avoided two approaches come into consideration the active or passive treatment of water. The active treatment usually includes a water processing plant and processes as neutralization, metal removal, clarification, etc. Meanwhile, passive treatments refer to systems relaying on natural processes for the removal of contaminants. Moreover, considering the complexity of site-specific conditions hybrid treatments are commonly implemented.

As outlined in Fig. 8, in addition to mine closure and dismantling of the facilities, backfilling and flooding of the mines may be required and these are part of the measurements that guarantee physical stability of a mine site and are non-compliant. In the case of mine flooding, the operative mine drainage systems are stopped in order to allow natural surface and groundwater to enter the mine cavities through shafts and adits or to fill an open cast excavation. This process can take several years or even decades and large volumes of water are necessary that would be unavailable for other purposes during the flooding period. Flooding measures specifics are closely connected to geotechnical measures, geomechanically stability and chemical stabilization of runoffs might be required. Depending on the mine's infrastructure and the mine sites hydrogeology, groundwater recovery management and water treatment might be needed as in the case of underground mines, in many cases, perpetual water discharges from the underground mine openings cannot be avoided.

Keeping a close eye on water management during closure and setting up an efficient monitoring system is vital for the transition to Post-Mining. The new created environment sets the base for the regional development. Depending on the area, this might be more civilization- or nature-driven. All scenarios rely on a "healthy" water system that is not harming the basis of existence. For human and wildlife, access to fresh water often increases the quality of life. The closure as transitioning stage can last decades as water system are (re)established. For the beginning, an active water treatment is considering to reduce pollution peaks. Covering and sealing waste and tailings is isolating the pollutants and must be initially done for reclaiming these areas. Groundwater management is usually required for a longer period. Finally, self-stable, passive system are preferable to keep the high stability without high operational efforts.

## 4.2 *(Historic) Post-mining Areas*

Water management in former mining regions is particularly challenging as the natural storage function of soils and vegetation has been affected. In historic post mining areas, where mining dates back several centuries, water suppliers and communes have to deal with insufficient data on the spatial extent, geometry and the present condition of the historic mine workings and water pathways which leads to uncertainties in the planning and distribution of water resources. Moreover, climate change exacerbates these uncertainties. Heavy rainfall events can not only lead to unexpected water releases but can also lead to the deterioration of water quality. Many post-mining regions have to deal with subsidence which requires artificial drainage and results in high energy demand for pumping systems (Fig. 10).

In addition to the various challenges and planning uncertainties caused by an insufficient data basis or data structure in the past, there are also many advantages of former mining regions. Figure 7 shows a good overview of the different potentials of the former mining regions. The potential of post-mining landscapes can be roughly divided into natural and cultural potential. The cultural potential includes not only the traditions and ways of life that have often evolved over centuries as a result of mining, but also the potential for subsequent use of the existing infrastructure. The natural potentials include on the one hand the after-use of landscapes for research and recreation but also the opportunity to store natural resources such as water. To store sudden water volumes in a controlled manner and to make them available again when needed, dynamic reservoirs are required. Dynamic water storage systems provide a flexible way for a region to adapt to different climatic events. However, the construction of a dam is subject to the morphology of a suitable region, has high investment costs, and is a significant interference with the land surface areas, its natural surroundings and monuments. In the case of water storage at the surface, losses due to evaporation are increased by rising temperatures. The water is also exposed to temperature fluctuations, leading to increased algae and bacteria growth on warm summer days. Additionally, not all former mining regions can be reforested fast enough to improve their natural storage capacity. Underground water reservoirs represent therefore one sustainable solution with the possibility of synergies between water storage and energy supply systems. Former mining areas have the advantage that they already host underground cavities so these would not have to be excavated separately. The projects "Energy and Water Storage Harz (EWAZ)" and "Water Storage Harz 2050 (WSH2050)" present approaches for the after-use potentials of historic mining infrastructure in the Harz Mountains. Climatically, the Harz Region is one of the areas with the highest precipitation rates in Germany with its annual precipitation of 1326 mm on the Clausthal plateau. Furthermore, its storage capacity based on six dams in the western Harz and four groundwater works from the interconnected system of the Harzwasserwerke GmbH (HWW) supplies around two million people and numerous important industrial companies from the Harz Mountains to Bremen with drinking water every day.[48] The low mountain range not only has a special status due to the UNESCO World Heritage the "Upper Harz Water Regale", but the water quality also stands out in comparison to many other areas in northern Germany[49] and is therefore not only of special water management relevance for areas with nitrate-polluted aquifers but also

48 Harzwasserwerke GmbH (2022).

49 Niedersächsischer Landesbetrieb für Wasserwirtschaft Küsten- und Naturschutz (2019).

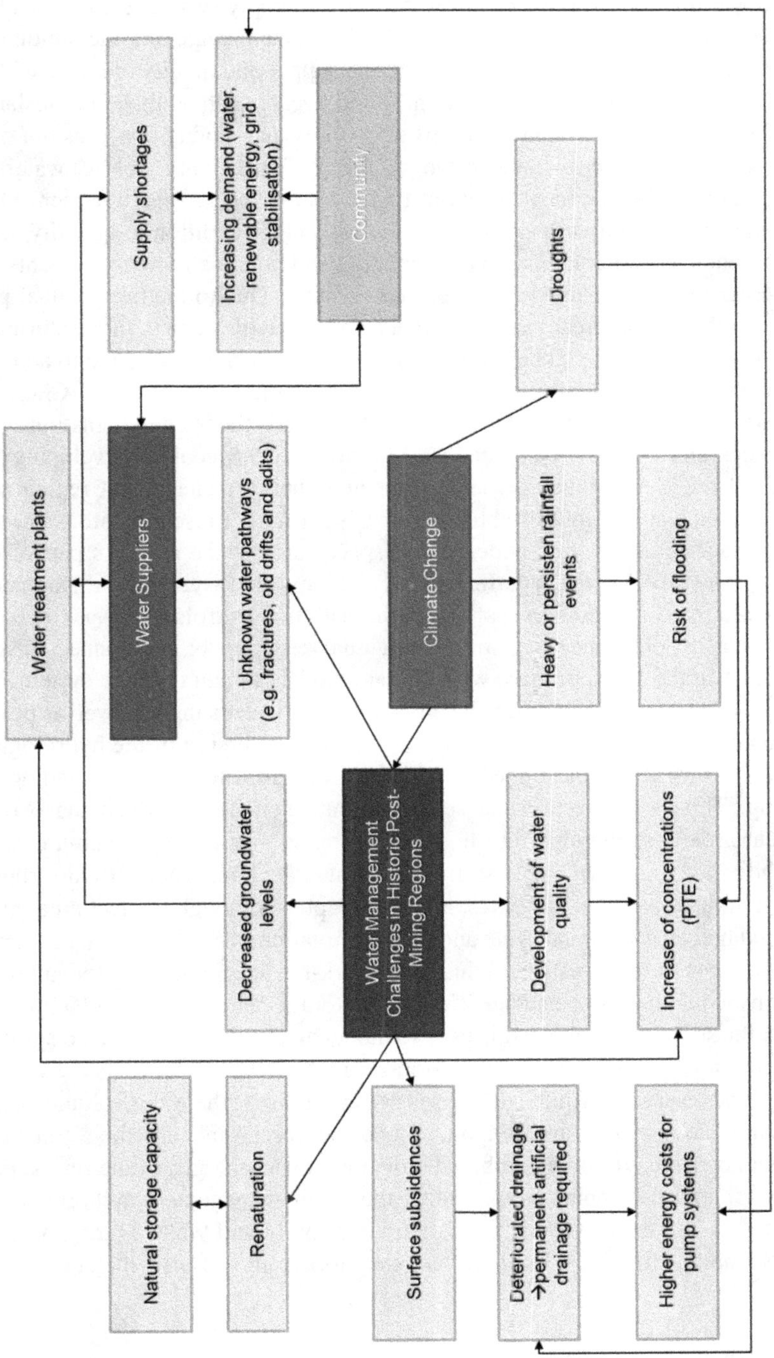


**Fig. 10** Interconnections of water management challenges in (historic) post-mining areas

for regions in the rain shadow of the Harz Mountains with an annual average of only 481 mm precipitation over 50 years.[50] If the current supply system is not expanded, the HWW see the challenge that as a result of climate change and the simultaneously increasing demand of the user region, the supply and quality of water will be limited in the medium and long term and flood events such as those in Goslar in 2017 will increase. The demand in terms of quantity and quality concerns not only the (drinking) water supply but also the provision of sufficient (cooling) water for industry, such as the chemical plants in Langelsheim, the sewage treatment plant pipelines, and agricultural operations in the foreland. In addition to quantity, temperature trends, microbiological parameters, pH and potentially toxic elements are of particular interest for regional water management. Due to the high annual precipitation rate and the already existing underground cavities due to the centuries of mining history, their potential as water reservoirs should be investigated to actively integrate them into the current water supply network. Therefore, the Clausthal University of Technology (CUT), the HWW, and the Federal Institute for Geosciences and Natural Resources (BGR) have set the goal of developing the potential of the Harz Mountains as a water and energy management region in a holistic, digital way and thus establishing a basis for further research into water and energy storage in above- and underground spaces around the area. As part of the previous project EWAZ, the department of Hydrology and River Basin Management of LWI has already combined a spatially high-resolution hydrological water balance model with a model for the operation of individual reservoirs or interconnected storage systems in the Western Harz with climate models to simulate the system services: Flood protection, low water elevation, drinking water supply as well as power supply using energy storage to simulate. The planning version of the hydrological software PANTA RHEI developed at HYWAG was used for the hydrogeological modelling. This is a deterministic semi-distributive model in which the overall catchment area is subdivided into high-resolution sub-areas. These sub-areas are also subdivided into the smallest hydrological calculation units (hydrotropes) according to land use, soil attributes, and topography. Although the leaf area index is included here in the annual cycle and the crop rotation of the long-term economy, current changes in the forest stand due to the bark-beetle and the last dry summers and storms would have to be added. The linked storage operation model IGOmod is based on the software Gecko and is used to model arbitrary interconnected systems and operation rules for multi-purpose storage. PANTA RHEI provides the natural inflows of the reservoirs, while IGOmod is used to simulate the water volume operation of individual reservoirs and interconnected reservoirs and their transfers. Parameters considered in the combined model are meteorological time series, processed as spatial data fields from data of the water management authorities, the HWW, and results of the project Climate Impacts on Inland Water Management in Lower Saxony (KliBiW).[51] Furthermore, the storage model also includes single

[50] Harzwasserwerke GmbH (2022).

[51] Nistahl et al. (2021), pp. 711–718.

areas of underground mining cavities. In the WSH 2050 project, this already existing combined model is to be extended to include the entire underground system. This expanded model is intended to serve as a management tool that enables targeted forecasts about holistic, adaptive water management of the region while reducing conflicts of use by generating various options for action for different issues and simulating and comparing their influence on water quality and quantity so that the current water demand by the Harz can also be ensured in the future.

Another after-use concept includes the geothermic energy use of mine water in abandoned flooded mines. Oppelt et al. (2022) investigate the efficiencies and possibilities of geothermal energy production as mine water has almost constant temperatures all year around and is ideal for heating and cooling. Based on a literature research Oppelt et al. (2022) state that 44 active mine water geothermal plants exist. 10 out of 18 plants in Germany are located in the Ore Mountains. Mine water geothermal energy competes with other renewable and fossil energy sources, but has become much more attractive in the context of recent geopolitical developments. The economic viability of the plants is also studied and presented by Oppelt et al. (2022). A challenge is the elevated load (e.g. bacteria, dissolved/undissolved substances and other suspended matter) of the mine water, which leads to precipitation processes on heat exchangers and thus to a reduction in the efficiency of the mine water geothermal plants. Various optimization options are currently being researched to minimize the fouling process. Where appropriate quantities and potential gradients are available, the mine water can also be used to generate electricity in hydropower plants. In Addition, synergy effects can be created in water treatment, as shown by the SULFAMOS joint project. In this project funded by the BMBF, the sulphate-contaminated mine water is treated as drinking and irrigation water using forward osmosis. A by-product of this process is gypsum for the building materials industry, which is generated by precipitating the sulphate.[52] Other exciting concepts and research projects have been presented at the Kassel22 "Let's talk about Grubenwasser" symposium.

Unfortunately, the operating costs of treatment are currently higher than the revenues generated by the consumption of the treated water by various end users. While there are already various solutions to the obstacles of water management for active mines, the barriers for closed, historic mines are higher and the market opportunities are lower.

Post-use and water management strategies that are resonant to climate variability are needed. The interactions between former mining activities and water resources are site-specific and exhibit complex system behavior characterized by feedback effects. Disturbances such as extreme weather events and a change in management strategy can lead to unexpected system changes. The basis for sustainable post-use is therefore an understanding of the dynamics of heterogeneous mine water systems and the identification of the system-determining factors and is part of the current research at the institute of mining at the Clausthal University of Technology. It is,

[52] BMBF (2022).

therefore, necessary to first identify the key parameters needed for responsible and sound water management and to collect the best available knowledge, including local knowledge, and where appropriate traditional knowledge, and define the best approaches on possible ways, positive impacts and challenges, strengths and weaknesses, in improving the water management of post-mining areas.

## 5 Integration into Academic Teaching

As shown by the water aspect during the lifecycle of the mine, it is an omnipresent topic in Mining. Hence, all persons working in the mining industry are somehow affected and need skills within this framework. The base for these skills needs to be set during the studies and water needs to be addressed as a topic. Similar to Safety and Sustainability, water is an overarching topic, and approaches for integration in the curricula can be adapted from those. Concepts for the integration of these overarching themes are presented and implemented by Binder et al. (2018)[53] and Hutwalker et al. (2021)[54] using the approach initially introduced by Shields et al. (2014)[55] for petroleum and mining engineering. So also, for water, four approaches of integration can be used:

- Coverage of some water matters in existing courses
- Specific courses on water
- Intertwining the field of water in regular courses, and
- Specialization in the field of water.

As presented in the previous sections, water influences all stages of the lifecycles of mines. Hence, at least the first approach needs to be used in modern curricula by addressing the relevance of water for the different areas. Addressing the topic sensitizes students to the topic. Specific courses in water need these sensitized students. Cooperation with other study programs allows the integration of specific courses from other curricula such as hydrogeology, but required previous knowledge might be challenging. On the other hand, specific courses for the discipline can be developed when the target group is large enough. To include the topic deeper in existing curricula, intertwined courses e.g. connecting mine planning and water can be introduced as compulsory courses. During the course completion, the students develop a complex and integrated concept of mine water connected to their existing skills. Hence, this approach is promising but also challenging in conception and implementation. Creating specialized study fields on mine water might be difficult to manage as enough interested students must be available. Partially, this concept is followed when students choose electives in the field of water and address the topic

[53] Binder et al. (2018), pp. 435–441.

[54] Hutwalker et al. (2021), pp. 175–183.

[55] Shields et al. (2014), pp. 390–403.

in theses and projects. Student research projects and theses can be linked with larger research topics and specialized skill sets for urgent topics can be developed.

Regardless of the concept, the course objectives should be formulated first following Cowan and Harding (1986)[56] as well as Biggs (1996)[57] while considering the chosen concept integrating the topic of water in the overall context, and addressing target-group-specific skills in the area of water. For experts in different fields, understanding the future needs of graduates is both challenging and essential for effective course design.

## 6 Conclusion and Outlook

Water is a very valuable resource that is required and utilized by mines throughout the mine life cycle. Conflicts of use between mining operations and local water supplies will continue to increase due to growing demand and global warming. Therefore, sustainable water management in mining and the rethinking of current processes is critical for the mining industry to achieve a 'social license to operate' and long-term profitability. By proactive water management best introduced in the earliest planning stage, challenges are minimized and synergies might occur, moving away from a reactive to a proactive approach. The approach Blue Mining representing an academic approach so far is ready to implemented in an active project. It offers the possibility of more synergies and increased resource efficiency of a mining project in the sense of the circular economy through the targeted identification, promotion and optimization of utilization cycles. Water in particular offers many potentials for process cycles and multiple uses. In particular, the use of hydropower and energy storage can support local sustainable energy supply and thus the achievement of a region's climate protection goals. Modern dynamic modelling approaches and online monitoring are gaining in importance in water management in mining as these enable the understanding of complex system interrelationships and the consideration of unforeseen events. Thorough planning and implementation of a water management plan from the blueprint of the mine combined with dynamic system investigation supports and accelerates decision-making during mining operations, saves resources, creates additional value and safety by being able to consider long-term impacts of the water balance of a region from the beginning. Until now, only a few mines use digital technologies and dynamic water balance models, as the barriers to digitalization are higher for already active and small-scale mines. However, new communication systems offer enormous potential to reduce these barriers. Partial approaches can also be implemented in existing mines and post-mining region but a retroactive socio-economic and biophysical rehabilitation of abandoned mines is far costlier than preventing these impacts through intelligent mine design

[56] Cowan and Harding (1986), pp. 103–109.

[57] Biggs (1996), pp. 347–364.

and technologies and ongoing monitoring and maintenance. The introduction of modern, holistic, intelligent water management systems and models after the exploration and planning stage is more challenging as the base from exploration and planning, as important background data is missing at times. Nevertheless, the historical post-mining regions also offer enormous potential for reuse, as illustrated in the EWAZ and WSH 2050 projects. Current research is distancing itself from the idea of considering mine water only as waste from a mining operation, but is developing new recycling and utilization potentials. For all those synergetic activities, the collaboration with local partners especially local water providers is essential to act in favor of a regional development in one synergetic water system. Research projects present models for those actions and demonstrate by the high number of collaborators the complexity as well. For a sustainable integration as a state of the art, water aspects must be incorporated in university education in different ways to train the needed experts.

## References


APEC (2018) Mine Closure Check List for Government, p 104

Biggs JB (1996) Enhancing teaching through constructive alignment. High Educ 32(3):347–364

Binder A, Langefeld O, Hitch M (2018) Integration of Sustainability in Mining Engineering Education. Integration von Nachhaltigkeitsaspekten in die Bergbauausbildung. In Mining Report Glückauf 154(5):435–441

BMBF (2022) SULFAMOS. https://bmbf-wave.de/Verbundprojekte+nach+Themenfeldern/Salzhaltiges+Wasser/SULFAMOS.html. Accessed 22 May 2023

Cowan J, Harding AG (1986) A logical model for curriculum development. Br J Educ Technol 17(2):103–109

CSIRO (2013) Helping miners find the right water balance. https://csiropedia.csiro.au/helping-miners-find-the-right-water-balance/. Accessed 22 May 2023

Department of Water Affairs and Forestry (2008) Best Practice Guideline A6: Water Management for Underground Mines

Gao L, Bryan BA, Liu J, Li W, Chen Y, Liu R, Barrett D (2017) Managing too little and too much water: robust mine-water management strategies under variable climate and mine conditions. J Clean Prod 162:1009–1020. https://doi.org/10.1016/j.jclepro.2017.06.101. Accessed 22 May 2023

Harzwasserwerke GmbH (2022) https://www.harzwasserwerke.de/ueber-uns/portrait/. Accessed 22 May 2023

Hutwalker A, Binder A, Langefeld O (2021) Arbeitsschutz im Bergbau und die Berücksichtigung in der Hochschulausbildung. Occupational safety and health and its consideration in higher education. In Mining Report Glückauf 157(2):175–183

ICMM (2019) Integrated mine closure: good practice guide, 2nd edn. February 2019, p 132

Kauppila P, Räisänen ML, Myllyoja S (2011) Best Environmental Practices in Metal Ore Mining. Finnish Environment 29en, p 222

Kellner M, Langefeld O (2013) "Blue Mining" - the future of mining. In: 6th International Conference on Sustainable Development in the Minerals Industry. Milos, Greece

Kretschmann J (2020) Post-mining—a holistic approach. Min Metall Explor 37:1401–1409. https://doi.org/10.1007/s42461-020-00265-y. Accessed 22 May 2023

Kunz NC, Moran CJ (2014) Sharing the benefits from water as a new approach to regional water targets for mining companies. J Clean Prod 84:469–474

McKinley DC, Miller-Rushing AJ, Ballard HL, Bonney R, Brown H, Cook-Patton SC, Phillips TB (2017) Citizen science can improve conservation science, natural resource management, and environmental protection. Biol Conserv 208:15–28

Melchers C, Goerke-Mallet P, Kleineberg K (2020) Elements and aspects of the post-mining era. In: Done for Good 2.0 - results in post-mining research, p 346. ISBN: 978-3-949115-00-4

More KS, Wolkersdorfer C, Kang N, Elmaghraby A (2020) Automated measurement systems in mine water management and mine workings – a review of potential methods. In: Water Resources and Industry Bd. 24

Mueller S (2015) What is Goldsim? Boliden Mining, Sweden. WaterSmart Seminar on Management of water balance in mining areas, 28th of August 2015. Finnish Environment Institute, Helsinki

New Zealand Petroleum & Minerals (2017) Minerals guidelines: scoping, pre-feasibility and feasibility studies; MB13243_1748

Niedersächsischer Landesbetrieb für Wasserwirtschaft Küsten- und Naturschutz (ed) (2019) Grundwasserbericht Nidersachsen. Sammlung-Parameterblätter-Datenbestand 2018. With assistance of Annette Kayser. 1st edn. www.nlwkn.niedersachsen.de. Accessed 22 May 2023

Nistahl P, Müller T, Lange A, Meon G (2021) Modellierung von vernetzten Speichern unter Mehrfachnutzung und Klimaänderung im westlichen Harz. In KW Korrespondenz Wasserwirtschaft 14(11):711–718

Northey SA, Mudd GM, Saarivuori E, Wessman-Jääskeläinen H, Haque N (2016) Water footprinting and mining: where are the limitations and opportunities? J Clean Prod 135:1098–1116

Oppelt L, Grab T, Wunderlich T, Storch T, Fieback T (2022) Mine water geothermal energy -abandoned mines as a green energy source. 10.34641/clima.2022.55. Accessed 22 May 2023

Punkkinen H, Räsänen L, Mroueh UM, Korkealaakso J, Luoma S, Kaipainen T, Backnäs S, Turunen K, Hentinen K, Pasanen A, Kauppi S, Vehviläinen B, Krogerus K (2016a) Guidelines for mine water management. VTT Technol 266. ISBN 978-951-38-8443-7

Punkkinen H, Räsänen L, Mroueh UM, Korkealaakso J, Luoma S, Kaipainen T, Backnäs S, Turunen K, Hentinen K, Pasanen A, Kauppi S, Vehviläinen B, Krogerus K (2016b) Guidelines for mine water management. https://cris.vtt.fi/en/publications/guidelines-for-mine-water-management. Accessed 22 May 2023

Shields D, Verga F, Andrea Blengini G (2014) Incorporating sustainability in engineering education: adapting current practices to mining and petroleum engineering education. Int J Sustain High Educ 15(4):390–403. https://doi.org/10.1108/IJSHE-02-2013-0014. Accessed 22 May 2023

State of Victoria, Department of Primary Industries (2014) Management of water in mines and quarries, Victorian legal and policy requirements. Environ mental Guidelines

Water Research Commission (2015) Water and mining, developing risk criteria for water management aspects of mine closure. https://www.wrc.org.za/wp-content/uploads/mdocs/PB_2127_Water%20management%20aspects%20of%20mine%20closure.pdf. Accessed 22 May 2023

Weeser B, Kroese JS, Jacobs S, Njue N, Kemboi Z, Ran A, Breuer L (2018) Citizen science pioneers in Kenya - a crowdsourced approach for hydrological monitoring. Sci Total Environ 631:1590–1599

# The German Federal Constitutional Court's Climate-Decision and Consequences for Raw Materials Supply

**Walter Frenz**

## Contents



---

This text is based on the author's book "Grundzüge des Klimaschutzrechts" (3 edn 2023).

---

W. Frenz (✉)
LFG Berg-, Umwelt- und Europarecht, RWTH Aachen University,
Aachen, Nordrhein-Westfalen, Germany
e-mail: frenz@bur.rwth-aachen.de


W. Frenz, A. Preuße (eds.), *Yearbook of Sustainable Smart Mining and Energy 2022*, Yearbook of Sustainable Smart Mining and Energy - Technical, Economic and Legal Framework 2, https://doi.org/10.1007/978-3-031-41873-0_5

**Abstract** The German Federal Constitutional Court considers the civil rights of the younger generations to be violated by the fact that the climate protection provided for by law to date is not sufficient to protect them from excessive climate burdens in the future. To put it simply: If those living today do not reduce greenhouse gas emissions sufficiently, the young will have to do so even more and will then be overburdened.

The German Federal Constitutional Court's climate decision thus offers a good opportunity for more effective climate protection in Germany, which must, however, be reconciled with the more ambitious EU targets set by the EU Climate Act. Here, Germany has to make a weighty, as required solidary contribution according to its strong economic power. This is the real challenge, which the German Federal Constitutional Court unfortunately did not take into account. Climate protection is not only international, but also European, as Section 4 (1) Sentence 5 of the Climate Protection Act shows.

The Russia-Ukraine war shows that climate protection is also about securing raw materials. This must therefore be ensured efficiently through international efforts. This is the only way to achieve the energy transition. Climate protection must also be conceived on a global scale. This means that all circumstances that are indispensable for climate protection must be considered globally. This also applies to the necessary raw materials. The supply of these raw materials must be secured across national borders.

# 1 Fundamentals

## *1.1 Result*

As requested, the German Federal Constitutional Court declared the German Climate Protection Act from 2019 as partly unconstitutional. The legislator therefore had to deliver significant improvements. This applied at any rate to the period after 2030, because more detailed specifications on how to achieve the goal of climate neutrality by 2050, as envisioned under Section 1 Sentence 3 of the German Climate Protection Act, were completely lacking for this period of time. However, the German Federal Constitutional Court calls for a fair distribution of the burdens between older and younger generations and wants to avoid a situation where, due to insufficient efforts up to and including 2030, heavy burdens fall on the younger generations from 2031 onwards.[1] As a result, the Climate Protection Act actually had to be improved for the period up to 2030 as well. Otherwise, intergenerational equity may not be achieved. This is exactly what the legislator did by raising the interim reduction target from 55% to 65% by 2030.

However, the German Federal Constitutional Court considered the provisions of sections 3 (1) sentence 2, 4 (1) sentence 3 German Climate Protection Act in conjunction with Annex 2 to be constitutional until 2030.[2] It only provides the legislator with some key points from a very differentiated interpretation of environmental state objectives, without specifying the measures in detail. Only the long-term path to reduce $CO_2$ emissions for the period after 2030 must be clearly defined and mapped out in the German Climate Protection Act, so that those affected can prepare themselves for the significant restrictions to be expected.[3] In other words, it is a matter of legitimate expectations for the future.

The legislator continues to have the freedom to define the details as long as it ensures a continuous reduction of $CO_2$ emissions in accordance with the constitutional requirements. In this respect, the German Federal Constitutional Court is more specific than the Court of Justice in The Hague,[4] which only placed the Dutch state under the obligation to do more than it had done so far to avert the consequences of global warming. Before the German Federal Constitutional Court's decision, on the other hand, it had been predicted that the Climate Protection Act suits would not be successful.[5] At most, a declaratory decision by the Federal Constitutional Court like the one in The Hague was considered.[6] In Karlsruhe, however, this was ultimately achieved by not merely declaring "more climate protection", but by declaring the contested sections 3 (1) sentence 2, 4 (1) sentence 3 German Climate Protection Act 2019 in conjunction with Annex 2 as incompatible with the constitutional rights, because the reduction targets were not extended beyond 2030.[7] They instead remained applicable in order to avoid a temporary loophole; the legislator had to have acted by December 31, 2022.[8] However, he acted immediately, so the amended Climate Protection Act applies.

[1] German Federal Constitutional Court (GFCC), Cases 1 BvR 2656/18 inter alia, ECLI:DE:BVerfG:2021:rs20210324.1bvr265618, margin no. 182. In addition Frenz (2021a), pp. 810 et seq. regarding the following as well.

[2] GFCC, Cases 1 BvR 2656/18 inter alia, ECLI:DE:BVerfG:2021:rs20210324.1bvr265618, margin no. 231.

[3] GFCC, Cases 1 BvR 2656/18 inter alia, ECLI:DE:BVerfG:2021:rs20210324.1bvr265618, margin no. 243 et seq.

[4] Supreme Court Netherlands, Case 200.178.245/01, ECLI:NL:GHDHA:2018:2591, Case 19/00135, ECLI:NL:HR:2019:2007, No. 5.7.7.

[5] E.g. Groß (2020), pp. 337 et seq.; Meyer (2020), pp. 894 et seq.; Frenz (2020), pp. 1 et sqq.

[6] Groß (2020), p. 337 (342).

[7] GFCC, Cases 1 BvR 2656/18 inter alia, ECLI:DE:BVerfG:2021:rs20210324.1bvr265618, margin no. 267.

[8] GFCC, Cases 1 BvR 2656/18 inter alia, ECLI:DE:BVerfG:2021:rs20210324.1bvr265618, margin no. 268.

Although this meant that the German Federal Constitutional Court only required legislative activity for the period after 2030, the German Climate Protection Act was also amended for the period before that, and in particular with an increased target reduction mark for $CO_2$ emissions of 65% instead of the previous 55%. This is not a contradiction, however, as a logical consequence it is inherent in the German Federal Constitutional Court's climate decision: The sooner the current generations develop increased efforts for climate protection, the sooner the burdens will remain bearable for the younger generations, so that a proportionate distribution takes place.

## 1.2 *Justifications*

The discussion so far has been focused on the question of the enforceability of state protection obligations arising from fundamental rights.[9] The Court of Justice in The Hague therefore emphasizes the protection aspect towards all citizens of the Netherlands. In contrast, the German Federal Constitutional Court, in accordance with its previous dogmatics,[10] does not allow the protection aspect to prevail, but rather the aspects of freedom, which, in this respect, have been redefined. It emphasizes that the protection of life and physical integrity under the first sentence of Article 2 (2) German Constitutional Law includes protection against impairments of fundamental rights by environmental pollution, regardless of by whom or by what circumstances they are threatened, and affirms an objective-legal obligation to protect also with respect to future generations.[11] However, the court found no violation of the obligation to protect under fundamental law because the German legislator has taken protective measures that are not obviously unsuitable.[12] However, a concept for protection aimed at the goal of climate neutrality is indispensable. In this respect, the German Climate Protection Act 2019 was already sufficient.[13]

However, the German Federal Constitutional Court found the rights of freedom of the younger generations to be violated because the legislator had not taken sufficient precautions to cope with the emission reduction obligations: Due to insufficient efforts and thus permitted emissions by 2030, very high emission reduction obligations, which have an advance interference-like effect on the freedom of the young complainants and are comprehensively protected by the constitutional rights,

---

[9] GFCC, Cases 1 BvR 2656/18 inter alia, ECLI:DE:BVerfG:2021:rs20210324.1bvr265618, margin no. 268.

[10] E.g. GFCC, Case 1 BvR 2203/95 Ozon NJW (1996), p. 651, leading case: GFCC, Case BvR 1025/82 inter alia, BVerfGE 85, 191 (212 et seq.), Case 1 BvR 1301/84, BVerfGE 79, 174 (202).

[11] GFCC, Cases 1 BvR 2656/18 inter alia, ECLI:DE:BVerfG:2021:rs20210324.1bvr265618, headnote 1.

[12] GFCC, Cases 1 BvR 2656/18 inter alia, ECLI:DE:BVerfG:2021:rs20210324.1bvr265618, margin no. 154.

[13] GFCC, Cases 1 BvR 2656/18 inter alia, ECLI:DE:BVerfG:2021:rs20210324.1bvr265618, margin no. 155 et seq.

may arise in later periods.[14] The established reduction targets under Section 3 (1) Sentence 2 and Section 4 (1) Sentence 3 German Climate Protection Act 2019 in conjunction with Annex 2 were unconstitutional to the extent that they gave rise to disproportionate risks of interference with future civil liberties.[15] This approach constitutes the further development of the points of view asserted in the constitutional complaints, above all by adolescents, young adults and minors.[16]

In other words, the German Federal Constitutional Court strives to prevent an excessive encroachment on future liberties. In doing so, the court calls upon fundamental rights as defensive rights against climate protection measures and fears all too great restrictions due to the climate protection standard set by Article 20a German Constitutional Law. Thus, its foundation is based on the constitutional defensive rights of future generations. It is therefore a matter of the comprehensive preservation of freedom—an aspect that could also come into play in the case of Covid-19 protection measures, if they served precisely to enable those affected to regain the comprehensive development of freedom.[17]

The German Federal Constitutional Court has chosen an approach based on civil liberties rather than on the duty to protect, which has profound consequences, especially since only the fundamental rights of the younger generations are considered to be violated. Moreover, this approach based on civil liberties is accompanied by sizable limitations.

Thus, the Federal Constitutional Court has created a completely new approach, which, however, only lets the younger and future generations be affected by a lack of current climate protection efforts. At the same time, the freedom of the middle and older generations is necessarily limited. In this respect, freedom is necessarily narrowed and allocated within the framework of climate protection. At the same time, these generations cannot demand more climate protection, since the German Climate Protection Act insofar is sufficient. In this context, they are equally affected by the phenomena of climate change, albeit possibly in later years.

In addition, the German Federal Constitutional Court did not take into account that although the Climate Protection Act has so far set more detailed targets only up to 2030, Germany will have to make greater efforts in the process of implementing the more detailed targets of the EU Climate Act, as already presupposed in Section 4 Climate Protection Act. In this respect, the interaction with European climate protection must be taken into consideration.[18] For this reason, it would be insufficient to derive national requirements from the environmental state objective.[19] In

[14] GFCC, Cases 1 BvR 2656/18 inter alia, ECLI:DE:BVerfG:2021:rs20210324.1bvr265618, margin no. 182 et seq.

[15] GFCC, Cases 1 BvR 2656/18 inter alia, ECLI:DE:BVerfG:2021:rs20210324.1bvr265618, margin no. 183.

[16] GFCC, Cases 1 BvR 2656/18 inter alia, ECLI:DE:BVerfG:2021:rs20210324.1bvr265618, margin no. 60 et seq.

[17] Frenz (2021b), p. 543 (550).

[18] For more details on this topic Frenz (2021c), p. 86 (88).

[19] See Frenz (2021d), pp. 201 et seq.

addition, if the level of protection specified by the EU is not met, stronger climate protection measures must be generally enforceable for all generations and thus for all persons whose health and property may be affected, as the recent massive flood in July 2021 in Germany underscored with well over 100 deaths and incalculable material damage.[20]

## 2 Violations of Fundamental Rights of Young and Future Generations Due to Inadequate Burden-Sharing Regulations

### *2.1 Approach Based on the Civil Liberties with Far-Reaching Limitation of Freedom*

#### (a) Approach of the German Federal Constitutional Court

The basis for the unconstitutionality of the German Climate Protection Act is thus not the duties to protect under constitutional law, but rather the "comprehensively protected freedom of the complainants"[21] and thus the fundamental rights in their entirety. Which ones are affected is not specified. However, reference is made to the protection of all human activities of freedom by the constitutional rights in special fundamental rights of freedom and "in any case by those in Article 2 (1) German Constitutional Law as the fundamental general right of freedom"[22] and thus by the general freedom of action.[23] Thus, it concerns a violation of all freedom activities protected by fundamental rights of young and future generations because of too low emission reductions today and thus too sharp ones in the future.

The German Federal Constitutional Court highlights that currently there are "numerous forms of private, professional and economic activity […] that still directly or indirectly cause $CO_2$ to be released into the Earth's atmosphere".[24] This already indicates the comprehensive course correction in social and economic life that the German Federal Constitutional Court predetermines, resulting from ensuring "that freedom is respected when making a transition to climate neutrality".[25]

[20] Already Frenz (2021e), p. 123 (128 et seq.) as well as more detailed and in view of the catastrophic flood in Germany 2021, Frenz (2021f), pp. 715 et seq.

[21] GFCC, Cases 1 BvR 2656/18 inter alia, ECLI:DE:BVerfG:2021:rs20210324.1bvr265618, margin no. 183.

[22] GFCC, Cases 1 BvR 2656/18 inter alia, ECLI:DE:BVerfG:2021:rs20210324.1bvr265618, margin no. 184.

[23] GFCC, Case 1 BvR 253/56, BVerfGE 6, 32 (36 et seq.).

[24] GFCC, Cases 1 BvR 2656/18 inter alia, ECLI:DE:BVerfG:2021:rs20210324.1bvr265618, margin no. 184.

[25] GFCC, Cases 1 BvR 2656/18 inter alia, ECLI:DE:BVerfG:2021:rs20210324.1bvr265618, margin no. 183.

This is because the Federal Constitutional Court considers the remaining emissions budget for Germany as essentially depleted by 2030[26] and therefore envisages a comprehensive, far-reaching change in private and economic activity in the interest of $CO_2$ neutrality for the period thereafter. Drastic Greenhouse gas reduction burdens are therefore imminent, thereby risking "unacceptable impairments of fundamental rights from today's perspective".[27]

It is therefore even more positive that the legislator has now also tightened the obligations up to 2030 in the German Climate Protection Act. In this way, $CO_2$ reduction starts as early as possible in order to give younger and future generations as much leeway as possible and to prevent the depletion of their $CO_2$ budget as early as 2031.

### (b) Allocated Freedom: climate protection-related and shared instead of natural

The protection of freedom to be preserved today follows from the threat of a narrowly restricted freedom in the future that should be mitigated. This is the present protection of freedom for the young and future generations, which conversely requires a stronger restriction of freedom for those living now, so that the restrictions in the future remain bearable. Thus, in climate protection, only a certain amount of freedom is conceded from the beginning, which is distributed fairly and intergenerationally. In total, the $CO_2$-relevant freedom is limited by the $CO_2$ emissions conceded by the state; this measure of freedom must be distributed: More freedom for some means less freedom for others.

This is freedom according to the necessary climate protection—a model that is far away from the foundations of original natural freedom or, depending on the point of view, returns it to its natural basis by including climate protection and the preservation of national livelihoods. It guarantees the necessary protection of fundamental rights.[28] The extent of this protection, however, is not automatically predetermined, but must be ensured by the legislator—in balance with the not allocated but natural rights of defense, also in the form of $CO_2$-relevant activities. Climate protection must not have an absolute effect.[29] However, according to the German Federal Constitutional Court, this is largely achieved;[30] it is primarily a matter of burden sharing. The German Federal Constitutional Court balances it more in terms of fundamental rights, not so much in terms of climate protection as such.

---

[26] GFCC, Cases 1 BvR 2656/18 inter alia, ECLI:DE:BVerfG:2021:rs20210324.1bvr265618, margin no. 233, 246.

[27] GFCC, Cases 1 BvR 2656/18 inter alia, ECLI:DE:BVerfG:2021:rs20210324.1bvr265618, margin no. 245.

[28] Based on the obligations to protect fundamental rights Frenz (2021e), p. 123 (128 et seq.).

[29] GFCC, Cases 1 BvR 2656/18 inter alia, ECLI:DE:BVerfG:2021:rs20210324.1bvr265618, margin no. 246.

[30] GFCC, Cases 1 BvR 2656/18 inter alia, ECLI:DE:BVerfG:2021:rs20210324.1bvr265618, margin no. 185 et seq.

### 2.2 *No Restriction of the Scope of Fundamental Rights Protection, But Less and Weaker Protection*

After all, the German Federal Constitutional Court still protects activities involving $CO_2$ emissions under fundamental rights. Thus, climate change does not constitute an external circumstance that already restricts the scope of protection of the occupational liberty or the freedom of property, as has been assumed in some cases for coal-fired power generation because of its harmful greenhouse gas emission.[31] However, this permission of $CO_2$ emissions is associated with a reduction in the fundamental liberties of future generations and thus has an advance interference-like effect on their freedom.[32] Every exercise of freedom, and thus that of young and future generations, is limited by the necessary protection of the climate, as demanded by the climate protection requirement of Article 20a German Constitutional Law.[33] In this context, the possible exercise of fundamental rights becomes even more limited the sooner the available $CO_2$ budget runs out: "As climate change intensifies, such exercise of freedom will be accorded ever less weight within the balancing process due to its ever greater impact on the environment."[34]

This means that freedom is less and less able to assert itself against the need for climate protection. The present generations must even more ensure the progress of climate protection in order to facilitate any freedom activities at all for future generations. At some point, the use of $CO_2$-related freedom will have to be essentially prevented anyway, since this is the only way to stop human-induced global warming.[35] It is true that $CO_2$ activities are still protected by fundamental rights. However, they are increasingly easier to restrict. Climate protection is becoming more and more prevalent.

## 3 Constitutional Obligation to Take Climate Action of Article 20a German Constitutional Law

### 3.1 *Absoluteness?*

The constitutional limit for the $CO_2$-relevant exercise of freedom results from the constitutional obligation to take climate action of Article 20a German Constitutional Law. At the same time, this limits the constitutional justification of the risk of the

[31] Schomerus and Franßen (2018), pp. 117 et seq., differently Kreuter-Kirchhof (2018), pp. 64 et seq.

[32] GFCC, Cases 1 BvR 2656/18 inter alia, ECLI:DE:BVerfG:2021:rs20210324.1bvr265618, margin no. 184.

[33] GFCC, Cases 1 BvR 2656/18 inter alia, ECLI:DE:BVerfG:2021:rs20210324.1bvr265618, margin no. 185.

[34] GFCC, Cases 1 BvR 2656/18 inter alia, ECLI:DE:BVerfG:2021:rs20210324.1bvr265618, margin no. 185.

[35] GFCC, Cases 1 BvR 2656/18 inter alia, ECLI:DE:BVerfG:2021:rs20210324.1bvr265618, margin no. 186.

loss of freedom in the future due to the emitted $CO_2$ until 2030. The regulations of the German Climate Protection Act that make it possible (Section 3 (1) sentence 2, Section 4 (1) sentence 3 in conjunction with Annex 2) must therefore be compatible with decisions regarding elementary fundamental rights, especially those of Article 20a German Constitutional Law, and they must not lead to disproportionate burdens on the complainants' liberties in the future.[36] Article 20a German Constitutional Law obliges the realization of climate protection even after 2030 to the extent required by the constitution and therefore does not allow emissions to be permitted until 2030 that prevent this realization.[37]

As a result, concrete constitutional obligations in the area of climate protection do follow from the environmental state objective, which only needs to be fleshed out at the outset. It is actually up to the legislator to lay down a requirement in this regard. However, the German Federal Constitutional Court considers climate protection so elementary that it derives a directly effective constitutional obligation to take climate action.

## 3.2 *Priority of Climate Protection*

It is true that Article 20a German Constitutional Law "does not take absolute precedence over other interests", but must be balanced with other constitutional goods and principles.[38] In this respect, economic and social aspects must also be taken into account, in accordance with the principle of sustainable development. However, this is apparently based on a relative priority. After all, "any overshoot of the critical temperature [...] would only be justifiable under strict conditions – such as for the purpose of protecting fundamental rights. Within the balancing process, the obligation to take climate action is accorded increasing weight as climate change intensifies."[39]

It is indeed weighed up, but with increasing climate change practically always at the expense of the $CO_2$-relevant freedom and thus the concerns of the economy as well as the private activities in form of car traffic, air travel, meat consumption, etc. In contrast, the European Court of Justice emphasized that the needs of the European economy were taken into account to the greatest possible extent in EU emissions trading, although its targets also arose from international agreements.[40]

---

[36] GFCC, Cases 1 BvR 2656/18 inter alia, ECLI:DE:BVerfG:2021:rs20210324.1bvr265618, margin no. 188.

[37] GFCC, Cases 1 BvR 2656/18 inter alia, ECLI:DE:BVerfG:2021:rs20210324.1bvr265618, margin no. 190.

[38] GFCC, Cases 1 BvR 2656/18 inter alia, ECLI:DE:BVerfG:2021:rs20210324.1bvr265618, margin no. 198, 246.

[39] GFCC, Cases 1 BvR 2656/18 inter alia, ECLI:DE:BVerfG:2021:rs20210324.1bvr265618, margin no. 198.

[40] ECJ, Case T-178/05 United Kingdom/Commission ECLI:EU:T:2005:412, margin no. 60.

### 3.3 Definition by the Paris Target

The German Federal Constitutional Court sees the necessary balance of interests materialized in the definition of the Paris climate target in Section 1 sentence 3 German Climate Protection Act.[41] In this way, it sees the constitutional obligation to take climate action substantiated and an interaction established at the same time: The temperature benchmark used as a basis in Section 1 Sentence 3 German Climate Protection Act provides orientation of its own under constitutional law.[42] This means that the Paris target of keeping global warming well below 2 degrees Celsius and, if possible, limiting it to 1.5 degrees Celsius is practically constitutional. This defines the constitutional obligation to take climate action, and does so as a basis,[43] as formulated in the third sentence of Section 1 of the German Climate Protection Act.

However, this means that the Paris Climate Agreement, which is based on cooperation between states and does not create any direct obligations for them, becomes binding for Germany with regard to its objective via the environmental state target. At the subsequent conference in Katowice, the 1.5-degree target was increasingly targeted, but not regarded as a firm obligation. It remains up to each individual state how they treat and pursue this target. This construction is maintained in the following climate conferences.

### 3.4 International Climate Protection

This also builds a bridge to the international climate protection target. In view of the global effects and causes, climate protection is in any case necessarily international and can therefore only be achieved through international cooperation.[44] This "international dimension"[45] requires the state to act internationally and thus to make a substantial contribution to global climate neutrality.[46] Therefore, the state cannot refer to the fact that there are also greenhouse gas emissions in other states.[47] Rather, it follows that there is "a constitutional necessity to actually implement one's own

[41] GFCC, Cases 1 BvR 2656/18 inter alia, ECLI:DE:BVerfG:2021:rs20210324.1bvr265618, margin no. 213.

[42] GFCC, Cases 1 BvR 2656/18 inter alia, ECLI:DE:BVerfG:2021:rs20210324.1bvr265618, margin no. 213.

[43] GFCC, Cases 1 BvR 2656/18 inter alia, ECLI:DE:BVerfG:2021:rs20210324.1bvr265618, margin no. 210.

[44] GFCC, Cases 1 BvR 2656/18 inter alia, ECLI:DE:BVerfG:2021:rs20210324.1bvr265618, margin no. 200.

[45] Schulze-Fielitz (2018), Art 20a margin no. 11.

[46] GFCC, Cases 1 BvR 2656/18 inter alia, ECLI:DE:BVerfG:2021:rs20210324.1bvr265618, margin no. 201 et seq.

[47] GFCC, Cases 1 BvR 2656/18 inter alia, ECLI:DE:BVerfG:2021:rs20210324.1bvr265618, margin no. 203.

climate action measures at the national level – in international agreement wherever possible". The state may therefore "avoid creating incentives for other states to undermine this cooperation. Its own activities should serve to strengthen international confidence in the fact that climate action – particularly the pursuit of treaty-based climate targets – can be successful while safeguarding decent living conditions, including in terms of fundamental freedoms."[48]

The aim is to make Germany a positive role model for global climate protection.[49] In this context, the legislator determines which sectors must contribute to (internationally exemplary) climate protection and to what extent. Germany's climate-related exemplary function under Article 20a German Constitutional Law does not necessarily imply that all $CO_2$-relevant sectors and activities must serve as role models. This requires a normative definition of standards, for example for certain private, professional or economic activity or for $CO_2$ reduction obligations. In this context, $CO_2$-related percentages and budgets could be determined for recycling and waste management in the future, even if the quota for reuse and recycling in Section 14 German Circular Economy Act (KrWG) also indirectly reduces $CO_2$ emissions.

If the Paris Climate Agreement also provides for a voluntary mechanism, thus practically obliging Germany to implement this agreement in the most climate-protective way possible. The climate target is defined by law in Section 1 Sentence 3 German Climate Protection Act and is thus constitutional by virtue of the interaction and as an expression of the environmental state objective. It must be implemented in such a way that other states develop confidence in Germany's climate protection efforts and thus are encouraged to take their own measures as well. In this respect, Germany can and should be a role model.

Through this constitutionally mandated role model function, the Paris Climate Agreement, which is based on voluntary mechanisms and a climate target that is by no means firmly agreed upon, becomes nationally binding for Germany. At the same time, the German legislator still has to monitor whether the climate protection target may need to be made more ambitious, and thus has an obligation to adapt to the latest developments and knowledge in science, comparable to risk prevention with regard to nuclear energy.[50] However, a lowering of the climate target can hardly be justified.[51]

The environmental state goal therefore has an effect nationally without further elaboration via an international treaty that Germany entered and the international scope of the climate protection requirement and forces the legislator to take

48 GFCC, Cases 1 BvR 2656/18 inter alia, ECLI:DE:BVerfG:2021:rs20210324.1bvr265618, margin no. 203.

49 See already Frenz (2021g), para 1 KSG marginal no. 14.

50 GFCC, Cases 1 BvR 2656/18 inter alia, ECLI:DE:BVerfG:2021:rs20210324.1bvr265618, margin no. 212, referencing Case 2 BvL 8/77, BVerfGE 49, 89 (130, 132) regarding Article 1 (1) sentence 1 German Constitutional Law.

51 GFCC, Cases 1 BvR 2656/18 inter alia, ECLI:DE:BVerfG:2021:rs20210324.1bvr265618, marginal no. 212.

measures in accordance with firm requirements of the German Federal Constitutional Court with an intensity typical of health-related risk prevention.

## 4 Concrete National Residual Budget of 6.7 Gigatons?

### *4.1 Recourse to IPCC and German Advisory Council on the Environment (Sachverständigenrat für Umweltfragen, SRU)*

It is true that Article 20a German Constitutional Law itself does not specify a precise distribution key and, above all, for reasons of justice it does not specify Germany's particular share of the burden for $CO_2$ reduction. Nevertheless, according to the German Federal Constitutional Court, the obligation to realize the climate protection target in international cooperation, which follows from the environmental state objective, requires a German contribution that clearly "promotes mutual trust in the willingness of the Parties to take action, and does not create incentives to undermine it".[52] There are uncertainties in the more detailed specification, which is why a binding basis for constitutional review is not (yet) possible.[53] However, the required temperature increase threshold of well below 2 degrees Celsius and, if possible, of 1.5 degrees Celsius must be converted into a corresponding $CO_2$ emission quantity. For this conversion, the German Federal Constitutional Court uses the figures of the IPCC, which has estimated various calculation models with a global $CO_2$ residual budget of between 420 and 1170 gigatons from 2018.[54]

On this basis, in turn, the German Advisory Council on the Environment (SRU) has derived a concrete national residual budget of 6.7 gigatons remaining from 2020 for the goal of limiting the rise in average Earth temperatures to 1.75 degrees Celsius with a probability of 67%.[55]

---

[52] GFCC, Cases 1 BvR 2656/18 inter alia, ECLI:DE:BVerfG:2021:rs20210324.1bvr265618, margin no. 225.

[53] GFCC, Cases 1 BvR 2656/18 inter alia, ECLI:DE:BVerfG:2021:rs20210324.1bvr265618, margin no. 236.

[54] IPCC, "Special Report on Global Warming of 1,5 °C" (2018) chapter 2, p. 108, chart 2.2.

[55] GFCC, Case 1 BvR 2656/18 inter alia, ECLI:DE:BVerfG:2021:rs20210324.0324.1bvr265618, margin no. 219, German Advisory Council on the Environment, "Umweltgutachten 2020: Für eine entschlossene Umweltpolitik in Deutschland und Europa", p. 52, 88, margin no. 111.

## 4.2 No Fixed Figure

Although the German Federal Constitutional Court also recognizes the dependence of the German Advisory Council on the Environment on valuations and also on uncertainties in the calculation, which is why this does not yet result in a "exact numerical benchmark for review by the Federal Constitutional Court" and no objection to the statutory regulations up to 2030,[56] it nevertheless bases its more detailed investigations on this calculation;[57] the statutory reduction targets must take this into account.[58] Based on the residual budget value of 6.7 gigatons determined by the German Advisory Council on the Environment, the Federal Constitutional Court did consider Section 3 (1) Sentence 2 and Section 4 (1) Sentence 3 German Climate Protection Act in conjunction with Annex 2 to be constitutional with regard to the $CO_2$ quantities permitted up to 2030, although it emphasized that this already largely consumes the underlying residual budget by 2030.[59]

However, the existing budget is practically depleted, leaving future generations no more room for $CO_2$-relevant exercises of freedom soon after 2030 (barely one year).[60] This fact would argue in favour of considering even greater reduction burdens before 2030 as mandatory. However, the German Federal Constitutional Court does not go so far. The calculation that the $CO_2$ budget will be used up by 2031 can therefore only be of an exemplary nature. The final determination is up to the legislator, as the German Federal Constitutional Court also emphasizes.[61] The latter reacted promptly with an increased $CO_2$ change target of 65% by 2030, but without setting an overall maximum emissions budget. Nevertheless, there is no explicit obligation to do so either. On April 21, 2021, the European Union agreed on a reduction target of 55%, which was laid down in the EU climate law that has since been passed, to be achieved through a set of measures outlined by the EU Commission in its "Fit for 55" climate package of July 14, 2021.

---

[56] GFCC, Cases 1 BvR 2656/18 inter alia, ECLI:DE:BVerfG:2021:rs20210324.1bvr265618, margin no. 236 et seq.

[57] GFCC, Cases 1 BvR 2656/18 inter alia, ECLI:DE:BVerfG:2021:rs20210324.1bvr265618, explicitly margin no. 231.

[58] GFCC, Cases 1 BvR 2656/18 inter alia, ECLI:DE:BVerfG:2021:rs20210324.1bvr265618, margin no. 229, 237.

[59] GFCC, Cases 1 BvR 2656/18 inter alia, ECLI:DE:BVerfG:2021:rs20210324.1bvr265618, margin no. 230 et seq.

[60] GFCC, Cases 1 BvR 2656/18 inter alia, ECLI:DE:BVerfG:2021:rs20210324.1bvr265618, margin no. 246.

[61] GFCC, Cases 1 BvR 2656/18 inter alia, ECLI:DE:BVerfG:2021:rs20210324.1bvr265618, margin no. 236.

### 4.3 Nonetheless Pre-determining Importance Despite Other Calculation Possibilities Also According to International Law

Due to the lack of determinations for the period after 2030, the German Federal Constitutional Court sees the legislator's obligation of distributing over time the reductions in $CO_2$ emissions as required under Article 20a German Constitutional Law up to climate neutrality in a forward-looking manner and in a manner that does not violate fundamental rights.[62] This obligation follows from the requirement of proportionality. However, this view is based on the assumption of the residual budget according to the German Advisory Council on the Environment (Sachverständigenrat für Umweltfragen), although it involves considerable uncertainties. In addition, the IPCC report on climate figures was welcomed at the COP24 in Katowice, but was not taken as formally binding. Thus, the German Federal Constitutional Court, like the German Advisory Council on the Environment, uses a calculation that is not firmly recognized under international law as basis. Article 4 (4) of the Paris Climate Agreement only provides for a commitment to absolute economy-wide emission reduction targets, not to fixed budgets.

Even more serious is the calculation of the residual budget for Germany according to its share of the world population of 1.1%.[63] At the same time, its $CO_2$ emissions are currently just under 2%.[64] It would therefore also be conceivable to require all countries to reduce their emissions equally, thus granting Germany a residual budget that is almost twice as high. Admittedly, there is the objection that many countries still have to master a development in order to reach a standard such as that of the industrialized countries, which already emitted $CO_2$ to a considerable extent in the past. Article 2 (2) of the Paris Climate Agreement, which the German Federal Constitutional Court refers to,[65] provides for implementation according to the principle of common but differentiated responsibilities and respective capabilities in the face of different national circumstances. Therefore, economic and social aspects in accordance with the sustainable development referred to in Article 2 (1) of the Paris Climate Agreement can also be implemented. Therefore, it is not required to focus on this part of the world population, which will hardly be possible without a profound deindustrialization and reduction of prosperity, unless it is possible to manage

[62] GFCC, Cases 1 BvR 2656/18 inter alia, ECLI:DE:BVerfG:2021:rs20210324.1bvr265618, margin no. 273.

[63] GFCC, Cases 1 BvR 2656/18 inter alia, ECLI:DE:BVerfG:2021:rs20210324.1bvr265618, margin no. 225.

[64] GFCC, Cases 1 BvR 2656/18 inter alia, ECLI:DE:BVerfG:2021:rs20210324.1bvr265618, margin no. 202.

[65] GFCC, Cases 1 BvR 2656/18 inter alia, ECLI:DE:BVerfG:2021:rs20210324.1bvr265618, margin no. 225.

economic growth in a climate-neutral way very quickly, as propagated by the EU Commission in its Green Deal.[66]

The Paris Climate Agreement allows other calculations for a national $CO_2$ residual budget than what the German Federal Constitutional Court, following the Council of Environmental Experts, has factually based it on, and does not oblige to determine it. The German Federal Constitutional Court derives an obligation to calculate it from the international scope of the climate protection requirement under Article 20a German Constitutional Law, in order to promote the confidence of other states in Germany's will to implement climate protection.

## 4.4 *Contradiction with the Theory of Materiality*

Elsewhere, the German Federal Constitutional Court emphasizes that the legislature must make the decisive decisions. Thus, it points out that the legislative process provides the necessary balance of interests with the required legitimacy.[67] If this is the case, however, only the content of the legislative process can be taken as a basis. This provision does not specify the parameters according to which the residual budget attributable to Germany is to be measured in order to achieve the global climate target set out in section 1 sentence 3 German Climate Protection Act. The extent to which reduction commitments actually have to be met by Germany depends largely on these parameters. The IPCC's assumptions for the global reduction requirements from 2018 for a 67% probability of limiting global warming to 1.5 degrees Celsius were already 420 gigatons worldwide, but with regard to a 2-degree Celsius target they were 1170 gigatons.[68]

With such a range that is decisive for all further impairments of fundamental rights in order to reduce $CO_2$-emission, a legal decision would have been necessary regarding the manner of calculating the German residual budget. The German Federal Constitutional Court demands this for the additional measures of $CO_2$-reduction with a long-term perspective until long after 2030. On this depends how fundamental the changes in economic activities and private activity (consuming meat, travelling by plane, driving by car, etc.) are. The limitation to 1.5 degrees Celsius and even 1.75 degrees Celsius cannot just be taken as basis, because this requirement is not binding under international law, as shown by the COP24 in

[66] No 1 (Introduction) to the Communication from the Commission to the European Parliament, the Council, the European Economic and Social Committee and the Committee of the Regions, "Sustainable Europe Investment Plan, European Green Deal Investment Plan", COM(2020) 21 final.

[67] GFCC, Cases 1 BvR 2656/18 inter alia, ECLI:DE:BVerfG:2021:rs20210324.1bvr265618, margin no. 213.

[68] GFCC, Cases 1 BvR 2656/18 inter alia, ECLI:DE:BVerfG:2021:rs20210324.1bvr265618, margin no. 219, IPCC, "Special Report on Global Warming of 1,5 °C" (2018) chapter 2, p. 108 chart 2.2.

Katowice.[69] The in section 1 sentence 3 of the Climate Protection Act in accordance to the Paris Climate Treaty determined maximum is significantly under 2 degrees Celsius without a more specific definition of the value.

# 5 Raw Materials for Climate Protection

## *5.1 Repercussions for Energy Transition and Climate Protection*

Energy transition and climate protection are further more crucial challenges for policymakers. The German Federal Constitutional Court's climate decision called for the early initiation of the necessary development and implementation processes also for the period after 2030, and cited the mobility transition as an example.[70] Precisely through this transition, $CO_2$ emissions can also be limited to a large extent. In its wake, a considerable amount of oil, which also comes from Russia, will be saved. However, this also increases the demand for electricity that therefore must come all the more from renewable energies if $CO_2$ emissions are actually to be reduced.

The EU Commission is also targeting electromobility in its "Fit for 55" climate package, according to which $CO_2$ emissions from passenger cars are to be reduced by 100% by 2035.[71] The European Parliament and the Council of the European Union approved this aim.[72] The German government plans in its coalition agreement with at least 15 million electric cars in 2030.[73] The electromobility thus envisioned depends not only on available electricity, but also on lithium-ion batteries. Eastern Ukraine, meanwhile, is home to lithium reserves, as well as numerous other critical raw materials, such as cobalt, manganese and rare earths.[74]

[69] GFCC, Cases 1 BvR 2656/18 inter alia, ECLI:DE:BVerfG:2021:rs20210324.1bvr265618, margin no. 235: "fails to satisfy the requirement of pursuing efforts to limit the temperature increase to 1.5 °C".

[70] GFCC, Cases 1 BvR 2821/11 inter alia, ECLI:DE:BVerfG:2016:rs20161220.1bvr282111, margin no. 249, Cases 1 BvR 2656/18 inter alia, ECLI:DE:BVerfG:2021:rs20210324.1bvr265618, margin no. 231.

[71] European Commission, European Green Deal: Commission proposes transformation of EU economy and society to meet climate ambitions. Press release from 14.7.2021, p. 2. https://ec.europa.eu/commission/presscorner/detail/en/ip_21_3541. Accessed 22.5.2023.

[72] European Parliament, 'Deal confirms zero-emissions target for new cars and vans in 2035'. Press release from 27.10.2022. https://www.europarl.europa.eu/news/de/press-room/20221024IPR45734/deal-confirms-zero-emissions-target-for-new-cars-and-vans-in-2035. Accessed 22.5.2023.

[73] Coalition agreement of SPD, Bündnis 90 Die Grünen and FDP 2021. https://www.spd.de/fileadmin/Dokumente/Koalitionsvertrag/Koalitionsvertrag_2021-2025.pdf. Accessed 22.5.2023, p. 51.

[74] 'EU, Ukraine sign "strategic partnership" on raw materials' (Euractiv, 13.7.2021). https://www.euractiv.com/section/circular-economy/news/eu-ukraine-to-sign-strategic-partnership-on-raw-

On a deeper level, the Russia-Ukraine war affects the switch to renewable energies itself. The energy transition is intended to reduce $CO_2$ emissions enormously by replacing conventional electricity with green electricity. In its "Fit for 55" climate package, the EU Commission states that the energy sector accounts for 75% of $CO_2$ emissions.[75] At the heart of this energy transition is an increase in power generation from renewable energies. The EU climate package envisages such an increase from 32% to 40% by 2030.[76] Now the European Parliament strengthened the aim to 45%.[77] Section 1 of the Renewable Energy Sources Act sets an expansion target of 80% (previously 65%), which means that Germany, in line with its strong economic performance, will contribute its share within the framework of the solidarity-based burden sharing that applies in the energy sector in particular,[78] in accordance with Article 2 (2) of the EU Climate Act.

## 5.2 *Linkage with Raw Material Security*

However, in Germany the raw materials needed for the additional solar modules, wind turbines and power lines—for example, for transporting electricity from northern to southern Germany—must be available all the sooner. If this is not the case, there is a risk that the energy transition will fail. This applies regardless of the problems of e.g. protection of species, which the coalition agreement new German government intends to give relative priority to the expansion of wind power for a limited period of time in the consideration of protected goods[79]—an approach that is critical under EU law in this generality.[80]

materials/?_ga=2.48976099.491578310.1649246206-1665917058.1649246206. Accessed 22.5.2023.

[75] European Commission, Communication from the Commission to the European Parliament, the Council, the European Social and Economic Committee and the Committee of the Regions. 'Fit for 55': delivering the EU's 2030 Climate Target on the way to climate neutrality. COM(2021) 550 final, p. 11.

[76] European Commission, Communication from the Commission to the European Parliament, the Council, the European Social and Economic Committee and the Committee of the Regions. 'Fit for 55': delivering the EU's 2030 Climate Target on the way to climate neutrality. COM(2021) 550 final, p. 11.

[77] European Parliament, 'Parliament backs boost for renewables use and energy savings'. Press release from 14.9.2022. https://www.europarl.europa.eu/news/en/press-room/20220909IPR40134/parliament-backs-boost-for-renewables-use-and-energy-savings. Accessed 22.5.2023.

[78] ECJ, Case C-848/19 P Germany/Poland ECLI:EU:C:2021:598, margin no. 38 et seq.

[79] Coalition agreement of SPD, Bündnis 90 Die Grünen and FDP 2021. https://www.spd.de/fileadmin/Dokumente/Koalitionsvertrag/Koalitionsvertrag_2021-2025.pdf. Accessed 22.5.2023, p. 56.

[80] Frenz (2022), p. 87 (88 et seq).

## 5.3 Securing Raw Materials as Part of the Energy Transition

However, the continued operation of coal-fired or even nuclear power plants will only be politically feasible for a short time at most. Therefore, the only way forward is to step up the expansion of renewable energies. And this requires the raw materials needed for this expansion. The energy transition as an essential part of climate protection hence is based to a large extent on the supply of raw materials. Against the backdrop of the Russia-Ukraine war, climate protection as a whole is also about securing raw materials. This is a necessary component of the climate protection requirement developed by the German Federal Constitutional Court based on the environmental state objective provision in Article 20a German Constitutional Law.[81] Germany must therefore ensure that its supply of raw materials is secure. Only in this way can it assume the exemplary function for climate protection under conditions worth living in,[82] assigned to it by the German Federal Constitutional Court. However, this also applies to the EU, which, in accordance with Article 191 (1), 4th indent, TFEU (Treaty on the Functioning of the European Union), also intends to play an important role at international level under its climate package.[83]

Against the background of limited own reserves and the cutting off of access to numerous raw material deposits by the Russia-Ukraine war, the EU's climate goals, like the climate protection requirement from Article 20a German Constitutional Law, require an expansion of the perspective for securing raw material supplies to include international reserves. The best solution would be to develop a set of rules for raw materials that grants access to all states and at the same time ensures global compliance with minimum standards for extraction. The German coalition agreement of 2021 places particular emphasis on this.[84]

Efforts could be made to ensure that the availability of raw materials is recognized as a component of climate protection in a follow-up climate agreement and that the signatory states guarantee access that ensures this availability. The loss of key raw materials from Russia and Ukraine makes such an approach mandatory. Otherwise, climate protection will come to a standstill and neither the European nor the national climate goals can be achieved.

---

[81] GFCC, Cases 1 BvR 2821/11 inter alia, ECLI:DE:BVerfG:2016:rs20161220.1bvr282111, margin no. 197 et seq.

[82] GFCC, Cases 1 BvR 2821/11 inter alia, ECLI:DE:BVerfG:2016:rs20161220.1bvr282111, margin no. 203 et seq.

[83] European Commission, Proposal for a directive of the European Parliament and of the Council amending Directive 2003/87/EC establishing a system for greenhouse gas emission allowance trading within the Union, Decision (EU) 2015/1814 concerning the establishment and operation of a market stability reserve for the Union greenhouse gas emission trading scheme and Regulation (EU) 2015/757, COM(2021) 551 final, p. 16.

[84] Coalition agreement of SPD, Bündnis 90 Die Grünen and FDP 2021. https://www.spd.de/fileadmin/Dokumente/Koalitionsvertrag/Koalitionsvertrag_2021-2025.pdf. Accessed 22.5.2023, p. 56.

## 6 Conclusions

1. Climate protection lawsuits have not yet been successful in Germany. In the Netherlands, the state was not ordered to take concrete measures, but to do more than it has done so far to avert the consequences of global warming.
2. The German Federal Constitutional Court considers the civil rights of the younger generations to be violated by the fact that the climate protection provided for by law to date is not sufficient to protect them from excessive climate burdens in the future. To put it simply: If those living today do not reduce greenhouse gas emissions sufficiently, the young will have to do so even more and will then be overburdened.
3. By 2030, Germany's residual $CO_2$ budget threatens to be largely exhausted. Therefore, countermeasures must be taken as soon as possible to ensure that the restrictions on the liberties of the younger generation that result from climate protection are kept to a minimum and thus do not infringe on fundamental rights. The basis for this is the expected impairment of freedom in order to meet the environmental standards that follow from the fulfilment of the climate protection requirement under Article 20a of the German Constitutional Law.
4. The environmental state objective results in a climate protection requirement that demands an effective limitation of $CO_2$ emissions. This is the reason for current and future restrictions on freedom. At the same time, it limits the development of freedom granted by the German Federal Constitutional Court in $CO_2$-relevant areas. In this respect, freedom exists only in accordance with climate protection. This overall degree of freedom must then be distributed between the present and future generations or today's young generations. This is required by the law of freedom and is carried out by means of proportionality that is also future-oriented.
5. The German Federal Constitutional Court mentions measures for the transport sector—but only as a possible example. It is essential to establish transparent and far-reaching benchmarks for long-term behavioral changes. In concrete terms, there is a need for continuous $CO_2$ reductions that are directed far into the future. This requires clear specifications by the legislature, which cannot leave the field to the legislator without a clear reduction path, as has been the case up to now under Section 4 (6) of the Climate Protection Act. It must define the essential contours itself, as is now the case.
6. The German Federal Constitutional Court's climate decision thus offers a good opportunity for more effective climate protection in Germany, which must, however, be reconciled with the more ambitious EU targets set by the EU Climate Act. Here, Germany has to make a weighty, as required solidary contribution according to its strong economic power. This is the real challenge, which the German Federal Constitutional Court unfortunately did not take into account. Climate protection is not only international, but also European, as Section 4 (1) Sentence 5 of the Climate Protection Act shows.

7. The Russia-Ukraine war shows that climate protection is also about securing raw materials. This must therefore be ensured efficiently through international efforts. This is the only way to achieve the energy transition. Climate protection must also be conceived on a global scale. This means that all circumstances that are indispensable for climate protection must be considered globally. This also applies to the necessary raw materials. The supply of these raw materials must be secured across national borders.

## References

Frenz W (2020) Klimaschutz und Menschenwürde, UPR, pp 1 et seqq

Frenz W (2021a) Anmerkung zu einer Entscheidung des BVerfG, Beschluss vom 24.03.2021 (1 BvR 2656/18, 1 BvR 78/20, 1 BvR 96/20, 1 BvR 288/20), DVBl, pp 810 et seqq

Frenz W (2021b) Anmerkung zu einer Entscheidung des VGH Baden-Württemberg, Decision of 5.2.2021 (1 S 321/21), DVBl, pp 543 et seqq

Frenz W (2021c) Anpassung des Klimaschutzes an Corona und Green Deal, DVBl, pp 86 et seqq

Frenz W (2021d) Klimaschutz nach BVerfG-Beschluss und EU-Klimagesetz, EnWZ, pp 201 et seqq

Frenz W (2021e) Einklagbarer Anspruch auf mehr Klimaschutz, ZNER, pp 123 et seqq

Frenz W (2021f) Klimagrundrecht – BVerfG, Beschl. v. 24.3.2021, 1 BvR 2656/18 DÖV, pp 715 et seqq

Frenz W (2021g) Gesamtkommentar Klimaschutzrecht, 1. ed. Erich Schmidt Verlag, Berlin

Frenz W (2022) Klimaschutz i.e.S. nach dem Ampel-Koalitionsvertrag: Ökostromausbau und früherer Kohleausstieg, NuR, pp 87 et seqq

Groß T (2020) Die Ableitung von Klimaschutzmaßnahmen aus grundrechtlichen Schutzpflichten, NVwZ, pp 337 et seqq

Kreuter-Kirchhof C (2018) Rechtliches Gutachten zur Positionierung des Wirtschafts- und Energieministeri-ums NRW im Hinblick auf die Kommission Wachstum, Strukturwandel und Beschäftigung

Meyer S (2020) Grundrechtsschutz in Sachen Klimawandel? NJW, pp 894 et seqq

Schomerus T, Franßen G (2018) Klimaschutz und die rechtliche Zulässigkeit der Stilllegung von Braun- und Steinkohlekraftwerken

Schulze-Fielitz H (2018) In: Dreier H (ed) GG, 3rd edn. Mohr Siebeck Verlag, Tübingen

# Climate Change and the Extraterritorial Reach of the State Duty to Protect

**Chiara Gentile**

## Contents



**Abstract** The constitutional debate on climate change has been growing as a result of both the introduction of constitutional clauses on the protection of the environment and the increase of climate litigation against states' governments. National courts have delivered important decisions on the obligation for states to take climate action, namely on mitigation and adaptation measures. A key and yet unclear issue is whether states have a constitutional duty of protection also vis-à-vis the environ-

This text is based on the presentation held by the author during the World Congress of Constitutional Law in Johannesburg, 5–9 December 2022.

C. Gentile (✉)
University of Ferrara, Ferrara, Italy
e-mail: chiara.gentile@unife.it



W. Frenz, A. Preuße (eds.), *Yearbook of Sustainable Smart Mining and Energy 2022*, Yearbook of Sustainable Smart Mining and Energy - Technical, Economic and Legal Framework 2, https://doi.org/10.1007/978-3-031-41873-0_6

ment outside their jurisdiction. The German Constitutional Court had a chance to address it in the Neubauer case, but evidently chose not to take a stand.

This chapter investigates the (extra)territorial scope of the state duty of protection under the constitution against environmental impairments caused by climate change. Drawing from the Neubauer case, this chapter outlines the different implications of mitigation and adaptation measures abroad and argues that an extraterritorial duty to protect seems conceivable only in respect to mitigation measure. In the last decades, the progress in the environmental protection under international law prompted constitutional changes.

## 1 Introduction

This chapter investigates the (extra)territorial scope of the state duty of protection under the constitution against environmental impairments caused by climate change. It draws on the assumption that due to its cross-border nature, climate change affects the environment and people worldwide; accordingly, state measures aiming at hindering climate change might impact also on the environment beyond national borders.

Climate change and the issue of state responsibility for transboundary environmental damage have long been confronted with at the international level. The Declaration of Stockholm of 1972 is considered the "seminal point" for the the public recognition of worldwide environmental degradation and the inadequacy of state responses.[1] From the 1970s the debate on environmental protection and climate change has improved and it has led to the creation of forms of international cooperations, such as the Conference of Parties under the UNFCCC.

A major step forward was reached with the Paris Agreement of 2015, whose aim is to "strengthen the global response to the threat of climate change, in the context of sustainable development and efforts to eradicate poverty", by "holding the increase in the global average temperature to well below 2 °C above pre-industrial levels and pursuing efforts to limit the temperature increase to 1.5 °C above pre-industrial levels", and by "increasing the ability to adapt to the adverse impacts of climate change and foster climate resilience and low greenhouse gas emissions development".[2]

In the last decades, the progress in the environmental protection under international law prompted constitutional changes. The constitutional debate on the topic has been growing thanks to the introduction of constitutional clauses on the protection of the environment and to the increase of climate litigation against states' governments.[3] Individuals and organizations acting on behalf of individuals have been

---

[1] Boyd (2012), pp. 3 and 89.

[2] Article 1(1) Paris Agreement.

[3] Setzer and Higham (2022), p. 11.

invoking the state responsibility for not complying with the objectives set at the international level, namely in the Paris Agreement, as well as at the national level, primarily in the constitution and the laws implementing the constitution.

Climate change derives from both natural and human activities. Intuitively, the law can only regulate the latter. States regulate polluting activities that take place abroad or within national borders but produce effects (also) abroad, in particular due to greenhouse gas (GHG) emissions. Mining, and especially fossil fuel extraction, and mineral processing are, among others, activities that highly contribute to GHG emissions. As a consequence, their impact on climate change may damage the environment all over the world.

In the case of transboundary environmental damage due to state climate (in) action, can individuals residing in other territories claim the (in)acting state's responsibility? Accordingly, does the state duty of protection under the constitution have extraterritorial effects?

Whether states have a constitutional duty of protection also vis-à-vis the environment outside their jurisdiction is still debated. In the *Neubauer* case of 24 March 2021, the German Constitutional Court (GCC) addressed the issue of the extraterritorial reach of the state duty of protection under the Basic Law against the environmental impairments caused by climate change. Even though it declared such a duty conceivable in principle, the GCC did not take a resolute position on the issue (paras. 101, 173–181). In other words, for the time being, it is not yet clear whether the state duty to protect reaches beyond German borders and people residing abroad can claim the state responsibility for its violation.

As it emerges from *Neubauer*, the issue of extraterritoriality has different implications depending on whether mitigation or adaptation measures are considered. In respect to the environment inside the state perimeter, states shall implement both mitigation and adaptation measures. By contrast, as far as the environment located abroad is concerned, states can only be conceivably subject to a duty of mitigation. This is because adaptation would require a more 'pervasive' action of the state in other states' territories, which may collide with the sovereignty principle under international law. Accordingly, as will be exposed, an extraterritorial state duty to protect seems conceivable only with respect to mitigation measures.

## 2 Climate Action

State action against climate change is highly invoked by the international community, as shown especially within the United Nations' network. Under Article 2 UNFCCC, states aim to the "stabilization of greenhouse gas concentrations in the atmosphere at a level that would prevent dangerous anthropogenic interference with the climate system". The Paris Agreement has recalled and further specified such objective in Arts. 2 and 4. Furthermore, according to the 13th goal of the 17 UN Sustainable Development Goals, states shall take urgent action to combat climate change and its impacts.

Climate is heating rapidly. The average global temperatures have increased by over 1 °C, particularly in the late twentieth century: in 2022 the average global temperature was about 1.15° above the pre-industrial (1850–1900) level.[4] According to the World Meteorological Organization, years between 2015 and 2022 have been the warmest since records began, fueled by ever-rising GHG concentrations and accumulated heat.[5] Among all GHG, $CO_2$ alone contributes more than 2/3 to global warming.[6] The mining sector is one of the major emitters of GHG,[7] and its environmental impacts will become even more significant in the future.[8]

In order to keep the raise of temperature to well below 2, preferably 1.5°, compared to pre-industrial levels, countries have set specific (binding) targets under the Paris Agreement of 2015, providing for effective and equitable adaptation and mitigation measures to significantly reduce vulnerability and contribute to climate resilience. Global warming levels get close to the Paris Agreement temperature goal only if full implementation of the net-zero pledges is assumed, since policies currently in place with no additional action would result in global warming of 2.8° over the twenty-first century.[9]

To combat this destructive trend, the pricing of carbon pollution and investment in low-carbon technologies are critical to staying within the total carbon budget to limit global warming to 2 °C, and some industries, such as the mining industry, "will play a critical role in transitioning to a low-carbon future".[10]

The mining sector will confront with climate change also on a different level. Even though little knowledge exists on how climate change may impact mining operations and the extractives sectors, changing climatic conditions will have both direct (operational and performance-based) and indirect (securing of supplies and rising energy costs) impacts on the mining sector.[11] Since mining is particularly vulnerable to climate change,[12] it will need to evolve and adapt to the process of climate change.

Mitigation measures, such as taxes on carbon emissions—further analyzed in para 6—have been adopted in order to contain GHG emissions. Due to the extraterritorial spread of GHG, this will also influence the environmental impacts of emissions in states other than the one of origin, in such a way as to reduce transboundary environmental damage.

---

[4] WMO, Press Release Number: 12012023.

[5] WMO, Press Release Number: 06112022.

[6] IPCC, Climate Change 2022: Mitigation of Climate Change, Summary for Policymakers, 2022, p. 7.

[7] Rüttinger and Sharma (2016), p. 1.

[8] Norgate and Haque (2010), pp. 266 ff.

[9] UNEP, Emissions Gap Report 2022, 27 October 2022, 21.

[10] Cox et al. (2022), p. 17.

[11] Rüttinger and Sharma (2016).

[12] Rüttinger and Sharma (2016); Arent et al. (2014), pp. 368 ff.

## 3 The State Responsibility for Transboundary Environmental Damage Under Human Rights Treaties

Under international law, states must ensure that activities carried out within their jurisdiction do not cause damage to the environment in areas beyond their jurisdiction[13] (*no-harm rule*, which is considered the only customary principle with respect to international environmental law[14]). When assessing the state failure to comply with its obligation to take climate action, human rights treaties—such as the Interamerican Convention on Human Rights and the European Convention on Human rights—play a key role before international courts and bodies, but also before national courts.

Several ground-breaking cases have been rendered on the protection of the environment, and, in more recent years, on climate change itself.[15] With respect to transboundary environmental damages and the (in)acting state responsibility, at the international level two major cases are *Sacchi et al. v. Argentina et al.* of the UN Committee on the Rights of the Child and the 2017 Advisory opinion of the Inter-American Court of Human Rights.

In *Sacchi et al. v. Argentina et al.*, sixteen children filed a petition against Argentina, Brazil, France, Germany, and Turkey, claiming that those states had not adopted the necessary measures to reduce GHG emissions and hinder climate change, in order to protect their present and future rights to life, health, and cultural heritage. On 8 October 2021, the UN Committee dismissed the case on procedural grounds, as petitioners had not previously exhausted internal remedies. Nonetheless, it found that states shall adopt measures for the prevention of transboundary environmental damage and the violation of human rights abroad, and that they have extraterritorial responsibilities for the carbon pollution coming from their territory, since its production falls under their effective control.[16]

The UN Committee recalled and further developed what the Inter-American Court of Human Rights had stated in its 2017 Advisory opinion, namely that states can be held responsible for any significant damage to people abroad caused by activities led within their borders, and thus under their effective control.[17] Therefore, states "must take measures to prevent significant environmental harm to individuals inside – and outside – their territory".[18] The exercise of effective control was

[13] ICJ, Legality of the threat or use of nuclear weapons. Advisory Opinion of July 8, 1996, para 29; Philippines Commission on Human Rights, Final Report – National Inquiry on Climate Change, 6 May 2022, 73 and 77; ICJ, Case of Pulp Mills on the River Uruguay (Argentina v. Uruguay), Judgment of April 20, 2010, paras. 101 and 204.

[14] Kotzé and Muzangaza (2018), pp. 291–292.

[15] Setzer and Higham (2022), p. 11.

[16] UN Committee on the Rights of the Child, 8 October 2021, CRC/C/88/D/104-105-106-107-108/2019, Sacchi et al.

[17] Inter-American Court of Human Rights, OC-23/17, paras. 101–103.

[18] Ibid., para 140.

extended to internal *activities* with external effects on the basis of an extensive interpretation of the concept of "jurisdiction" under Article 1 of the Inter-American Convention on Human Rights.[19]

In Europe, it is worth notice that even though the European Court of Human Rights has not yet delved into the extraterritoriality of the ECHR with respect to climate change, there are now at least two pending climate cases that might lead to a "landmark judgment"[20] on the issue. The first is *Duarte Agostinho et al.*,[21] in which the applicants claim the responsibility of 33 Contracting states due to the transboundary effects of GHG emissions coming from their territory, which contribute to global warming and result in the failure to comply with their undertakings under the Paris Agreement. The second is *Greenpeace Nordic et al.*,[22] which originates from a Judgment in which the Supreme Court of Norway interpreted Article 112 of the Norwegian constitution (right to a healthy environment) as it protects the environment only within national borders.[23]

At the state level, *Urgenda* represents a paradigmatic case of the role of human rights treaties in climate litigation before national judges. The Dutch Supreme court held the state responsible for the violation of the positive obligations deriving from the rights to life and private life (Arts. 2 and 8 ECHR) due to the excessive amount of GHG that were allowed under the Dutch law. *Urgenda* is considered a 'model' of national climate litigation, as it inspired other cases in the European context, such as the *Climate Case Ireland* of 31 July 2020, the *Affaire du siècle* of 3 February 2021, and the *Klimaatzaak* of 17 June 2021.[24] However, neither *Urgenda* or any of these Judgments addressed the issue of the (extra)territorial reach of the state duty of protection and its responsibility for transboundary environmental damage.

## 4 The State Duty to Protect in National Constitutions

As anticipated, public recognition of environmental degradation prompted constitutional amendments on the protection of the environment.

[19] Müller (2022), pp. 119–120; Quattrocchi (2018), pp. 686–687.

[20] Heri (2022), p. 926.

[21] ECtHR, no. 39371/20, Duarte Agostinho et al. v. Portugal et al. Following the Duarte model, on 3 march 2021, two applications were filed before the ECtHR by two Italian citizens, who argue that the greenhouse gas emissions of several states have caused global warming, affecting their living conditions and mental health. The two applications are no. 14615/21, Uricchio v. Italy and 31 other States, and no. 14620/21, De Conto v. Italy and 32 other states.

[22] ECtHR, no. 34068/21, Greenpeace Nordic et al. v. Norway.

[23] Supreme Court of Norway, 22 December 2020, no. 20-051052SIV-HRET, para 155, A. Gociu, S. Roy, Extraterritoriality of Oil Constitutionalism in People v Arctic Oil, in EJIL: TALK!, 16 February 2021.

[24] Bagni (2021), p. 1036.

To this day, the most common constitutional provision is the state duty of protection, which figures in at least 144 constitutions. [25] Even if phrased in many different ways, such duty imposes upon the state an obligation to conserve or protect the environment.[26] A paradigmatic example is Article 20a of the German Basic Law (*Grundgesetz*, GG), which affirms that "mindful also of its responsibility towards future generations, the state shall protect the natural foundations of life and animals…".[27] Following the same wording, Article 9(3) of the Italian constitution—as amended by constitutional law no. 1/2022—affirms that "[the Republic] shall protect the environment, biodiversity and ecosystems, also in the interest of future generations…".[28]

In some constitutions, there is no provision directly articulating a state duty to protect the environment, but there is recognition of a subjective right to live in a healthy environment. According to Boyd, in these cases the "basic human rights concept that there can be no right in the absence of a corresponding duty can be used to infer that the government has an obligation to protect the environment".[29] However, not all duties are in a correlative relationship with a subjective right, and therefore not all duties are corresponded with subjective rights.[30] Duties of this kind provide other persons (individuals, public organs, etc.) with a *power* to demand for the fulfillment of the duty—a power that is just a "guarantee of the compliance with the duty".[31] The state duty to protect against climate change falls within this category, since it does not, as such, entail a specular right to live in a healthy environment.

Nonetheless, a duty to protect may behave as a constitutional parameter to assess the violation of other fundamental rights: [32] if the state has not taken adequate climate action and its laxity causes an environmental catastrophe (such as a flood that submerges the houses and farms of a community of farmers), the violation of the affected individuals' fundamental rights is ascertained in the light of the violation of the state duty of protection. For example, even though Article 20a GG provides for a state objective (*Staatsziel*)[33] and from it no subjective right can be derived,[34] it operates as a parameter (*Maßstab*) for the evaluation of the state violations of individual fundamental rights and liberties.[35]

---

[25] Boyd (2013), p. 11. Most recently, see Amirante (2019); Amirante (2022), pp. 75 and 127 ff.

[26] Boyd (2013), p. 11.

[27] GCC, 24 March 2021, 1 BvR 2656/18, 1 BvR 288/20, 1 BvR 96/20, 1 BvR 78/20, Neubauer, para 196.

[28] In the European context, other examples are the Dutch Constitution, Article 21; the Greek constitution, Article 24; the Latvian Constitution, Article 115; the Slovenian Constitution, Article 5.

[29] Boyd (2013), p. 11.

[30] Romano (1947), p. 96.

[31] Romano (1947), p. 97.

[32] Rodi (2022), p. 102.

[33] Di Fabio (2004), margin no. 95.

[34] Callies (2022) margin no. 30; Weinrich (2022), pp. 47 and 127; Schulze-Fielitz (2013), margin no. 82; Neubauer, para 112.

[35] Rodi (2022), p. 102.

Generally, constitutional clauses on the duty of protection do not clarify whether the duty's efficacy is confined within state borders.[36] Yet the broad object of the protection—"natural foundations of life and animals" in Article 20a GG or "the environment, biodiversity and ecosystems" in Article 9(3) Italian constitution—can hardly be framed within the state, as climate change knows no boundaries:[37] if the state shall protect the environment, and climate change causes environmental impairments also far away in space, the state climate action shall protect or at least not jeopardize the environment beyond its borders. An "international dimension" shall therefore be accorded to this duty of protection and to the state responsibility.[38]

Due to the silence of the constitutional charters, the determination of the spatial reach of the state duty of protection seems to be referred to national (constitutional) courts.

Whereas national courts have delivered important decisions on the obligation to take climate action, they have not clarified whether the state has a constitutional duty to protect also the environment beyond its territory. The GCC had a chance in *Neubauer*, but evidently chose not to take position.

## 5 The *Neubauer* Case

### 5.1 Overview

*Neubauer* is considered a ground-breaking case in climate change litigation,[39] also from a constitutional perspective for (re)affirming[40] that the federal legislator has an obligation to take climate action (*Klimaschutzgebot*) deriving from Article 20a GG.[41]

The facts are known: the plaintiffs, some residing in Germany and some residing in Bangladesh and Nepal, filed four constitutional complaints against several provisions of the Federal Climate Change Act of 12 December 2019 (*Bundes-Klimaschutzgesetz*, KSG) and the state failure to take further measures to reduce GHG emissions after 2030. Sections 3, § 1, and e 4, § 1, KSG together with Annex 2 provided for the reduction of GHG emissions of at least 55% within 2030 compared to the 1990 levels, without specifying the emission amounts for the following

[36] In the European context, the Slovenian Constitution is an exception, as in Article 5 it explicitly limits the protection of fundamental rights and freedom to the Slovenian territory.

[37] Rodi (2022), p. 101; Callies (2022), margin no. 50.

[38] In analogous terms, with respect to the responsibility towards future generations, Grasso, L'espansione della categoria dei doveri costituzionali nella riforma costituzionale sull'ambiente, in eticaeconomia.it, 4 April 2022.

[39] Kotzé (2022), p. 1423; Rodi (2022), p. 129; Bifulco (2021), pp. 421 ff.

[40] GCC, 24 November 2010, 1 BvF 2/05, para 135.

[41] Neubauer, para 196, Aust, Klimaschutz aus Karlsruhe: Was verlangt der Beschluss vom Gesetzgeber?, in VerfBlog, 5 May 2021.

years. The plaintiffs argued that the state climate action was not sufficient to limit the raising of the temperature to 2 °C, preferably 1.5 °C, as established in the Paris Agreement, especially from 2031 on. This would result in the violation of the state duty to protect arising from Article 20a GG and the plaintiffs' rights to life and physical integrity (Article 2(2) first sentence GG), to property (Article 14(1) first sentence GG), to a minimum ecological standard of living (Article 2(1) in combination with Article 20a GG), and to a future coherent with human dignity (Article 2(1) in combination with Article 1(1) GG).

In the light of the scientific evidence on anthropogenic climate change, its consequences, and the associated risks described in the reports published by the Intergovernmental Panel on Climate Change (IPCC), the GCC held the KSG provisions governing national climate targets and the annual emission amounts allowed until 2030 to be incompatible with the objectives set in the Paris Agreement, as well as inadequate for the protection of the fundamental rights of future generations, insofar as they lacked sufficient specifications for further emission reductions from 2031 onwards.[42] Accordingly, the GCC found a violation of the intergenerational responsibility and the obligation to take climate action enshrined in Article 20a GG.[43] In all other respects, the constitutional complaints were rejected.[44]

## 5.2 *The Issue of Extraterritoriality*

### 5.2.1 Claims

For what herein concerned, the GCC rejected the complaints raised by several children residing in Bangladesh and Nepal,[45] who claimed that Germany had violated its duties of protection arising from Article 2(2) first sentence GG and Article 14(1) first sentence GG—respectively the right to life and physical integrity and the right to property—due to insufficient climate action efforts.[46] They argued that their health and property were and continue to be exposed to the environmental impairments caused by climate change, such as heavy rainfall, landslides, extreme heat, floods, cyclones, droughts, fires, and drought—events that impact on their houses, businesses, and farms.[47]

Alleging that the GHG emissions from the German territory were partly responsible for the impairments experienced in their home countries, the plaintiffs invoked the responsibility of the German state under Article 1(3) GG, which binds the legislature, the executive, and the judiciary to the basic rights enshrined in the Basic

[42] Neubauer, paras. 185–190.

[43] Ibid., paras. 198 ff.

[44] Ibid., para 229.

[45] 1 BvR 78/20.

[46] Neubauer, para 78.

[47] Ibid., paras. 83 and 78.

Law, and Article 25 GG, which incorporates the general rules of international law in federal law. In relation to the former, since it does not provide for a spatial restriction to its efficacy, they argued that the protection afforded by Article 1(3) GG to fundamental rights should reach also people living abroad; in relation to the latter, they claimed that Germany violated customary international law because state authorities failed to meet their customary international law obligations to avoid damage to the environment in other states (*no-harm rule*). The plaintiffs relied on the legislature's inability to set proper climate protection targets and adopt adequate adaptation and (especially) mitigation measures.[48]

Opposing to such claims, the Federal Government and Parliament objected the lack of a "sufficiently close connection" of the plaintiffs to Germany, proved by the fact that climate change is a global phenomenon and by the impossibility to establish a direct causality link between the German legislation and the environmental damage in Nepal and Bangladesh. Besides this, fundamental rights would not imply the content of an "effective and appropriate" protection even towards people abroad, due to the principle of non-intervention under international law and the factual limits resulting from the limited resources of the state.[49]

### 5.2.2 The Decision

The GCC recognized standing to the plaintiffs,[50] but rejected the claim, finding no violation of the duty of protection (*Schutzpflicht*) arising from fundamental rights.[51]

When *Neubauer* was decided it had not hitherto been clarified under which conditions the Basic Law impose on the German state a duty to protect the rights of people abroad against impairments caused by climate change.[52] Albeit defining it "conceivable in principle", the GCC chose not to delve into the question of whether such duty to protect places the German state under an obligation with respect to the plaintiffs, finding it of no relevance for the case.[53] As it emerges from the Judgment, this seems a deliberate choice.

Firstly, the GCC focused on Article 1(3) GG, and recalled its Judgment of 19 May 2020 on the surveillance led by Federal Intelligence Service on the telecommunications abroad (so-called *BND* Judgment), in which it affirmed—for the first time—that the binding effect of Article 1(3) GG is comprehensive and not restricted to the German territory. Such a conclusion was based on three main considerations: *1*) no reference is made by the Basic Law to national borders as a limit to the binding effect of fundamental rights; *2*) at the time in which the Basic Law was approved,

[48] Ibid., para 84.

[49] Ibid., paras. 85–89.

[50] Ibid., para 101.

[51] Ibid., para 173.

[52] Ibid., paras. 101 and 175.

[53] Ibid., para 174. Critically on the GCC choice not to develop the issue, see Jahn (2022), p. 71.

there was no unspoken consensus on the fact that fundamental rights are not applicable to the actions of German state organs abroad; and *3*) particularly in response to the violence of Nazi tyranny, Article 1(3) GG aimed at achieving a comprehensive binding effect of fundamental rights rooted in human dignity.[54]

However, in the *BND* Judgment the GCC confined this principle to fundamental rights in their defensive dimension (*Abwehrrechte*), i.e. as rights against state interference such as in the case of surveillance measures carried out by the Federal Secret Service, and clarified that the situation might be different for fundamental rights in their dimension as positive obligations of the state (*Leistungsrechte*) or—like in *Neubauer*—as the basis for objective duties of protection (*Schutzpflichten*).[55]

Rights in their defensive dimensions refrain the state—in all its organs—from interferences into the private sphere of individuals;[56] an example is the right to privacy against mass surveillance. Rights in their positive dimension require a *facere*, a positive intervention of the state for their realization;[57] an example is the right to health when it requires the state to perform medical assistance. Rights as basis for duties of protection require the state to protect the legal good of the individual from external attacks; originally, they were conceived as protecting the individual only from attacks by third private parties, but currently the majority of scholars and the constitutional case-law include interferences caused by natural disasters or other phenomena, such as climate change.[58]

In *Neubauer* the *BND*-principle was recalled without being further analyzed. The GCC only affirmed that one possible factor capable of establishing a constitutional duty of protection would be that the severe impairments already or potentially faced by people abroad due to climate change are caused to some extent by GHG emissions emanating from Germany[59]—which actually seems the situation claimed by the plaintiffs.

Secondly, the GCC stated that even if it existed, a constitutional duty of protection vis-à-vis people abroad would not have the same content that it has vis-à-vis people residing in Germany. The main obstacle to the extraterritorial extension of such duty is the principle of sovereignty under international law, which prohibits the German state to act in territories out of its jurisdiction.[60] With regard to this matter, the reasoning of the GCC builds on the measures that must be undertaken to hinder climate change.

In order to comply with the temperature targets set in the Paris Agreement, Germany is obliged to: *1*) adopt measures that help to slow down global warming

---

[54] BND, paras. 89–92.

[55] Ibid., paras. 88 and 104. Critically, see Bifulco and Gentile (2022), p. 147; Reinke, Rights reaching beyond Borders, in VerfBlof, 30 May 2020.

[56] Jellinek (1892), pp. 98 and 105 ff.; Dreier (2013), p. 3. margin no. 84; Isensee (1992), margin no. 2.

[57] Jellinek (1892), pp. 98 and 127 ff.; Dreier (2013), margin no. 89.

[58] Hendler (2009), p. 401; Unruh (1996), p. 17; contra. Dreier (2013), margin no. 104.

[59] Neubauer, para 175.

[60] Ibid., para 176.

(i.e. mitigation measures); 2) implement measures that, while not actually mitigating climate change, do alleviate its adverse impacts (i.e. adaptation measures).[61] These measures include the creation of open spaces, the restoring, unsealing, renaturing and reforesting of suitable areas, and building houses with low and aerodynamic design. The unilateral implementation of such measures in foreign territories would result in the violation of the sovereignty of the foreign state. Therefore, due to the sovereignty principle, "it is practically impossible for the German state to afford protection to people living abroad by implementing adaptation measures there".[62]

However, in its reasoning the GCC only focused on the legal and factual obstacles of external *adaptation*, but ignored *mitigation*, which was instead referred to by the plaintiffs. As a consequence, the decision seems quite evasive insofar it omits deliberately to deal with the extensions of a constitutional duty of protection with respect to mitigation measures, which, as will be stressed, do not interfere with other states' sovereignty. In alternative, the GCC could have said that the state has a duty to protect against climate also vis-à-vis people living abroad, even though such a duty differs from the one that it has vis-à-vis its residents as it would not have the same options at its disposal due to the limits deriving from the sovereignty principle.

*Last but not least*, the GCC stated that even if the German state were obliged under Arts. 2(2) first sentence and 14(1) GG to afford protection to the plaintiffs in Bangladesh and Nepal by taking action to hinder the rise in temperature, such a duty of protection would not be violated by the challenged provisions, because the legislator had undertaken climate measures not manifestly unsuitable or completely inadequate for achieving the protection goal, by ratifying the Paris Agreement and promulgating the KSG.[63]

Two final remarks on *Neubauer*. First, the GCC assimilates mitigation or adaptation measures, as it treats them as if they were of the same nature and gave the same implementation problems; as will be argued, this argument does not seem to be embraceable. Second, the GCC does not build its reasoning on the objective duty to protect enshrined in Article 20a GG, but rather focuses on the duties of protection deriving from fundamental rights (Arts. 2 and 14 GG). It is difficult to assess whether the outcome would have been different on the basis of Article 20a.

---

[61] Ibid., para 177.

[62] Ibid., para 178. Nonetheless, Germany shall undertake positive steps to protect people in poorer and harder-hit countries, by providing these countries with financial resources (para 179).

[63] Ibid., para 180.

## 6 The Extraterritorial Duty to Protect with Respect to Adaptation and Mitigation Measures

*Neubauer* sheds light on the fact that adaptation and mitigation measures have different implications on the forementioned state duty to protect the environment. The following analysis aims at showing that an extraterritorial duty to protect is conceivable only in relation to mitigation.

Climate change adaptation consists in the "process of adjustment to actual or expected climate and its effects in order to moderate harm or take advantage of beneficial opportunities";[64] in other words, it aims at "reducing climate risks and vulnerability mostly via adjustment of existing systems".[65] Adaptation takes many shapes and forms, ranging from building flood defenses, setting up early warning systems for cyclones and switching to drought-resistant crops, to redesigning communication systems, business operations and government policies.[66] It embraces all aspects of human society,[67] as it impacts on several fields: property rights, freedom of economic enterprise, public and private health, working conditions, etc.

Whereas the implementation of adaptation measures within state borders does not encounter limits under the sovereignty principle, the same cannot be affirmed with respect to their implementation beyond the state borders. As adaptation requires a positive *facere* of the state, its extraterritorial implementation faces an insurmountable obstacle under international law: states cannot unilaterally apply or enforce rules of conduct in respect of persons, property or events beyond their territory, because they would otherwise infringe the other state's sovereignty.[68] Exemplifying, a state could not build modern train lines or irrigation systems in territories out of its jurisdiction. For this reason alone, a duty of protection can hardly be argued with respect to adaptation measures. Nonetheless, states must abide by the no-harm rule,[69] and they are politically responsible for ensuring that positive steps are taken to protect people in poorer and harder-hit countries, as Article 9 of the Paris Agreement affirms that "developed country Parties shall provide financial resources to assist developing country Parties with respect to both mitigation and adaptation".[70]

With respect to mitigation measures, the above considerations on extraterritoriality do not apply.

[64] IPCC (2022), Summary for Policymakers (available at: https://bit.ly/3OqmjMZ), recalling the definition of adaptation give in IPCC's Fifth Assessment Report.

[65] Ibid., 20.

[66] Further measures are enlisted in I.R. Noble et al. (2014). Adaptation needs and options. https://bit.ly/3tQ8le5. Accessed 22 May 2023.

[67] Craig (2010), p. 29.

[68] Kamminga (2020).

[69] Adelman (2019), pp. 123 ff., spec. 127.

[70] In analogous terms, see Article 4(3) of the UNFCCC, Neubauer, para 179. So far, however, states have failed to provide such an economical assistance to developing countries.

Climate change mitigation aims at reducing the sources or enhancing the sinks of GHG.[71] Under the UNFCCC, every state is required to adopt "measures on the mitigation of climate change, by limiting its anthropogenic emissions of greenhouse gases and protecting and enhancing its greenhouse gas sinks and reservoirs".[72] The major sources of GHG emissions are indeed human activities:[73] the burning of fossil fuels—including coal, gas and oil—was responsible for 86% CO2 emissions between 2011 and 2021.[74] In the field of climate mitigation, the role of the mining industry is "contentious", since, on the one hand, it is an intensive energy user and greenhouse gas emitter and is therefore considered environmentally and socially destructive, and yet, on the other hand, it is the primary supplier of the infrastructure needed to mitigate the destructive effects of climate change.[75]

In order to mitigate climate change and contain emissions, states have been adopting a wide range of instruments, such as carbon taxes, i.e. taxes on fossil fuels or other substances containing carbon (which impact directly on the mining industry);[76] cap-and-trade systems, setting emissions cap and issuing a quantity of emission allowances consistent with that cap;[77] command-and-control regulations on multiple economic sectors,[78] etc.[79] The implementation of such instruments takes place within the state territory (and under its jurisdiction), without requiring a positive intervention in other states, and therefore without conflicting with the principle of sovereignty.

Nonetheless, since GHG emissions produced in a specific country do not stop at the state border,[80] the effects of mitigation strategies travel beyond the state territory and affect the environment abroad.[81]

The transborder impacts of mitigation measures is efficaciously pointed out in *Neubauer*, where the GCC acknowledges that "by reducing the GHG emission produced in Germany, the German state could protect people living abroad against the consequences of climate change just as it could protect those living in Germany".[82]

---

[71] IPCC, Summary for Policymakers, in O. Edenhofer et al., Climate Change 2014: Mitigation of Climate Change, 2014, 4. https://bit.ly/2HihPHR. Accessed 22 May 2023.

[72] Article 4(2) lett. a) UNFCCC.

[73] Fawzy et al. (2020), p. 2070; Yue and Gao (2018), p. 243 ss.

[74] https://www.ipcc.ch/report/ar6/wg1/downloads/report/IPCC_AR6_WGI_Full_Report_smaller.pdf. Accessed 22 May 2023.

[75] Cox et al. (2022), p. 5.

[76] On green taxes (i.e. taxes on GHG emissions), see F. Castañeda Ibarra v. Government of Zacatecas. https://bit.ly/3hVmX9f. Accessed 22 May 2023.

[77] Van Asselt (2016), p. 332.

[78] Mayer (2021), p. 143.

[79] IPCC (2022), Summary for Policymakers, in Climate Change 2022: Mitigation of Climate Change, 29.

[80] Weinrich (2022), p. 106.

[81] Weinrich (2022), p. 106.

[82] Neubauer, para 178.

To conclude, with regard to mitigation measures, unlike for adaptation, there seem to be minor obstacles to the configuration of an extraterritorial state duty to protect and, accordingly, to the recognition of the state responsibility in the case of the violation of such duty.

## 7 Conclusion

That states have a duty of protection against the impairments caused by climate change is widely acknowledged. As the UN High Commissioner for Human Rights pointed out, "states (duty-bearers) have an affirmative obligation to take effective measures to prevent and redress these climate impacts, and therefore, to mitigate climate change, and to ensure that all human beings (rights-holders) have the necessary capacity to adapt to the climate crisis".[83] This statement condenses the state duty to adopt mitigation and adaptation measures. As it refers to all human beings as rights-holders, no difference should be made between citizens and non-citizens; as it does not mention state territory as a limit, it could be assumed that such a state duty reaches beyond national borders (in compliance with the limits set under international law).

As climate change crosses boundaries, so do the effects of state measures adopted and implemented to hinder and contain the phenomenon; accordingly, the territorial scope of application of the state duty to protect should not be confined to national borders, and states should be held responsible for insufficient or inadequate climate action. In other words, "if pollution can travel across the border, so can legal responsibility".[84]

With respect to adaptation measures, such a duty is in practice unattainable due to the limit of the sovereignty principle. The scenario changes with respect to mitigation measures, whose implementation takes place within the national perimeter but whose effects extend also beyond it. In the light of the above considerations on mitigation and mining, it is clear that the extraterritorial duty to protect would also concern the mining sector and (the regulation of) its GHG emissions.

If an extraterritorial duty to protect is conceivable with regard to mitigation measures, and accordingly the state can be held responsible for violating it due to insufficient or inadequate climate mitigation action, i.e. for its GHG emissions, how can inhabitants of other states claim and prove such a violation?

On procedural grounds, the criterion of extraterritorial adjudicative jurisdiction must be satisfied for a state to have authority to decide competing claims in respect to events beyond its territory, such as climate change, in which the cause is internal (GHG emissions) and the event is external (environmental damage). The exercise of extraterritorial adjudicative jurisdiction is permitted only if there is a sufficient

---

[83] OHCHR (2016), Key Messages on Human Rights, Climate Change and Migration.

[84] Banda (2018), p. 1.

connection between the state and the external event.[85] To this purpose, interesting cues are offered by the effective control test developed by the Inter-American Court on Human Rights in OC 23/17 (and recalled in *Sacchi et al.*), according to which the state has jurisdiction insofar it exercises effective control on the activities of production of GHG within its territory.

This model of extraterritorial jurisdiction would not solve all the issues related to the assessment of the transboundary damage, as for example the establishment of the causality link between the damage and the GHG emissions hailing from a specific state. It is nearly impossible to precisely prove such link,[86] since it is extremely arduous to trace the "direct and certain causal nexus" between a climate damage and the emissions from a particular state or group of states.[87] Likewise, proving how the damage impairs human rights might be difficult, but in this case a key role would be played by the extensive scientific evidence documenting the risk of harm due to climate change, which may be used to establish a causal link between environmental harm and the violation of human rights. [88]

Yet, the suggested model of extraterritorial jurisdiction would allow persons from other countries, whose environment has been seriously damaged by the effects of GHG emissions coming from another state, to invoke the (in)acting state responsibility for its insufficient or inadequate action of mitigation. In order to assess the quality of mitigation measures, and if the state complied with the temperature targets set at the international level, the Paris Agreement represents an essential parameter. As an international binding act, it guides the action of the legislature and the interpretation and application of the law by the judiciary. Depending on the constitutional structure of states, such Agreement may have 'supra-legal' rank[89] or function as an "interposed norm" between the constitution and the law,[90] guiding the interpretation of legal provision(s) in a way that is consistent with the aims set in the Agreement, and therefore contributing to the process of "constitutionalization of international environmental law".[91]

---

[85] Kamminga (2020).

[86] An exception is Luciano Lliuya c. RWE AG, in which judges and experts flew to Peru to ascertain environmental damages caused by the pollution of a German enterprise in the area of residence of the plaintiff, cfr. Steininger and Herrera (2022), Travelling Courts and Strategic Visitation, in VerfBlog, 1 June 2022.

[87] Maljean-Dubois (2018). To answer the question "whether and to what extent anthropogenic climate change has altered the likelihood and intensity of an individual extreme weather event to occur", scientists adopt a probabilistic event attribution, see Minnerop and Friederike (2020), p. 55. On the topic, see Quirico (2018), pp. 187 ff.

[88] Keller and Heri (2022).

[89] See, on this, the Judgment of the Brazilian Supreme Court, 1 July 2022, PSB et al v. Brazil.

[90] As in Italy, according to Article 117(1) of the constitution.

[91] Kotzé and Muzangaza (2018), p. 27.

## References

Adelman S (2019) Sovereignty and environmental human rights. In: Elgar Encyclopedia of Environmental Law, Cheltenham, n. 7.8, pp 123 ff., spec. 127

Amirante D (2019) L'ambiente preso sul serio. Il percorso accidentato del costituzionalismo ambientale, in DPCE, no. spec., pp1 ss

Amirante D (2022) Costituzionalismo ambientale. Atlantico giuridico per l'Antroprocene, Bologna

Arent D, Pless J, Mai T, Wiser R et al (2014) Implications of high renewable electricity penetration in the U.S. for water use, greenhouse gas emission, land-use, and materials supply. Appl Energy 123, pp 368 ff

Aust (2021) Klimaschutz aus Karlsruhe: Was verlangt der Beschluss vom Gesetzgeber?, in VerfBlog, 5 May 2021

Bagni S (2021) La costruzione di un nuovo "eco-sistema giuridico" attraverso i formanti giudiziale e forense, in DPCE online, spec. n., 1036

Banda ML (2018) Inter-American Court of Human Rights' Advisory Opinion on the Environment and Human Rights. Am. Soc'y Int'l L, p 1

Bifulco R (2021) Cambiamento climatico, generazioni future (e sovranità) in una stori-ca sentenza del Tribunale costituzionale federale tedesco. Astrid Rassegna 12, pp 421 ff

Bifulco R, Gentile C (2022) I fragili confini del territorio. Federalismi.it 4:147

Boyd DR (2012) The environmental rights revolution. A global study of constitutions, human rights, and the environment. Vancouver-Toronto, pp 3 and 89

Boyd DR (2013) The Status of Constitutional Protection for the Environment in Other Nations, David Suzuki Foundation, available at: https://bit.ly/3Fto5KD

Callies C (2022) Article 20a. In: Maunz T, Dürig G (eds) Grundgesetz Kommentar, XCVIII, München, margin no. 30

Cox B, Innis S, Kunz NC et al (2022) The mining industry as a net beneficiary of a global tax on carbon emissions. Commun Earth Environ 3:17

Craig RK (2010) Stationarity is dead. Harv Environ Law Rev 34:29

Di Fabio U (2004) Article 2 Abs. 2 Satz 1 GG. In: Maunz T, Dürig G (eds) Grundgesetz Kommentar, XLIII, München, margin no. 95

Dreier H (2013) Vorbemerkungen vor Artikel 1. In: Dreier H (ed) Grundgesetz-Kommentar, Tübingen, 3, margin no. 84

Fawzy S et al (2020) Strategies for mitigation of climate change. Environ Chem Lett 18:2070

Gociu A, Roy S (2021) Extraterritoriality of Oil Constitutionalism in People v Arctic Oil, in EJIL: TALK!, 16 February 2021

Grasso (2022) L'espansione della categoria dei doveri costituzionali nella riforma costituzionale sull'ambiente, in eticaeconomia.it, 4 April 2022

Hendler R (2009) Zur verfassungsrechtlichen Pflicht des Staates zum Schutz vor Bodenwertminderungen, in UPR, n. 11+12, 401

Heri C (2022) Climate change before the European Court of Human Rights: capturing risk, ill-treatment and vulnerability. EJIL 33:926

Isensee J (1992) Das Grundrecht als Abwehrrecht und als staatliche Schutzpflicht. In: Isensee J, Kirchhof P (eds) Handbuch des Staatsrechts der Bundesrepublik Deutschland, 2003, 9, margin no. 2

Jahn J (2022) Internationaler Klimaschutz mithilfe nationaler Verfassungsgerichte. ZaöRV 82(1):71

Jellinek G (1892) System der subjektiven öffentlichen Rechte, Milan, 1912, pp 98 and 105 ff

Kamminga MT (2020) Extraterritoriality. In: Wolfrum R (ed) The Max Planck encyclopedia of public international law. Oxford

Keller H, Heri C (2022) The Future is Now: Climate Cases Before the ECtHR, in Nordic Journal of Human Rights, n, 40.1, pp. 153 ff

Kotzé (2022) Neubauer et al. versus Germany. German Law J 22(8):1423

Kotzé LJ, Muzangaza W (2018) Constitutional international environmental law for the Anthopocene? RECIEL, pp 291–292
Maljean-Dubois S (2018) Climate change litigation. In: The Max Planck encyclopedia of international law. Oxford
Mayer B (2021) Climate change adaptation and the law. Va Environ Law J 39:143
Minnerop, Friederike (2020) Climate change and causation. Buff Environ Law J 27:55
Müller A (2022) Justifying extraterritorial human rights obligations. In: Gibney M et al (eds) The Routledge handbook on extraterritorial human rights obligations. Abingdon-New York, pp 119–120
Norgate T, Haque N (2010) Energy and greenhouse gas impacts of mining and mineral processing operations. J Clean Prod 18(3), pp 266 ff
Quattrocchi G (2018) Applicabilità extraterritoriale della Convenzione americana dei diritti umani e prevenzione del danno ambientale transfrontaliero. Diritti umani e diritto internazionale 12(3):686–687
Quirico O (2018) Climate change and state responsibility for human rights violations: causation and imputation. NILR 65(2), pp 187 ff
Reinke (2020) Rights reaching beyond Borders, in VerfBlof, 30 May 2020
Rodi M (2022) Handbuch Klimaschutzrecht, München, p 102
Romano S (1947) Doveri. Obblighi, in Id., Frammenti di un dizionario giuridico, Milano, 1947, p 96
Rüttinger L, Sharma V (2016) Climate change and mining. A foreign policy perspective. https://bit.ly/3kZXxbU 1. Accessed 22 May 2023
Schulze-Fielitz H (2013) Article 20a. In: Dreier H (ed) Grundgesetz-Kommentar, Tübingen, n. 3, margin no. 82
Setzer J, Higham C (2022) Global Trends in Climate Change Litigation: 2022 Snapshot. Grantham Research Institute on Climate Change and the Environment and Centre for Climate Change Economics and Policy, London, p 11
Steininger, Herrera (2022) Travelling Courts and Strategic Visitation, in VerfBlog, 1 June 2022
Unruh P (1996) Zur Dogmatik der grundrechtlichen Schutzpflichten, Berlin, p 17
van Asselt H (2016) The design and implementation of greenhouse gas emissions trading. In: Gray KR, Tarasofsky R, Carlarne CP (eds) The Oxford handbook of international climate change law, p 332
Weinrich M (2022) Die grundrechtlichen Umweltschutzpflichten des Staates, Berlin, pp 47 and 127
Yue XL, Gao QX (2018) Contributions of natural systems and human activity to greenhouse gas emissions. Adv Clim Change Res, p 243 ss

# Corporate Climate Liability

## After Sharm El-Sheikh, Shell and Daimler Judgement

**Walter Frenz**

### Contents



W. Frenz (✉)
LFG Berg-, Umwelt- und Europarecht, RWTH Aachen University,
Aachen, Nordrhein-Westfalen, Germany
e-mail: frenz@bur.rwth-aachen.de


W. Frenz, A. Preuße (eds.), *Yearbook of Sustainable Smart Mining and Energy 2022*, Yearbook of Sustainable Smart Mining and Energy - Technical, Economic and Legal Framework 2, https://doi.org/10.1007/978-3-031-41873-0_7

**Abstract** The ruling of the Higher Regional Court of Hamm of November 30, 2017, on RWE's responsibility for (imminent) climate damage in Peru is now almost a classic. It has been joined by the Shell judgment of the District Court of The Hague, which obliged the oil company to make $CO_2$ reductions, and conversely the judgment of the County Court Stuttgart, which dismissed climate claims against Daimler. The climate conference in Sharm El-Sheikh 2022 decided on a climate damage fund. Does this result in new aspects for civil law dogmatics?

## 1 Initial Facts

Are German companies liable for global climate damage? At the Sharm El-Sheikh conference, a climate damage fund for the benefit of the countries most affected by climate change was launched. The industrialized countries are to provide it with capital, however to what extent and by whom specifically is not yet certain. The outcome of the proceedings before the Higher Regional Court of Hamm is also still unsure. In the Huarez case, a plot of land is to be protected against the effects of climate change. RWE is said to be obliged to contribute to the costs of protecting the plaintiff's property from a tidal wave that is threatened by the melting of a nearby glacier as a result of global warming and thus ultimately also because of RWE's $CO_2$ emissions. Here the plaintiff refers specifically to the protection of property according to Section 1004 German Civil Code. The Higher Regional Court of Hamm issued an order to take evidence on November 30, 2017[1] and did not dismiss the lawsuit as the previous instance.[2]

## 2 Location in the Current Climate Law Context

A lot has happened in terms of climate damage law since the evidence order by the OLG Hamm. At the level of international law, an instrument was created to compensate for climate damage. For the Glasgow climate summit in 2021, the countries of the Global South and island states were already calling for an independent financial institution to be set up to provide funds for this. After all, a pot of money was set up at this conference to provide aid for damage and losses in poor countries, thus fulfilling a long-standing demand. The industrialized countries, though refused to pledge concrete aid for damage already caused by climate change, which was

[1] OLG Hamm 30.11.2017 – 5 U 15/17, ZUR (2018), p. 118; more detailed in the general context Frenz (2022a), pp. 115 et seqq. also for the following.

[2] As stated by the previous instance LG Essen 15.12.2016 – 2 O 285/15, NVwZ (2017), p. 734.

criticized by the developing countries.[3] However, poor countries were to receive more financial aid for adapting to the often fatal consequences of the climate crisis. In this respect, an increase from the previous level of around 20 billion U.S. dollars to around 40 billion U.S. dollars was planned, thus doubling the amount.

It was not until the Sharm El-Sheikh climate conference in November 2022 that the agreement was reached on a compensation fund for damage already incurred for the benefit of particularly vulnerable countries. The fund is intended to cover the unavoidable consequences of global warming, such as increasingly frequent droughts, floods and storms, as well as rising sea levels and desertification. However, the climate conference in Sharm El-Sheikh has not yet decided on any sums for the new compensation fund, nor have the exact stakeholders been defined. A transitional commission is to be set up to draw up recommendations, which will then be discussed at the next climate conference in Dubai at the end of 2023. This transitional commission is to consist of ten representatives of the industrialized countries and thirteen of the developing countries.[4] There is also a lack of clear specifications for the supervision of this fund to date.

Nevertheless, this is a "milestone" and a "real success in the fight against climate change".[5] An independent institution has now been created so that developing countries do not have to act as petitioners in order to receive financial aid for the often-fatal consequences of the climate crisis in their countries or to adapt to it. The creation of the compensation fund is the logical counterpart to the increase in financial aid to developing countries. In any case, the industrialized countries implicitly recognize their responsibility for the damage in the countries particularly affected by climate change.

However, the development affects states and does not even justify concrete claims against them. Parallels could be drawn with a fund into which companies have to pay into that have caused high $CO_2$ emissions and considerable damage as a result. But such a fund is not up for discussion. It would also be difficult to establish under international law. At most, states could be obliged to require their companies to pay into such a fund.

---

[3] Tagesschau 13.11.2021, Klimagipfel erzielt Einigung. https://www.tagesschau.de/ausland/europa/klimakonferenz-einigung-abschlusserklaerung-101.html. Accessed 22.5.2023.

[4] Tagesschau 20.11.2022, COP27 einigt sich auf Abschlusserklärung. https://www.tagesschau.de/eilmeldung/klimakonferenz-cop27-eu-abschlusserklarung-klimawandel-101.html. Accessed 22.5.2023.

[5] As stated by the expert on climate Jan Kowalzig from Oxfam Deutschland, Tagesschau 20.11.2022, COP27 einigt sich auf Abschlusserklärung. https://www.tagesschau.de/eilmeldung/klimakonferenz-cop27-eu-abschlusserklarung-klimawandel-101.html. Accessed 22.5.2023.

# 3 Causality Affirmation Contrary to the Forest Damage Decision of the Federal Court of Justice?

## 3.1 *Full Proof Required*

The solution of the Huarez case is thus not shaped by developments in international law, but by civil law. In the multifaceted discussion about the prerequisites for a general liability for climate damage according to Section 823 BGB (German Civil Code) or the protection against future impairments according to Section 1004 para 1 BGB (German Civil Code), the focus is still on the causality of the respective process for the damage.[6] The full proof required for this according to general principles of civil law is not considered promising for climate damage.[7] Finally, liability for environmental damage is generally excluded if causation cannot be reliably proven.[8] Accordingly, only developments that can definitely be traced back to climate change are included.[9]

## 3.2 *German Federal Court of Justice on Forest Damage*

The case law of the Federal Court of Justice (BGH)[10] on forest damage remains dogmatic, for which an individual classification was rejected because of an incalculable number of emitters. This approach cannot be regarded as outdated even with reference to developments in international law. The fact that a compensation fund was set up means that the industrialized countries are jointly and not individually responsible. No industrialized country has yet recognized individual responsibility.

## 3.3 *No Fundamental Differences*

It is therefore not without reason that civil law arguments are used: Against an adoption of the BGH case law, it is argued that the $SO_2$ emissions dealt with in the BGH judgment should also be distinguished from the $CO_2$, which are the subject of the OLG Hamm proceedings. For the former as the cause of the acid rain over a certain piece of forest, it is not possible to determine which power plant they came from, as they remain in the near-earth air layers and are blown in different directions

[6] Pöttker (2014), pp. 62 et seqq.

[7] Chatzinerantzis and Appel (2019), p. 881 (882, 883 et seqq.).

[8] Hecker et al. (2013), p. 139 (158).

[9] Risse and Haller (2021), p. 3500 (3502); Frenz (2022a), pp. 115 et seqq. also for the following.

[10] BGH 10.12.1987 – III ZR 220/86, BGHZ 102, 350 (368).

depending on the wind direction.[11] $CO_2$ emissions, on the other hand, remain in the atmosphere for many years and, when added up, contribute to global warming increasing and causing climate damage.[12] However, this is also rejected with reference to the excessive complexity of the scientific causal chain.[13] In any case, the lack of precise allocation to individual power plants remains.

## 3.4 A Side Glance at the Federal Constitutional Court

Incidentally, the Federal Constitutional Court also dismissed a lawsuit because of acid rain: It did not see any violation of fundamental rights to protect, but the state had taken sufficient measures.[14] The Federal Constitutional Court did not see any violation of fundamental rights in its Climate Decision[15] either: in this respect, too, the Federal Constitutional Court emphasized the normative leeway. It only accepted obligations to reduce $CO_2$ based on the civil rights of subsequent generations. In this regard, a more detailed calculation was also made as to which reduction burdens Germany has to bear.

Thus, $CO_2$ emissions were allocated to a state, although these emissions, when added together, ultimately cause climate change, as the Federal Constitutional Court recently emphasized again: "At the same time, every emission of climate-damaging $CO_2$ from every state contributes equally to climate change."[16] However, the calculation of the German reduction burden, based on the IPCC's calculations of the still possible global $CO_2$ emissions by 2050 to meet the 1.5-degree target, in the climate resolution of March 24, 2021, was not used as a concrete constitutionally binding figure, but rather as an—albeit plausible—orientation mark. Ultimately, the concrete determination is left to the legislature, which must, however, adhere to the key points of the Federal Constitutional Court.[17]

Above all, this calculation concerned the reduction by the Federal Republic of Germany and not the liability of companies. It is true that governmental climate protection measures will regularly be able to assert themselves against the fundamental rights of defence of companies as well,[18] especially as climate change progresses; however, companies are not automatically liable for damages. Even the climate protection obligations that apply to them must first be defined by the

[11] Frank (2021), p. 305 (306).

[12] Frank (2021), p. 305 (306).

[13] Risse and Haller (2021), p. 3500 (3502).

[14] BVerfG 29.11.1995 – 1 BvR 2203/95, NJW (1996), p. 651.

[15] BVerfG, 24.3.2021 – 1 BvR 2656/18 i. a., BVerfGE 157, 30.

[16] BVerfG 23.3.2022 – 1 BvR 1187/17, NVwZ (2022), p. 861 (margin no. 121).

[17] BVerfG 24.3.2021 – 1 BvR 2656/18 i. a., BVerfGE 157, 30 (margin no. 236 f.).

[18] BVerfG 24.3.2021 – 1 BvR 2656/18 i. a., BVerfGE 157, 30 (margin no. 198 a. E.).

legislature. This was also emphasized by the Stuttgart Regional Court in its ruling of September 13, 2022.[19]

Therefore, it cannot be concluded from the Federal Constitutional Court's climate decision that $CO_2$ emissions must be allocated to companies for their liability and that these companies must be held responsible, even though the damage is cumulative. Such a question can at most arise in the case of avoidance efforts, and this was affirmed by the District Court of The Hague,[20] but denied by the Regional Court of Stuttgart.[21] First, however, the classical dogmatics must be consulted, before it is examined whether, despite the different perspective of avoidance efforts, conclusions for liability can be drawn from the two aforementioned civil court decisions.

# 4 Missing Classical Liability Requirements

## *4.1 Necessary Individualization*

In any case, even when considering the differences to $SO_2$ emissions, the individual damages still cannot be traced back to concrete $CO_2$ emissions from specific power plants.[22] In this respect, there is no individualization. However, this individualization is indispensable according to the case law of the Federal Court of Justice, although it is not required or possible to derive anything else from the Federal Constitutional Court's judgments on climate protection from a constitutional perspective, for example.

It is only in the aggregate and in interaction that the emitted $CO_2$ emissions have such an effect that they lead to damage. In relation to this, a percentage allocation is possible; for example, 0.47% of global $CO_2$ emissions since 1850 are attributed to RWE. However, this quantifiability can hardly suffice for an adequate damage-related individualization.[23] The link between the individual contribution and the concrete damage is missing.

It is true that RWE's 0.47% $CO_2$ emissions have contributed to global warming. However, increasing $CO_2$-emissions is an overall development. Only this then causes climate damage. Precisely 0.47% of this damage cannot be attributed to RWE and certainly not in total, which is also evidenced by Section 830, Paragraph 1, Sentence 2 of the German Civil Code (BGB), which, in the case of an

[19] LG Stuttgart 13.9.2022 – 17 O 789/21, NVwZ (2022), p. 1663 (margin no. 25). more detailed u. VII.1.

[20] See under VI.

[21] See under VII.

[22] See above III.3.

[23] As stated in contrast by Frank (2021), p. 305 (306 f.).

unascertainable concrete attribution, still presupposes a possible causation of the overall damage by each person against whom a claim is made.[24]

## 4.2 *Adequacy*

In addition, the 0.47% RWE share in the previous global $CO_2$ emissions lacks the necessary for the civil damage law adequacy.[25] It is also difficult to add up over the decades. The relevance only emerges today. It was not known or apparent in early times.[26] As a result, the predictability and recognisability of the impact of coal-fired power generation on the climate, which is required for adequacy, is also missing.[27] With regard to environmental liability, too, reference is made to the fact that liability risks for plant operators must be transparent, calculable and therefore insurable.[28]

## 4.3 *Equivalence*

Even the equivalence raises doubts. The problem is cumulative causality. It can only be affirmed if causality is also recognized in other states. Otherwise, the causality is excluded for some potentially liable parties and the other issuers have to bear the entire burden.

## 4.4 *Strict Liability as in Mining Law as a Way Out?*

Therefore, even strict liability, as provided for in mining liability in Section 114 Federal Mining Act (BBergG), cannot be considered. This, too, presupposes equivalence, although in this respect causation by a mining company is sufficient, which is, however, linked to the special features of mining, namely the necessary protection against the traditionally hazardous activity of mining.[29] The hazardous nature of $CO_2$ emissions only became apparent in more recent times and in summation as well as a long-distance effect—in contrast to mining.

---

[24] Risse and Haller (2021), p. 3500 (3502).

[25] Bereits LG Essen 15.12.2016 – 2 O 285/15, NVwZ (2017), p. 734 (736).

[26] Chatzinerantzis and Appel (2019), p. 881 (883).

[27] Differently Frank (2021), p. 305 (306 f.). More detailed Frenz (2022b), Einf. L: Klimaschadensrecht, margin no. 60, 66 et seqq.

[28] Risse and Haller (2021), p. 3500 (3502).

[29] Frenz (2019), para 114 margin no. 89 with further references.

The Meggen judgement of the German Federal Administrative Court requires that the attribution criterion still holds up, which is decisive according to the polluter-pays principle; a problem can arise at most from the fact that "an initially decisive contribution to causation recedes completely into the background, only the condition that now exists is to be evaluated, and responsibility is now assigned anew, for example according to the principle of the common burden on the general public."[30] Insofar as mining, with its special risks emphasized by the Federal Administrative Court (BVerwG), continues to have an effect, especially on the environment, the polluter responsibility continues to exist. This can also be seen—subject to the aforementioned differences—for $CO_2$ emissions that increase the earth's temperature. The fact that causal contributions recede into the background is a question of causality, and above all of overtaking causality through later causes such as (improper) construction activity or natural processes. It would be relevant if the farmer from Peru had caused any waterlogging near a melting glacier through his own construction defects.[31]

Above all, however, Section 114 Federal Mining Act (BBergG) provides for liability without fault, even in the case of legal conduct and in the case of aggregate damage if a company is certainly involved. Transferred to climate liability, this would also raise the question of whether all $CO_2$ emitters should really be liable—including, for example, car owners. The introduction of strict liability for climate damage would therefore also encounter practical difficulties.

Conversely, this view at mining law makes it clear that strict liability must be specifically ordered. The liability for climate damage is therefore also to be processed according to the purpose of the general civil law liability for damage regulations according to the classic model. Loosening up can only be considered selectively.

## 5 Global Litigation

### *5.1 Limited Forum Shopping*

In contrast to mining damage liability, which due to the necessary reference to mining does not extend to climate damage that only occurs through the burning of coal,[32] not through its mining, climate liability is international. If each individual contribution is causal, anyone affected could take action against any $CO_2$ emitter anywhere on earth, regardless of how small their contribution is. A worldwide

---

[30] BVerwG 18.12.2014 – 7 C 22.12, BVerwGE 151, 156 (margin no. 47) – Meggen.

[31] Frenz (2019) para 114 margin no. 70 as well as margin no. 111 et seqq., generally concerning building soil risks.

[32] More detailed Frenz (2019), para 114 margin no. 21 et seqq., as well as already Frenz (2018), pp. 99 et seqq.

network of various lawsuits could arise, which would hardly be transparent anymore. There is a risk of forum shopping, but it is limited by the fact that claims for damages in another country can only be considered if there has also been a damaging result in that country.

### 5.2 *Prevention of Multiple Substitutions*

In any case, an injured party may not be compensated for their loss or damage more than once. Either then only the small percentage that each emitter has contributed to global $CO_2$ emissions can be sued for. Then those affected would have to sue numerous emitters in order to have their damage compensated, even though this was caused by summation, i.e. it has an overall cause and can therefore hardly be individualized.

Or total damage could be claimed against each emitter. In that case, however, the courts would have to be informed or keep themselves informed as to where which damaged party has already received compensation and in what amount. In addition, the companies held jointly and severally liable could seek recourse from the co-polluters worldwide.

### 5.3 *Uniform Rules*

This suggests uniform rules in all countries to ensure that climate damage claims are dealt with in the same way. These could, for example, be introduced at a follow-up conference on climate change, so that they can then be established in the individual countries. Before that, however, corporate climate liability should be rejected.

## 6 Shell Judgement from The Hague

### 6.1 *First Direct Corporate Climate Obligations*

Nothing else emerges from the Shell judgment either: on May 26, 2021, Shell was sentenced by a Dutch court, albeit not to pay damages, but to reduce its $CO_2$ emissions by 45% until 2030, based on 2019.[33] $CO_2$ emissions from the production and supply chain were also attributed. The District Court of The Hague specifically refers to Shell's control and influence over the emissions released by end

[33] District Court The Hague 26.5.2021 – C/09/571932/HA ZA 19-379, ECLI:NL:RBDHA:2021:5337, no. 4.4.55.

consumers.[34] This judgment, however, refers to $CO_2$ reduction and not to liability for it. This issue is not dealt with, even though the serious and irreversible consequences of climate change are named.[35] Accordingly, the calculation of possible compensation for any consequences of climate change was left out.[36]

With its judgement, the District Court of The Hague made legal history by condemning companies to climate reduction obligations for the first time. Because of this reference, the court did not have to go into the percentage of the company's global $CO_2$ emissions, as is problematic in the climate action before the Hamm Higher Regional Court. This already shows the fundamental difference to claims for damages, which are based precisely on such a calculation. Nevertheless, corporate obligations for climate protection were affirmed. The question therefore arises as to whether the chosen legal basis also has repercussions on liability. The violation of primary obligations can lead to secondary obligations. Thus, if $CO_2$ reduction obligations are affirmed and breached, liability could come into consideration. However, the prerequisite for such an approach is that corporate $CO_2$ reduction obligations exist in the first place.

## 6.2 *Legal Basis*

### (a) **Dutch Civil Code**

The legal basis for the $CO_2$ reduction obligation is the unwritten standard of diligence in Book 6 Section 162 of the Dutch Civil Code: according to this, acting contrary to what is generally accepted under unwritten law is impermissible; therefore, Royal Dutch Shell must observe the standard of diligence customary in society when determining the Shell Group's corporate policy. In doing so, all the circumstances of each case must be taken into account.[37]

### (b) **ECHR**

The European Convention on Human Rights (ECHR), which only applies to states, is also included as a circumstance. The approach here is the right to life under Article 2 ECHR and the right to respect for private and family life under Article 8 ECHR. The district court recognizes the state-related nature of these rights; they cannot be asserted directly against companies. Indirectly, these human rights are said to play a role in the relationship between climate victims and a company, so the court takes human rights and the values they embody into account when interpreting

[34] District Court Den Haag, no. 4.4.25.

[35] District Court Den Haag, no. 4.4.10.

[36] Ekardt (2022), p. 15 (15).

[37] District Court The Hague 26.5.2021 – C/09/571932/HA ZA 19-379, ECLI:NL:RBDHA:2021:5337, no. 4.4.1.

the unwritten standard of diligence.[38] Indirectly, these rights are thus included after all—as this was the case with the Urgenda judgement[39] to which the district court explicitly refers,[40] and which preceded the Federal Constitutional Court climate decision of March 24, 2021,[41] which similarly called for stronger state reductions.

However, this reference to the ECHR because of $CO_2$ reduction obligations contradicts the ECHR itself: Indeed, the ECHR has not yet derived any climate or even environmental obligations from it, nor has it imposed any concrete climate protection measures.[42] So far, the ECHR has not led to any convictions for insufficient climate or even environmental protection efforts by states.[43] Rather, according to the European Court of Human Rights (ECtHR), they have a wide scope for design.[44] Parallel to Article 8 ECHR (respect for private life), a legislative and administrative framework for sufficient defence against dangers can also follow for the protection of life under Article 2 ECHR, to which the District Court of The Hague also referred,[45] without, however, the ECHR so far stating a deficient legislation; it limited itself to a deficiency in the application of the law.[46]

This scope of action can then hardly be narrower for companies. However, this is reflected in the judgment of the District Court of The Hague only to the extent that it merely establishes a reduction result without specifying concrete, consequential measures. According to the standards of the ECHR, however, there would not even have been such a numerical reduction obligation.

### (c) UN Guiding Principles parallel to the Federal Constitutional Court's Climate Decision

The District Court further places special emphasis on the more detailed fulfilment of the unwritten standard of diligence by international law. In doing so, it draws on the UN Guiding Principles (UNGP) as an authoritative and internationally recognized "soft law" instrument, which clearly differentiates between state and

---

[38] District Court The Hague 26.5.2021 – C/09/571932/HA ZA 19-379, ECLI:NL:RBDHA:2021:5337, no. 4.4.9.

[39] Hoge Raad of the Netherlands 20.12.2019 – 19/00135.

[40] District Court The Hague 26.5.2021 – C/09/571932/HA ZA 19-379, ECLI:NL:RBDHA:2021:5337, no. 4.4.10.

[41] Referring to this BVerfG, 24.3.2021 – 1 BvR 2656/18 i. a , BVerfGE 157, 30 (margin no. 157).

[42] Generally ECHR 29.6.1999, no. 29121/95, Asselbourg i. a./Luxemburg, Groß (2020), p. 337 (338).

[43] Comprehensively regarding the diverse judgements Braig (2013), pp. 7 et seqq.; Braig and Ehlers-Hofherr (2020), pp. 589 et seqq.

[44] ECHR 6.2.2018, no. 23225/05, Calancea i. a./Moldova, margin no. 31 et seqq., 20.3.2008, no. 15339/02 i. a., Budayeva/Russia, margin no. 128 et seqq., 147 et seqq.

[45] District Court The Hague 26.5.2021 – C/09/571932/HA ZA 19-379, ECLI:NL:RBDHA:2021:5337, no. 4.4.9 und 4.4.10.

[46] ECHR 30.1.2004, no. 48939/99, Öneryildiz/Turkey, margin no. 89, 132, Groß (2020), p. 337 (339).

entrepreneurial obligations, but without necessarily creating a tension. Rather, a balancing act is required to reconcile different needs.[47]

However, there are no corporate obligations under international law for climate protection. It is not for nothing that the UNGP distinguishes between the responsibility of states and the responsibility of companies; the former is more far-reaching and consists of an obligation to protect against human rights violations on their territory.[48] According to the UNGP, it is up to the states to weigh up the various societal interests. At this level, the balancing of the diverging interests takes place and, based on this, the companies may be required to do so. However, this does not exist without state regulation.

This is reminiscent of the Federal Constitutional Court's climate decision, which in this respect precisely emphasizes the normative scope for action.[49] It is true that climate protection measures will regularly prevail, especially as climate change progresses;[50] however, corporate climate protection obligations do not exist automatically and per se. Rather, their determination is left to the democratic process of weighing up not only the various normative concerns, but also the scientific and economic questions of fact, which are characterized by uncertainties.[51]

(d) **IPCC Budget**

This necessary normative specification also affects the available $CO_2$ budget. In contrast, the District Court of The Hague derives a $CO_2$ reduction obligation from the IPCC budget.[52] However, this budget is characterized by uncertainties and assumptions and leaves considerable leeway, which the Federal Constitutional Court specified with reference to an expert opinion by the German Advisory Council on the Environment (Sachverständigenrat für Umweltfragen – SRU).[53] Nevertheless, the Federal Constitutional Court does not require a specific numerical value for a German residual budget but instead makes the legislator responsible and demands that the SRU's calculations based on the IPCC estimate be taken into account as long as the uncertainty about the emission possibilities remaining to maintain the temperature threshold globally and nationally is still too great.[54] If there is a legislative prerogative, it is difficult for a Dutch district court to take certain numerical

---

[47] District Court The Hague 26.5.2021 – C/09/571932/HA ZA 19-379, ECLI:NL:RBDHA:2021:5337, no. 4.4.12.

[48] Item District Court The Hague 26.5.2021 – C/09/571932 / HA ZA 19-379, ECLI:NL:RBDHA:2021:5337, no. 4.4.12.

[49] BVerfG 24.3.2021 – 1 BvR 2656/18 i. a., BVerfGE 157, 30 (margin no. 249).

[50] BVerfG 24.3.2021 – 1 BvR 2656/18 i. a., BVerfGE 157, 30 (margin no. 198 at the End).

[51] Ekardt (2022), p. 15 (16) as well as more detailed Berkemann (2021), pp. 585 et seqq.; Ekardt and Heß (2021a), pp. 579 et seqq.; Ekardt and Heß (2021b), pp. 1421 et seqq.

[52] Abl. Ekardt (2022), p. 15 (16 f.).

[53] BVerfG 24.3.2021 – 1 BvR 2656/18 i. a., BVerfGE 157, 30 (margin no. 219 et seqq.).

[54] BVerfG, 24.3.2021 – 1 BvR 2656/18 i. a., BVerfGE 157, 30 (margin no. 236 f.).

values as a basis. This approach was also just emphasized by the Stuttgart Regional Court in its Daimler judgement.[55]

### 6.3 *Transfer to German Law?*

The District Court of The Hague affirms a corporate responsibility to respect human rights as a global standard of expected behavior regardless of states and beyond compliance with national laws and regulations protecting human rights.[56] Accordingly, there are separate entrepreneurial obligations as part of the unwritten standard of care under the Dutch Civil Code. This helps the components of the unwritten standard of care become binding, even though these components are non-binding[57]—at least in relation to companies. The starting point is therefore the Dutch Civil Code.

As a parallel, reference is made to Sections 1004, 823 Par. 2 German Civil Code (BGB) for violation of protective laws.[58] However, this does not refer to an unwritten standard of care formed from a synopsis of various legal principles, including UN soft law. Also, the obligation to tolerate according to Section 1004 para 2 BGB results from law (namely Section 906 BGB in the case of real property) or contractual permission, consent, etc. Therefore, in parallel to the ruling of the District Court of The Hague, direct corporate climate protection obligations cannot be assumed in this country.

## 7 Stuttgart District Court (Daimler)

### 7.1 *Primacy of the Legislator*

The judgment of the Stuttgart Regional Court of September 13, 2022, which dealt with claims for injunctive relief under Sections 1004 and 823 para 1 German Civil Code (BGB) against the marketing of combustion engine passenger cars and thus also with $CO_2$ reduction, also follows this line. According to this ruling, it is also primarily the responsibility of the legislator to set reduction targets. This must be done in compliance with constitutional requirements. In doing so, it has to take the climate protection requirement according to Article 20a German Constitutional Law (GG) as a basis, but has considerable creative freedom. In principle, it is not the

---

[55] P. forthwith VII.1.

[56] District Court The Hague 26.5.2021 – C/09/571932/HA ZA 19-379, ECLI:NL:RBDHA:2021:5337, no. 4.4.13.

[57] Insofar critically also Ekardt (2022), p. 15 (16 f.).

[58] Ekardt (2022), p. 15 (16).

task of the courts "to derive concretely quantifiable limits to global warming and corresponding emission quantities or reduction targets from the open formulation of Article 20a of the German Constitutional Law."[59]

The courts must apply the laws thus enacted, but may not assume the decisions reserved for the legislature. Otherwise, individuals would be entitled to use civil law to correct the legal regulations that he or she considers inadequate, and thus above all the production of combustion engines by the EU, which is permitted until 2035, in individual cases.[60] Therefore, an overall climate protection concept to be enacted by the democratically elected federal legislature is necessary, which extends to all areas of economic and social life in the entire Federal Republic of Germany, without individuals having the right to seek injunctive relief against individual companies by invoking their personal rights or the EGRC.[61] Even the individual German states could not be sued before the Federal Constitutional Court.[62]

## 7.2 Balancing Within the Framework of Section 823 German Civil Code

The most that can be considered in the context of Section 823 of the German Civil Code is a balancing of the personal rights of of those potentially affected in conjunction with Article 2 para 1 of the German Basic Law and the fundamental rights of the companies under Articles 12, 14 and 2 para 1 of the German Basic Law. Such a consideration of interests is assumed in the case of indirect effects on the personal rights of third parties.[63] However, in such a balancing of interests, the consequences for those affected from the behavior to be prohibited must at least be foreseeable.[64] However, that is not the case here because the restrictions on those potentially affected cannot be foreseen if Daimler continues to produce combustion engines. At most, this would be the case if emissions in the Federal Republic of Germany remained at the current level and there were no developments involving a reduction in the $CO_2$ content of the atmosphere or of other emitters.

Thus, there is no illegal violation of personal rights and fundamental rights.[65] This German climate protection lawsuit against companies was therefore rightly dismissed.[66]

---

[59] LG Stuttgart 13.9.2022 – 17 O 789/21, NVwZ (2022), p. 1663 (margin no. 25) referencing the Climate Decision by the Federal Constitutional Court.

[60] LG Stuttgart 13.9.2022 – 17 O 789/21, NVwZ (2022), p. 1663 (margin no. 25).

[61] LG Stuttgart 13.9.2022 – 17 O 789/21, NVwZ (2022), p. 1663 (margin no. 26).

[62] BVerfG 18.1.2022 – 1 BvR 1565/21, NJW (2022), p. 844 (margin no. 12 et seqq.).

[63] Grüneberg and Sprau (2022), para 823 margin no 95.

[64] LG Stuttgart 13.9.2022 – 17 O 789/21, NVwZ (2022), p. 1663 (margin no. 21).

[65] LG Stuttgart 13.9.2022 – 17 O 789/21, NVwZ (2022), p. 1663 (margin no. 22 f.).

[66] Likewise, Ruttloff (2022), editorial issue 21.

## 8 Results

The establishment of a compensation fund for climate damage at the conference in Sharm El-Sheikh affects states and does not (yet) establish any concrete obligations for them. For companies, the general principles of damage law remain. Corporate climate liability fails above all due to the necessary individualization of damage and the required adequate causality: the individual amounts are not significant and the (earlier) foreseeability of the damage is missing. Neither the Shell judgement of the District Court in The Hague, with its reach into the ECHR and the UN Goals, nor the Federal Constitutional Court climate decision change this. In the Daimler lawsuit, the Stuttgart Regional Court rightly rejects direct $CO_2$ reduction obligations on the part of companies. Therefore, there are also no primary obligations, the breach of which could trigger claims for damages.

## References

Berkemann J (2021) Die Entscheidungsformel des Klima-Beschlusses des BVerfG vom 24. März 2021 – Eine Analyse, ZUR, pp 585 et seqq
Braig K (2013) Umweltschutz durch die Europäische Menschenrechtskonvention. Helbing & Lichtenhahn Verlag, Basel
Braig K, Ehlers-Hofherr A (2020) Diese andere Potenzielle Katastrophe: Wie kann der EGMR dazu beitragen, die Klimakrise einzudämmen? NuR, pp 589 et seqq
Brüning C (2013) In: Hecker B, Hendler R, Proelß A, Reiff P (eds) Verantwortlichkeit und Haftung für Umweltschäden. Erich Schmidt Verlag, Berlin, pp 139 et seqq
Chatzinerantzis A, Appel M (2019) Haftung für den Klimawandel, NJW, pp 881 et seqq
Ekardt F (2022) Klimaklagen gegen Unternehmen – Das Den Haager Shell-Urteil, Klima und Recht, pp 15 et seqq
Ekardt F, Heß F (2021a) Intertemporaler Freiheitsschutz, Existenzminimum und Gewaltenteilung nach dem BVerfG-Klima-Beschluss – Freiheitsgefährdung durch Klimawandel oder durch Klimapolitik? ZUR, pp 579 et seqq
Ekardt F, Heß F (2021b) Bundesverfassungsgericht, neues EU-Klimaschutzrecht und das Klima-Ziel des Paris-Abkommens, NVwZ, pp 1421 et seqq
Frank W (2021) Klimahaftung der Energiekonzerne, RdE, pp 305 et seqq
Frenz W (2018) Bergschadenshaftung für den Klimawandel in Peru? JUTR, pp 99 et seqq
Frenz W (2019) BBergG. Erich Schmidt Verlag, Berlin
Frenz W (2022a) Haftung für Hochwasser- und Klimaschäden, ZNER, pp 115 et seqq
Frenz W (2022b) Gesamtkommentar Klimaschutzrecht, 2nd edn. Erich Schmidt Verlag, Berlin
Groß T (2020) Die Ableitung von Klimaschutzmaßnahmen aus grundrechtlichen Schutzpflichten, NVwZ, pp 337 et seqq
Grüneberg C, Sprau H (2022) BGB, 81st edn. Verlag C.H.Beck, München
Pöttker E (2014) Klimahaftungsrecht. Mohr Siebeck Verlag, Tübingen
Risse J, Haller H (2021) Klimaschutzklagen und Streitverkündung gegen den Staat, NJW, pp 3500 et seqq
Ruttloff M (2022) Klimaklagen – Klappe die Soundsovielte. NVwZ, editorial issue 21

# Clean Energy Transition in Africa's Mining Sector: A Rapid Reflection

**Angelina Anani and Kenneth J. Bansah**

**Contents**



**Abstract** Clean energy transition is crucial for reducing carbon footprint and ensuring environmental and mining sustainability. Contributing to the discourse on clean energy evolution, we focus on the continent of Africa and provide a quick reflection on the continent's clean energy transition in the mining sector. We explore the factors that affect Africa's transition and highlight the policies established to incentivize transition. We also discuss how the clean energy standards and initiatives affect global investment in Africa's mining industry and the readiness of the mining industry to transition. We observe that African countries' pledge to join the clean energy transition initiative has significant impacts on the continent's mining

A. Anani (✉)
Department of Mining and Geological Engineering, The University of Arizona, Tucson, AZ, USA
e-mail: angelinaanani@arizona.edu

K. J. Bansah
Department of Mining & Explosives Engineering, Missouri University of Science and Technology, Rolla, MO, USA


W. Frenz, A. Preuße (eds.), *Yearbook of Sustainable Smart Mining and Energy 2022*, Yearbook of Sustainable Smart Mining and Energy - Technical, Economic and Legal Framework 2, https://doi.org/10.1007/978-3-031-41873-0_8

industry to reduce carbon footprint and increase the production of critical minerals for clean energy technologies. Investment, legislation, infrastructure, and industry commitment are required to further drive the clean energy transition agenda on the continent.

## 1 Introduction

The campaign for clean energy is rapidly gaining momentum and acceptance across the globe due to increasing evidence of climate change and its attendant socioeconomic and environmental effects. The growing recognition of clean energy as an alternative source to fossil-based energy to ensure energy security has come with increased investment in clean energy initiatives and programs. Clean energy investment recorded 2% growth per year for five years after the Paris Agreement in 2015; but this growth in clean energy investment has increased to 12% since 2020.[1] Additionally, the enactments and adoption of new policies that aim to decarbonize the industrial economy have stimulated clean energy interests in the mining industry, and manufacturers and suppliers are presently deploying zero-carbon-operated mining equipment to the industry.

Africa is a major facilitator of the clean energy transition. The United Nations estimates that Africa hosts 30% of the world's mineral reserves, 40% of the world's gold, and nearly 90% of the world's chromium and platinum.[2] The massive reserve of natural resources, including critical minerals, which could help revolutionize the clean energy transition paradigm suggests that Africa's role in this sustainable mining initiative to decarbonize the industry is crucial. However, the question is, "Are mines in Africa prepared to transition from fossil fuels to zero-carbon energy sources, and how easy or difficult could the transition be?

Rising global demand for clean energy would require the extraction of more materials from the ground to advance and develop clean energy technologies. This situation would require the creation of new mines or the expansion of existing mining facilities, a change that would potentially alter the mining footprint. There are also concerns about the potential increase in the carbon footprint from mining operations if the industry continues to depend on fossil fuels to provide raw materials for clean energy technologies.[3] Admittedly, the mining industry to a large extent is ill-equipped for a clean energy transition. Many of the existing infrastructure in most mines depend on carbon-based fuels to function,[4] and large investments may be required to replace these facilities. Thus, industry players' spurring interest in clean energy sources to reduce carbon footprint could intensify opportunities for research

[1] International Energy Agency (2022).

[2] United Nations.

[3] Ali et al. (2022), p. 102984.

[4] Strazzabosco et al. (2022).

and innovations. Already, mines in countries such as Australia, Canada, and the United States have commenced the adoption of clean energy sources to decarbonize, and interest in these zero-carbon sources is rapidly growing in these countries. Mining firms such as Glencore and New Gold have ordered battery-electric mining equipment to augment their fleets, improve ventilation, and reduce greenhouse gas emissions.[5]

While Africa serves as an important source of raw materials for energy transition, there is limited information about the continent's preparedness to transition. To this end, our work explores the readiness of mines in Africa to transition from fossil fuels to clean energy sources and highlights the challenges and opportunities this energy transition offers. Governments in Africa largely control the mineral resources of the land and have a ubiquitous role in mineral governance. Typically, the mineral resources of the land are constitutionally vested in the central government, which makes the decision on mineral land concessions. Similar to the situation in other jurisdictions, governments in Africa promulgate mining and environmental regulations and ensure the implementation of those laws—suggesting that the government cannot be removed from decision-making related to clean energy transition in the mines. In the paper, we explore how African governments' actions can impede the global clean energy transition effort and provide a path toward promoting and sustaining the interest in energy transition on the continent. In the paper, we define renewable energy as a form of clean energy.

## 2 Clean Energy Transition in Africa

The clean energy transition initiative aims to transition from fossil-based energy like oil, natural gas, and coal to renewable energy like wind and solar. The incentive for this transition is to address the climate crisis and curtail a potentially catastrophic climate trajectory. To achieve this objective, a significant leap in technological advancement is required to produce a carbon-free energy sector.

The clean energy transition is currently propelled by the demand for energy storage solutions, electric transportation infrastructure, and the increasing use of technology. The long-term global target is to achieve net-zero greenhouse gas emissions by 2050. However, regulatory structures globally remain uneven on how this target will be achieved. On the African continent, about 12 countries (e.g., South Africa) have joined the pledge to meet this target but with no feasible strategic plan in place.

Twenty percent of the countries in the world are responsible for 78% of global greenhouse gas emissions.[6] Of the 54 countries in Africa, only South Africa is part of the 20%. Africa makes up a fifth of the world's population, contributing only 3%

---

[5] Sandvik (2023) and Green Car Congress (2022).

[6] Christensen and Olhoff (2019).

of global emissions and 6% of global energy use.[7] However, the continent is one of the most affected by climate change impacts such as drought, food and water scarcity, and human migration[8] but plays a significant role in achieving the clean energy transition with its diverse and significant potential for wind and solar energy. Globally, solar sites are designated to be 20% of the land area, and Africa holds 60% of this area.[9] The continent also houses some of the most significant minerals needed for clean energy technologies but has only 1% of the global solar PV installed. It is projected that as the cost of renewable energy decreases, more countries will increase the installation of solar PV and other sources of renewable energy.[10]

The strategic measure for clean energy transition must be tailored to Africa's ecosystem and not an adoption of what is employed by western countries. The reason is that 43% percent of Africa's population does not have access to electricity as of 2021,[11] providing an avenue for direct implementation of renewable energy without the need for large transitions of the energy system. In addition, the readily available resources (e.g., ample sunshine and critical minerals) offer opportunities for the continent to transition to clean energy. Current investments occur in countries such as Nigeria and Rwanda, which invested USD 620 million and USD 30 million for electrification using solar energy.[12] Solar PV installation rate prior to the COVID-19 pandemic equaled the global capacity increase in Africa. However, several challenges hinder the clean energy transition. For example, increasing national debt among many African countries limits their ability to invest in renewable energy. These challenges further limit investments in these countries, especially with stricter ESG standards, and between 2010 and 2020, the continent received less than 3% in clean energy investment.[13]

Africa lacks interconnection among countries resulting in uneven energy distribution and making the possibility of a single energy market difficult. The Africa Clean Energy Corridor initiative offers an opportunity for a shared vision on the continent and cross-border trade of renewable energy between the eastern and southern countries.[14] Technology readiness is also necessary for the long-term sustainability of renewable energy infrastructure. Yet, the current infrastructure (grid) for producing electricity is inadequate due to underinvestment, aging, theft, and natural disasters.

The continent's demand for electricity is set to increase to 75% by 2030.[15] This will increase the per capita electricity demand significantly, and the policies

---

[7] Senyagwa (2022).

[8] Schraven et al. (2020).

[9] Senyagwa (2022).

[10] Senyagwa (2022).

[11] Senyagwa (2022).

[12] Senyagwa (2022).

[13] Senyagwa (2022).

[14] IRENA (2019).

[15] Senyagwa (2022).

implemented by African leaders will dictate the successful transition to clean energy and the ability to meet energy demands. Africa has the potential to double its installed capacity of solar PV if the needed infrastructure can be successfully completed.

At the COP26 Climate Change Conference in 2021, participating African countries represented by the African Group of Negotiators on Climate Change (AGN) recognized the need to overcome existing barriers to clean energy investment in Africa. Of the 54 African countries, 53 of them have submitted climate action plans to mitigate greenhouse gas emissions. The plan included conditional and unconditional targets that hinged on the availability of financial and technological support (over USD 1.2 trillion) from countries with advanced economies like the United States. The successful implementation of the proposed plan will result in a 40% reduction in emissions by 2030. Such policies include the introduction of a solar water heating system in countries such as South Africa, Zimbabwe, Botswana, Lesotho, Mozambique, and Namibia. Egypt provides a tax exemption for raw materials needed for the manufacture of energy-efficient products. Other countries have also implemented policies to reduce energy consumption and regulate thermal efficiency in the building construction industry.

The clean energy transition also means a decline in hydrocarbons that generate substantial revenue for the continent. The current reduction in oil production is a result of political and economic turmoil and not due to the transition to clean energy. As many countries take the pledge to reduce their carbon emissions, oil production in Africa will continue to decrease. This situation may have economic repercussions on countries like Nigeria and Libya. On the other hand, natural gas production has increased due to the increasing demand for liquefied natural gas. Similar to coal and oil, the production of natural gas will decrease as the demand for clean energy transition grows. Domestic needs could affect demand for fossil fuel depending on the cost per unit compared to renewable energy sources. Most households use diesel-operated generators; however, when fuel prices are higher, the use of diesel no longer becomes affordable. Hence, a sustainable and cost-effective renewable energy alternative may be preferred in the future.

## 3 Policies for Energy Efficiency and Clean Energy Transition

The policies established by African countries play a crucial role in the success of energy transition and independence. But current policies fall short of the requirements needed to achieve sustainable development goals. If no modifications are made to current policies, 565 million people will still have no access to electricity. The number of people without electricity may increase as sub-Saharan Africa's population growth exceeds the number of people gaining access to electricity. Countries such as Kenya, Nigeria, and Ethiopia have implemented policies to ensure clean cooking in households. Egypt has provided incentives such as tax deductions, technical training, land allocation cost, and utility connection cost reduction for companies with projects in the areas of renewable energy, green hydrogen, carbon capture, and energy storage.

Ghana established a renewable energy act to deliver competitive market rates for electricity generation. The country also introduced a Renewable Energy Master Plan in 2019 to provide incentives such as tax reductions for materials, equipment, and machinery needed for the assembly and manufacture of renewable energy technology. Ghana also gives households access to financing for sustainable energy projects. Kenya has one of the cleanest electricity grids in the world due to its energy sector reform policies, including the Energy Act of 2019 and the 2021 Renewable Energy Auction Policy. Morocco adopted a bill to make renewable energy attractive for international and domestic investors by simplifying the permitting process.

Namibia's strategic plans include the Green Hydrogen Initiative, which aims to secure investment in the country. Kenya's initiatives already include the first large-scale green hydrogen project with a USD 10 billion investment. Nigeria has pledged a net-zero emission by 2060 and launched the Energy Transition Plan. The country's Climate Change Act of 2021 aims to provide a framework for achieving its net-zero emission goals. South Africa received a pledge of USD 8.5 billion for their clean energy transition from the US, the UK, and the European Union. As one of the largest emitters of $CO_2$, the country has developed an Integrated Energy Plan to shift from a coal-based energy sector to a diversified energy system that includes renewable energy. Also, the country has self-generation incentives for renewable energy for its citizens without needing a permit. This arrangement is essential for the country as it supplies electricity to neighboring Botswana and Namibia. Uganda's Climate Change Policy 2015 and Renewable Energy Policy aim to expand the nation's use of renewable energy such as wind, biomass, and solar.[16]

As many countries begin to pledge net-zero carbon in the next one to three decades, none of the policies will be sufficient to reach this goal. Of the 54 countries, 25 and 20 countries have clear strategic initiatives for electricity access and clean cooking, respectively. With the post-pandemic economic crisis, energy costs have become even less affordable to most African households, especially those living in extreme poverty. The continent's ability to provide electricity to half a billion people without access to electricity will hinge on reduced cost of energy and financial aid, although the desperation to access electricity and acquire a better life for its citizens may overshadow emission mitigation efforts.[17] Strategic policies to ensure that the continent meets its goal will require global technological and financial support.

[16] IRENA (2019).

[17] Adulugba (2021).

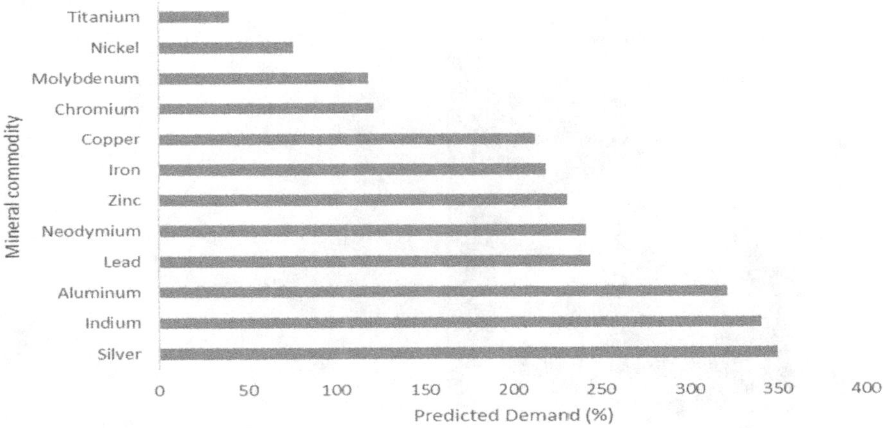


**Fig. 1** Demand for minerals due to clean energy transition (based on the roadmap to 2050)

## 4 Clean Energy Transition in Africa's Mining Industry

There are two main aspects pertaining to the role of the mineral industry in the clean energy transition initiative. First is the production of critical minerals needed for transition. Second is the energy use and emission levels of the industry itself.

**Critical Minerals**

The energy sector currently emits 72% of greenhouse gas worldwide.[18] The emission level is expected to rise as the demand for minerals increases. Clean energy technologies such as wind, solar PV, and concentrated solar power require about 10, 8, and 2 different types of minerals, respectively.[19] In Fig. 1, predicted changes in demand for minerals due to the clean energy transition are presented,[20] indicating that drastic changes are required in the mineral resource industry to meet such demands.

As the global demand for raw materials and fuel increases, it presents an opportunity for African countries to meet those demands. The continent is a rich source of critical minerals needed for clean energy technologies such as solar panels, batteries, and wind turbines. Africa is also rich in renewable energy resources such as wind, solar, and bioenergy. The minerals needed for clean energy technologies include cobalt, aluminum, graphite, copper, lithium, and nickel, and as shown in Fig. 2, the continent is currently a leading producer of manganese, platinum-group metals (PGMs), and cobalt.[21] For example, South Africa is the leading producer of

[18] World Resources Institute (2017).

[19] Hund et al. (2020).

[20] Gielen et al. (2019).

[21] Bovcon and Humphery-Smith (2022).

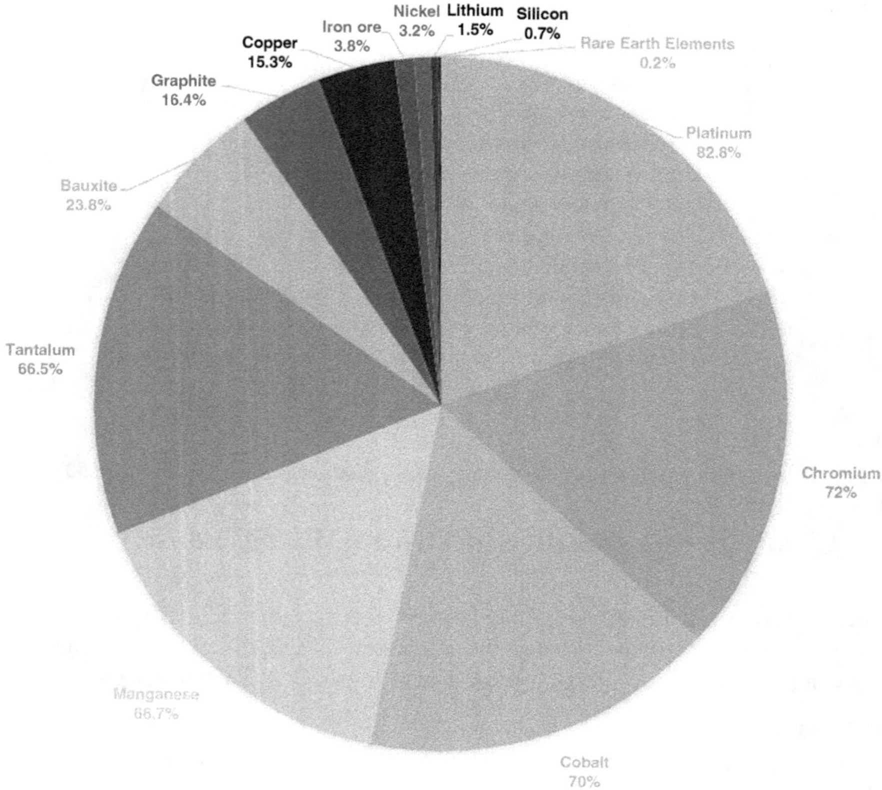


**Fig. 2** Sub-Saharan Africa's contribution to the world's critical minerals supply

PGMs, chromium, and manganese, while DRC produces 70% of the world's cobalt. The continent also bears a large share of bauxite, graphite, and copper.[22]

There are also other untapped resources, such as lithium and nickel. Current exploration and exploitation of these minerals are on a small scale. On average, the revenue generated by mining on the continent is 8%, and it is one of the sectors with significant direct foreign investment. The contribution to the continent's GDP can increase significantly with the demand for energy transition. The added benefit of exploiting critical mineral resources in Africa includes railways, roads, and power infrastructure for local communities and other sectors. One way to reduce the investment burden is shared infrastructure between the local government and the mining companies. Nevertheless, this arrangement is not a common practice as foreign mining companies prefer to control private infrastructure (e.g., energy and transportation) to curb unreliable local infrastructure and mitigate political risks. The transition to clean energy may present further challenges to foreign investors. There are several challenges that the continent must overcome to capitalize on its mineral

[22] Toledano et al. (2021).

resources. Such efforts include improved exploration to uncover mineral resources and improved characterization of mineral deposits.

Global exploration spending has decreased by at least 6% in the last decade. As part of the African Mining Vision developed by the African Union, African countries must improve resource development and improve their capacity to incentivize investment in new mines and facilities. There are several factors that attract investment in the mining sector. These factors include infrastructure support, robust governance, transparent regulations, and incentives. National and local governments must develop successful strategies to regulate and monitor the mineral resources sector.

**Energy Use and Emissions**

The mining industry consumes about 3.5% of the world's energy. Sixty percent of the energy used by the industry is consumed by equipment (e.g., haul trucks and excavators). The remaining energy consumption is distributed between comminution processes (36%) and other processes (4%). As the demand for mineral resources increases, energy use by the sector will also increase. The industry also contributes 4–7% of greenhouse gas (GHG) emissions globally. The adoption of renewable energy, alternative fuels, and efficient energy storage solutions can reduce the industry's carbon footprint. A push towards adopting renewable energy and alternative fuels and reducing greenhouse emissions includes carbon taxation in several countries. This levy has presented several challenges in the industry's ability to meet the demands of critical minerals. Countries such as South Africa have implemented a carbon tax of R120 + 2% per year/tonne of $CO_2e$.[23] The country's mineral industry, however, qualifies for tax-free exceptions of up to 85%. The majority of African countries do not explicitly have carbon tax disincentives. Incentives to reduce emissions and adopt renewable energy by the mining sector will hinge on local policies and levels of enforcement.

**Environmental, Social, and Governance (ESG) Requirements**

Current investments for new projects now require strict ESG compliance to attain capital. This ESG compliance impacts global investment in Africa's mining sector. There are particular aspects of ESG that are relevant to African countries but are not considered in developed countries, such as concerns surrounding child labor. Other concerns include waste and water management, air and water pollution, emission, corruption, and workers' safety. With the projected demand for critical minerals, revenue from the mineral industry is expected to double for the continent. This expectation is feasible if measures are implemented to meet ESG demands that attract investments.

Global climate change policies directly impact Africa's mineral extraction industry through increased production of minerals for clean energy applications and the industry's adoption of stringent ESG policies. ESG investment and carbon border taxes implemented in different countries could incentivize foreign investors to bring intermediate stages of production closer to reduce emissions and waste generation.

[23] Republic of South Africa (2019).

This plan could mean a large economic boost for the continent and increase foreign investment.[24]

## 5 Readiness of the Mining Industry to Transition

In the mineral resources sector, investors now demand prioritizing ESG factors as part of a clean energy transition and supporting decarbonization of the energy market. Mining companies will have to adapt quickly to meet the energy transition goals. The level of readiness to transition to clean energy will differ among countries. This will be impacted by the availability of technology, resources, investors' demand, the countries' policies, incentives, and disincentives.

Around the globe, several mines have commenced the adoption of renewable energy as part of being energy independent. Specifically in Africa, Newmont Mining has employed in its Akyem Gold Mine the Nomad mobile solar power array to generate electricity. The Bisha Mine in Eritrea has adopted a hybrid solar-diesel system to generate power for the mine. The Nampala Mine, operated by a Canadian mining company in Mali, is also commissioning a hybrid power plant to produce 3.9 MWp of electricity. The plant will also supply 40% of Guinea's Kiniero mine's energy needs. The Fekola Gold Mine, owned by B2Gold in Mali, has also deployed a 30 MW hybrid solar power plant. In 2018, IAMGOLD Essakane S.A. Gold Mine opened the world's largest hybrid solar/thermal plant (15 MWp) in Burkina Faso. Sedibelo Platinum in South Africa also plans to commission a 40 MW hybrid wind-solar plant and a 35 MW photovoltaic park in its power purchase agreement to provide energy for its mining sites. In 2022, Gold Fields' South Deep Mine in South Africa completed its solar PV plant to produce 50 MW of energy. The company is set to save $7 million per year on energy cost. Anglo-American and Engie developed an agreement in 2019 to commission a hydrogen power plant for the Mogalakwena platinum mine in South Africa.

Several factors hinder the industry's ability to transition, including land constraints for renewable energy infrastructure installation, lack of awareness and expertise in renewable energy, a need for proof of concept in order to be adopted by small to medium size companies, and conflicting business models between the energy and mining sector.[25] Currently, critical minerals produced on the continent consume on-grid energy fueled by coal. This option is attractive as it is less costly than a self-generating option. Therefore, cost plays a role in the industry's willingness to transition as energy costs up to 40% of the total operating cost.[26] However, as stated earlier, the grid is unreliable in many African countries, leaving many mining companies with backup systems that rely on fossil fuels. Many mines (e.g., Fekola Gold Mine) are increasing their adoption of a hybrid system (fossil and renewable energy) that ensures a reliable power supply. The readiness to transition

[24] Toledano et al. (2021).

[25] Igogo et al. (2020).

[26] Toledano et al. (2021).

**Table 1** Renewable energy initiatives in Africa

| Country | Initiatives |
|---|---|
| Mauritania | ● Resource mapping for renewable energy<br>● The assessment and revision of the national grid code.<br>● Providing support for renewable energy projects through the Sustainable Energy Marketplace and Project Navigator. |
| Niger | ● Development of a Rural Electrification Agency.<br>● Use of part of the electricity tax on rural electrification for renewable energy promotion.<br>● World Bank, African Development Bank and Power Africa to assist the rural electrification agency with mini-grid concessions. |
| Djibouti | ● Provided training on resource assessment and project development<br>● The incorporation of renewable energy policies in the Master Plan and Electricity Law.<br>● Tax exemption incentives for all renewable energy equipment.<br>● National strategy for energy conservation that focuses on renewables.<br>● USAID used RRA for geothermal intervention. |
| Senegal | ● Renewable energy law on electricity generated by renewable energy and their grid and purchase by the state of surplus renewable-based electricity from self-producers.<br>● World Bank action plan to develop renewables in Senegal. |
| Ghana | ● Developed a model for managing decentralized off-grid mini-grids |
| The Gambia | ● Enactment of a renewable energy law and established RE-fund.<br>● Provided capacity building for stakeholders within the renewable energy sector |
| Zambia | ● The Energy Regulation Authority is revising the Existing Grid Code for the incorporation of renewable energy.<br>● Received technical assistance from World Bank for Renewable Energy Resource Mapping<br>● Developed a Renewable Energy Strategy for the UNDP South-to-South project. |
| Eswatini | ● Integrated capacity building for renewables in their national power masterplan.<br>● Updated their energy sector Master Plan considering least-cost options |
| Mozambique | ● Developed the Mozambique Renewable Energy Investment Workshop for capacity building.<br>● Performed grid integration studies.<br>● Renewable energy resources mapping for the whole country.<br>● Adopted renewable energy feed-in-tariff.<br>● Revised the Rural Electrification Strategy based on renewable energy for the private sector. |

could be spurred on by reducing the cost of renewable energy and the cost penalty associated with stringent requirements by host countries.

The readiness of the mining industry to transition is also affected by the sustainability of renewable energy technologies and permitting requirements. This depends on the readiness of the countries themselves. The International Renewable Energy Agency (IRENA) performed an assessment of renewable energy readiness for several countries in Africa.[27] Their findings are summarized in Table 1. The agency provides capacity-building initiatives to increase the preparedness of African

[27] IRENA (2019).

countries. Evolving renewable energy companies (e.g., Chariot Ltd) are prepared to partner with the mining industry in Africa to develop and supply renewable energy.

## 6 Conclusion

We have highlighted the nuances of clean energy transition in Africa, specifically in the mining sector. We note Africa's potential to supply large quantities of critical minerals for clean energy technology to support the global clean energy revolution. Additionally, initial investments in the energy sector demonstrate the feasibility of Africa's mining industry to transition to clean energy. However, like other mines outside Africa, the continent's mining industry will continue to largely depend on fossil fuels until the industry replaces carbon-based fuel infrastructure with functional clean energy supply systems. Moreover, the continent must overcome challenges such as government disincentives, land constraints, lack of awareness and expertise, political risk, and capital investment to effectively transition from fossil fuels to clean energy. Host countries must employ standards and initiatives that attract investors to improve the mining industry's preparedness to transition to clean energy. Clean energy policies could be adopted as part of the requirements for responsible sourcing to engender interest and 'force' the industry to transition and minimize its carbon footprint.

## References

Adulugba C (2021) Sustainable energy transition in Sub-Saharan Africa. In: Asif M (ed) Energy and environmental security in developing countries. Advanced sciences and technologies for security applications. Springer, Cham. https://doi.org/10.1007/978-3-030-63654-8_8. Accessed 22 May 2023

Ali M, Tursoy T, Samour A, Moyo D, Konneh A (2022) Testing the impact of the gold price, oil price, and renewable energy on carbon emissions in South Africa: novel evidence from bootstrap ARDL and NARDL approaches. Resour Policy 79:102984. https://doi.org/10.1016/j.resourpol.2022.102984. Accessed 22 May 2023

Bovcon M, Humphery-Smith E (2022) ESG risks cast pall over quest for African critical minerals. https://www.maplecroft.com/insights/analysis/esg-risks-cast-pall-over-quest-for-african-critical-minerals/. Accessed 22 May 2023

Christensen JM, Olhoff A (2019) Emissions gap report 2019. United Nations Environment Programme (UNEP): Gigiri Nairobi, Kenya

Gielen D, Gorini R, Wagner N, Leme R, Gutierrez L, Prakash G et al (2019) Global energy transformation: a roadmap to 2050. https://www.h2knowledgecentre.com/content/researchpaper1605. Accessed 22 May 2023

Green Car Congress (2022) Glencore orders full fleet of Epiroc battery-electric mining equipment for Onaping Depth nickel and copper mine in Canada. https://www.greencarcongress.com/2022/07/20220709-epiroc.html. Accessed 22 May 2023

Hund K, La Porta D, Fabregas T, Laing T, Drexhage J (2020) Minerals for climate action: the mineral intensity of the clean energy transition. Climate-Smart Mining Facility

Igogo T, Lowder T, Engel-Cox J, Newman AM, Awuah-Offei K (2020) Integrating clean energy in mining operations: opportunities, challenges, and enabling approaches. https://doi.org/10.2172/1659921. Accessed 22 May 2023

International Energy Agency (2022) Record clean energy spending is set to help global energy investment grow by 8% in 2022. https://www.iea.org/news/record-clean-energy-spending-is-set-to-help-global-energy-investment-grow-by-8-in-2022. Accessed 22 May 2023

IRENA (2019) Scaling up renewable energy deployment in Africa - impact of Irena's engagement. https://www.irena.org/-/media/Files/IRENA/Agency/Publication/2019/Jan/IRENA_Africa_impact_2019.pdf?rev=120833913f8b4fe2bb30f92a6aba6ad5&hash=73AE71DE8524833B28ED66715BE24957. Accessed 22 May 2023

Republic of South Africa (2019) Carbon Tax Act 15 of 2019. https://www.gov.za/documents/carbon-tax-act-15-2019-english-afrikaans-23-may-2019-0000. Accessed 22 May 2023

Sandvik (2023) New Gold's New Afton mine grows Sandvik BEV fleet. https://www.rocktechnology.sandvik/en/news-and-media/news-archive/2023/01/new-golds-new-afton-mine-grows-sandvik-bev-fleet/. Accessed 22 May 2023

Schraven B, Adaawen S, Rademacher-Schulz C, Segadlo N (2020) Climate change impacts on human (im-) mobility in Sub-Saharan Africa: recent trends and options for policy responses. Deutsche Gesellschaft für Internationale Zusammenarbeit (GIZ) GmbH. https://www.adaptationcommunity.net/wp-content/uploads/2020/07/GIZ_Climate-impacts-on-human-mobility-Africa.pdf. Accessed 22 May 2023

Senyagwa J (2022) Africa Energy Outlook 2022. https://www.sei.org/publications/africa-energy-outlook-2022/. Accessed 22 May 2023

Strazzabosco A, Gruenhagen JH, Cox S (2022) A review of renewable energy practices in the Australian mining industry. Renew Energy. https://doi.org/10.1016/j.renene.2022.01.021. Accessed 22 May 2023

Toledano P, Dietrich Brauch M, Bhuwalka K, Busia K (2021) The case for a climate-smart update of the Africa Mining Vision. Karan and Busia, Kojo, The Case for a Climate-Smart Update of the Africa Mining Vision (April 1, 2021). https://iap.unido.org/articles/case-climate-smart-update-african-mining-vision. Accessed 22 May 2023

United Nations. https://www.unep.org/regions/africa/our-work-africa#. Accessed 22 May 2023

World Resources Institute (2017) Climate Analysis Indicators Tool. https://www.climatewatchdata.org/?source=cait. Accessed 22 May 2023

# Ventilation on Demand in Brazilian Underground Mines: Current Situation and Perspectives

**Anna Luiza Marques Ayres da Silva, Jonatas Michel de Cardoso Vieira, Welton Teixeira da Silva, and Sérgio Médici de Eston**

## Contents



**Abstract** Ventilation systems are estimated to represent 40 to 50% of a mine's total energy consumption. Therefore, ventilation on demand (VOD) is becoming increasingly widespread in underground mines. Added to more precise air quality and quantity controls and costs reduction, its great energy savings potential can also mean GHG emissions reductions, helping to mitigate anthropogenic global warming. Brazilian mines have been improving their ventilation systems. Despite the increasing number of current studies, there is a considerable research gap on VOD technology deployment. Thus, this work aims to determine its current scenario and adoption perspectives, presenting the mines that use it, details of their ventilation systems, and the implemented automation levels. Based on extensive bibliographic

A. L. M. Ayres da Silva (✉) · J. M. d. C. Vieira · S. M. de Eston
Mining and Petroleum Engineering Department of the University de São Paulo, São Paulo, Brazil
e-mail: alayres@usp.br; jonatas.vieira@usp.br; smeston@usp.br

W. T. da Silva
Nexa Resources, Vazante, Brazil


W. Frenz, A. Preuße (eds.), *Yearbook of Sustainable Smart Mining and Energy 2022*, Yearbook of Sustainable Smart Mining and Energy - Technical, Economic and Legal Framework 2, https://doi.org/10.1007/978-3-031-41873-0_9

research and results of a recent and thorough survey with Brazilian mining companies, approximately 20% of the mines have started automating their ventilation systems, reporting energy savings of up to 71%. All mines' main fans own at least the levels 1 and 2 automation strategies. 9% of mines have extended such automation to their auxiliary ventilation systems. The main challenges for VOD implementation range from lack of basic infrastructure to cultural issues. However, an expansion is expected, especially in new projects and medium- and large-scale metal mines.

## 1 Introduction

For underground mining to be possible, a safe, efficient, and economical ventilation system is essential to supply an adequate amount of oxygen in the workplace (for the workforce and combustion engines), for the dilution of contaminants in the underground, coming from the strata or the unitary mining operations, and for the maintenance of a suitable thermal environment for people and machines. Therefore, the ventilation circuit must be integrated into the mining method and follow its evolution throughout the life of the mine.[1]

As stated in the regulations on mine ventilation in countries with a tradition in the mining sector, the dimensioning of the ventilation system is generally based on determining a sufficient airflow to dilute these contaminants (especially from the combustion of diesel) at specific times in the mine's lifetime. According to Brazilian standards, its calculation, therefore, depends on the number of people and equipment with diesel engines in operation on the development and production openings (specifically on their rated power and utilization factors), on the consumption of explosives and the monthly tonnage blasted. There are also certain distinctions if the ore is coal or not, and clearly mention that ventilation must be maintained and operated regularly in underground activities. To ensure adequate air quantity and quality, frequent monitoring of air velocity and flow and the contents of the main contaminants at key points in the mine (in the intake air openings, working and development faces) must be carried out to verify if they remain below their respective occupational exposure limits (OELs), designated as tolerance limits (LTs) in Brazilian regulations.[2] In deep mines, on the other hand, the removal of heat coming mainly from strata, the air's auto-compression, and mine machinery, becomes a fundamental parameter. In this way, the volumes of air supplied are determined mainly by taking it into account.[3]

As mine ventilation systems are historically designed, therefore, based on peak air demand according to criteria based mainly on diesel engines or heat, it is usual

[1] McPherson (2018) and Hartman (1997).

[2] Ayres da Silva and De Eston (2021), pp. 348–365.

[3] Hardcastle and Kocsis (2004), pp. 51–57.

for them to operate at this highest level, despite the true air demand.[4] As a result, significant and costly airflow redundancies can occur in the mine. According to Hardcastle and Kocsis,[5] few areas of a mine need to be continuously ventilated, since each one houses discontinuous mining operations. Underground operations vary widely, and so does the exposure of personnel to contaminants.

Ventilation has therefore long been identified as a major energy consumer.[6] As reported by Kocsis, Hardcastle, and Hall,[7] in metal mines, auxiliary ventilation can be responsible for approximately 50% of the electrical energy consumed by the total mine ventilation system. Baiden et al.[8] even cite studies that state that ventilation is the unit operation with the greatest energy consumption in the underground mining industry according to most mine managers, with a total consumption from 60 to 80% of a mine's total electrical load. Consistent with more recent studies, however, it is estimated that ventilation systems may represent 40 to 50% of mine's total energy consumption[9] and that ventilation costs, depending on energy prices (usually increasing) can reach up to 50% of the mine's energy costs and 15% of its total costs.[10]

To deal with these challenges associated with mining ventilation and its dynamic characteristics, the use of ventilation on demand (VOD) began to be considered, mainly in the 1980s, in underground mines, especially in Canada and Sweden.[11] VOD in-mine trials in other regions also stand out at that time, such as those by Pickering and Robinson in 1984 for methane monitoring in British coal mines and by Hatakeyama in 1992 for carbon monoxide and methane gas monitoring, concentration prediction, and control in Japanese coal mines.[12] In Australia, Wu and Gillies[13] report research conducted in the early 2000s in coal and base metal mines to develop a computerized monitoring and simulation system to offer immediate or real-time information on each branch within an underground ventilation network, allowing interpretation of key system data and convenient operational changes in the mine ventilation system.

[4] Hardcastle et al. (2006), pp. 15–27.

[5] Hardcastle and Kocsis (2004), pp. 51–57.

[6] Udd et al. (1996).

[7] Kocsis (2004), p. 7.

[8] Baiden et al. (2005), pp. 457–462.

[9] De Souza (2018).

[10] Dicks and Clausen (2017) and Pacheco et al. (2018).

[11] Hardcastle and Kocsis (2004), pp. 51–57; Hardcastle et al. (2006), pp. 15–27; Udd et al. (1996); Kocsis (2004), p. 7; Baiden (2005), pp. 457–462; Nensen and Lundkvist (2005); Allen (2003), pp. 1–8; Hardcastle (1999), pp. 31–38; Tuck et al. (2006), pp. 11–14; Rustan (1979), pp. 223–229; Hardcastle et al. (1995), pp. 183–199; Rönnblom (1996), p. 5; Hardcastle et al. (1998), pp. 785–94; Hardcastle and Kocsis (2001), pp. 779–786; Kocsis et al. (2008), pp. 779–786.

[12] Acunã (2014), pp. 1–7.

[13] Wu and Gillies (2005), pp. 383–389.

As reported by Semin, Levin, and Maltsev,[14] this term was first mentioned in.[15] VOD can be understood as a technology for intelligently supplying airflows in the mine, with the least possible human intervention, so that variable amounts of airflow are provided in the different areas of the mine, according to the needs required in each sector, at a given time. In other words, it is the ability to supply the necessary volume of air with the necessary quality, to where it is needed, only when it is needed.[16] When the concept of VOD was proposed, the idea was, therefore, that the technology would go straight to the stage of air volume controlling based on the equipment/personnel location in a certain area of the mine. According to Acuña, Alvarez, and Hurtado,[17] as the technology evolved, it became clear that a more flexible approach was necessary to make a more practical solution. Providing an airflow that satisfies the mine's needs without surpluses, therefore, makes it possible to significantly reduce the mine's operating costs. In addition, VOD technology makes it possible to control whether the quantity and quality of airflow needed by underground labor are effectively provided, contributing to underground occupational health. Mine production can also be favored, as more accurate or even shorter re-entry times can be considered in the mining cycle after blasting. Acuña and Feliú[18] also mention the possibility of increasing the number of simultaneous working faces with the same initial available airflow, simply by better distributing the airflow remaining from inactive working faces to ongoing activities. On a larger scale, the great potential for energy savings employing the VOD strategy can also mean reductions in greenhouse gas (GHG) emissions, helping to mitigate anthropogenic global warming.[19] A study on the implementation of VOD software in certain Canadian and Swedish mines showed possible energy savings of 50% or more when considering energy for main and auxiliary fans and natural gas.[20]

As observed by Baiden et al.,[21] regardless the evident benefits of VOD, this technology was little adopted by the industry until the mid-2000s. The reasons for this are many, including the lack, at the time, of adequate, low-cost, comprehensive, and reliable supervision and control; and poor or overly simplistic cost-benefit analyses. Hardcastle and Kocsis[22] list the lack of the necessary infrastructure for control and information collection as the main reason. With the technological development of the last seventeen years, although it is not yet a standard, there is a greater tendency, worldwide, to consider in underground mines the adoption or improvement of a

[14] Semin et al. (2020), pp. 803–820.

[15] Hardcastle (1999), pp. 31–38; Hardcastle (1998), pp. 785–794.

[16] Tran-Valade and Allen (2013); Acuña et al. (2016), p. 12.

[17] Acuña et al. (2016), p. 12.

[18] Acuña and Feliú (2014), pp. 813–821.

[19] Kumar Katta et al. (2020); Moreau et al. (2021), pp. 1159–1167; Gonen (2021), pp. 72–82.

[20] Acuña et al. (2016), p. 12.

[21] Baiden et al. (2005), pp. 457–462.

[22] Hardcastle and Kocsis (2004), pp. 51–57.

VOD, or as it is usually identified in a generic way, a Ventilation optimization system,[23] Ventilation control system (VCS) or Ventilation management system (VMS),[24] with an increasing number of contributions around the world in this regard.[25]

It is worth mentioning that the term VOD is not used in its original concept in the present work, but to designate any ventilation control system.

In line with the world, Brazilian underground mines have been improving their ventilation systems and implementing or considering the use of VOD. Although in recent years there has been a significant increase in the number of studies and publications on underground mine ventilation in the country, their number is still quite small compared to other countries with a mining tradition. There is a considerable research gap, especially regarding the adoption of ventilation control systems in the country.

## 2 Objectives

Therefore, this work aims to review the main aspects of VOD technology to determine the current scenario of its adoption in Brazil, presenting the underground mines that use it, their ventilation systems, and the automation levels adopted. It also aims to highlight the main challenges currently faced for its implementation in the country.

## 3 Material and Methods

The study is based on an extensive and current bibliographic research and one of the results of a recent and thorough survey with Brazilian mining companies based on interviews about their ventilation systems. For the bibliographic research, reference textbooks on ventilation of underground mines, varied bibliographic databases, workshops, and annals of relevant congresses on mine ventilation were consulted.

---

[23] Griffith and Pinedo (2022).

[24] Acuña et al. (2016), p. 12; Kocsis (2009).

[25] Semin et al. (2020), pp. 803–820; Bartsch et al. (2010), p. 567; Gyamfi (2020); Kolobe et al. (2021), pp. 538–547; Shriwas and Pritchard (2020), pp. 1015–1021; Gutiérrez et al. (2021), pp. 262–271; Jacinto et al. (2021), pp. 71–87; Sjöström et al. (2020), pp. 533–539; Wallace et al. (2015), pp. 165–169; Chatterjee et al. (2015), pp. 65–73; Pandey et al. (2015), pp. 119–125; Ge et al. (2019), pp. 116–125; Lagowski (2013), pp. 58–60; Acuña and Allen (2017), pp. 89–95; Nel et al. (2019), pp. 379–390.

# 4 Results and Discussion

## 4.1 State of Art of VOD Technology

### 4.1.1 Ventilation Assets and Control Strategies

VOD systems basically comprise advanced software and electronically controlled hardware packages that allow the automated or semi-automated control of the various components of the mine ventilation system (ventilation assets): fans and airflow control devices, such as regulators and doors.[26] Control signals are usually sent through a telecommunications network available in the control rooms to on-site assets.[27] However, analyzing data and triggering commands can also be performed by a decentral control system.[28] Paramount, in both cases, is an adequate infrastructure to transmit information.[29]

There are different ways of classifying the VOD control strategies.[30] According to one of the most common manners, the control strategies are differentiated into the following types (referred to as implementation levels by software packages available on the market):[31]

- User or manual control (level 1);
- Time of day scheduling (level 2);
- Event-based (level 3);
- Tagging (level 4);
- Environmental control (level 5).

The manual or level 1 control strategy is the one carried out by the operator from a control room, defining different operating points for the various components of the ventilation system (remote-control system). For example, turning on, off, or changing the fan rotation frequency, opening, or closing doors. It is, therefore, the simplest level of control. It is observed that, before the manual control strategy, the "visualization" control strategy (or zero level according to some experts) is sometimes considered, in which the operator does not act directly on the fans (or another ventilation asset) controlling them remotely, however, is able to see everything that is happening in the mine, and in this way, take actions more quickly.

In the level 2 control strategy, "Time of day scheduling", the difference is that the on/off functioning or adjustment of fans, regulators, and doors is based on a given time input, from a certain schedule. This is normally done by coupling a timer to the control system to trigger individual actions or sequences of changes in the

[26] Acuña and Levesque (2021), pp. 442–457.

[27] Gutiérrez et al. (2021), pp. 262–271.

[28] Dicks and Clausen (2017).

[29] Dicks and Clausen (2017).

[30] Griffith and Pinedo (2022).

[31] Tran-Valade and Allen (2013); Acuña et al. (2016), p. 12.

ventilation asset set point at a certain time of day. In general, "Manual control" and "Time of day scheduling" are implemented together.[32]

The "Event-based" or level 3 control strategy is an "open" mode of control, that is, the automatic triggering of prescribed actions occurs in reaction to alternative events. Basically, it is any action that can be done in the mine and that does not have a predefined level of control through "Manual control", "Time of day scheduling", by "Tagging" or "Environmental control".[33] As the "Time of day scheduling", this control strategy uses the same principle as the "action trigger" but is based on any available variable that can be transmitted to the decision-making software. The typical example of an event-based control strategy is in case of fire or another emergency. In addition, it can also be used to trigger the cascade use of fans or optimize blast clearance. In these first control strategies there is no real-time knowledge of where the personnel and equipment are, but an "expected" location. [34]

The fourth control strategy is "by positioning" or "Tagging" (or level 4). It requires real-time knowledge of the location of the workforce and/or equipment so that the mine airflow can be distributed according to these parameters, as well as to the airflow calculation rules. Therefore, this type of control strategy is the one most associated with the original concept of VOD and requires a more complex infrastructure, usually composed of a tag and tracking system and a communication network integrated into the ventilation software so that convenient modifications can be made to the ventilation assets.[35]

The last control strategy is the "Environmental control" (level 5) and considers the automatic control of ventilation assets based upon real-time environmental data, obtained from sensors installed underground that measure, for example, the concentration of gas (CO, $CH_4$, $NO_x$, etc.), dust, DPM and wet and dry bulb temperatures. Although this control strategy is conceptually the most cost-effective for the ventilation system, depending on the country's ventilation regulations, it may not be fully implemented. Acuña, Alvarez, and Hurtado[36] mention that Ontario law and Chilean legislation, for example, by requiring the provision of a specific amount of airflow per hp of the diesel engine, prevent this environmental control strategy. As stated in Brazilian legislation, this specific amount of airflow is 3.5 (or 2.65 depending on certain characteristics of the diesel and engine) $m^3$/min/hp.[37] Nonetheless, onboard gas monitors can be used to indicate the quality of the air or if an equipment operator is in "clean" air.[38] Particularly in large mines that hold a huge hazard about unexpected events such as gas outbursts, monitoring with detection sensors and measuring devices can be useful for air quality analyses and react for example with

[32] Acuña et al. (2016), p. 12.

[33] Acuña and Allen (2017), pp. 89–95.

[34] Acuña et al. (2016), p. 12.

[35] Hardcastle and Kocsis (2004), pp. 51–57.

[36] Acuña et al. (2016), p. 12.

[37] Ayres da Silva and De Eston (2021), pp. 348–365.

[38] Hardcastle and Kocsis (2004), pp. 51–57.

higher airflow rates if some regulatory demands are not yet met.[39] This level is therefore usually used together with other control strategies as a failsafe system.[40] The ventilation system can be adjusted to provide additional airflow until the contaminant concentrations are below their acceptable OELs.

Control strategies in a VOD can, therefore, be adopted individually or in combination. According to Pacheco,[41] the VOD software and hardware packages available on the market usually provide two types of solutions:

- Complete solution (level 5 VOD);
- Partial solution (level 1 to 4 VOD).

In the complete or level 5 VOD solution, there is a combination of all control strategies (from level 1 to 5), resulting in a complex logical scheme as shown in Fig. 1, from Gyamfi, [42] in which there is a control of ventilation assets, ensuring continuous monitoring of air quality and adjustment of the necessary airflow consistent with to the real-time location information of personnel/equipment and environmental variables, obtained by the tags and mine sensors, respectively.

As advantages, the complete solution presents the provision of all the possibilities currently available for ventilation control: fully outsourced solution, from implementation, assembly to VOD system support; friendly use and great integration capacity, due to the implementation of extensive and somewhat standardized communication networks and automation systems. Disadvantages may include the need for a large infrastructure (depending on the mine size), more capital resources and a longer time to implement the system; little flexibility to purchase parts/support from another supplier; more complex development/applications or programming and possible conflict with current legislation on ventilation (due to the implementation of air quality sensors as a basis for the ventilation system adjustments).[43]

In the partial solution (level 1 to 4 VOD) there is any combination of VOD control strategies from level 1 to 4, therefore, it does not consider the measurement of environmental variables through sensors. As reported by Pacheco,[44] its advantages may include better suitability to the mine's needs, since the user can better choose the type of ventilation management desired; partially or fully outsourced solution; support facility (depending on the country); user-friendly, and great integration capability. The disadvantages are similar to those of the complete solution.

In addition to these two types of solutions for the mine ventilation system automation, there is also the "In-house development" in underground mines today. This is a current trend, in which the ventilation control system incorporates technology

[39] Dicks and Clausen (2017).

[40] Gyamfi (2020).

[41] Pacheco (2021).

[42] Gyamfi (2020).

[43] Pacheco (2021).

[44] Pacheco (2021).

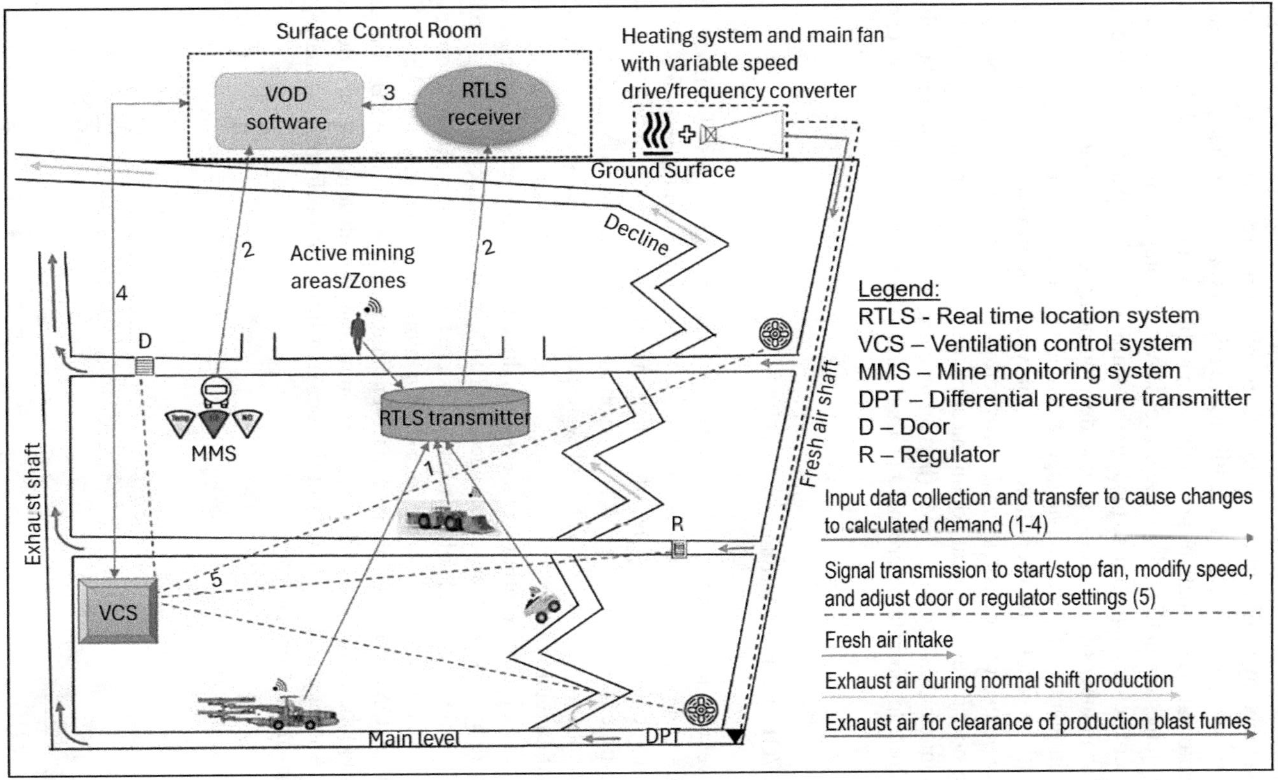


**Fig. 1** Schematic diagram of a complete or level 5 VOD solution (Gyamfi 2020)

developed internally in each mine over the years, and automation tools such as SCADA (Supervisory Control and Data Acquisition) or DCS (Distributed Control System) systems, or sometimes even simpler variants of these systems. It is not usually developed by third parties, but by the mine's workforce, including automation engineers, electrical engineers, ventilation engineers, among other professionals. These automation alternatives normally have a slightly more artisanal development but are quite adequate to the mine's needs. As advantages, the following can be mentioned: better adaptation to the mine's requirements; independent development; a smaller number of resources; simple development; less time and less infrastructure for implementation. Disadvantages may include: great ability to coordinate, not having all the benefits of a ventilation control system; unproven solution (depending on the expertise of the professionals involved in its development and implementation; on how robust the local communication networks are, among other factors); limited integration capacity; HMI (human machine interface) and modest environments (simple operation screens, but that fulfill their purpose) and independent support.[45]

### 4.1.2 VOD Main Components

Each level of implementation of a VOD system requires variations of software and hardware components. However, in general, the essential components for a VOD system implementation may include data acquisition system (for instance, sensors and real-time location systems, RTLS); central system process and ventilation control system (for example, the VOD software); controlled devices (ventilation assets) and communication facility.[46]

The control of the ventilation assets set is done through the programming of the VOD software. VOD software packages have undergone a great evolution in the last decade. Currently, several software packages are commercially available, however, the main ones available on the market (in alphabetical order) are: AbilityTM Ventilation Optimizer (formerly SmartVentilation/800xA) from ABB, Imagine from Meglab, NRG1-ECO (Energy consumption optimization) from Bestech, Serpent Automatic/Ventilation from Epiroc, VentSim Control (formerly SmartExec from Simsmart) from Howden.[47]

---

[45] Pacheco (2021).

[46] Gyamfi (2020); Pandey et al. (2015), pp. 119–125.

[47] Acuña et al. (2016), p. 12; Gyamfi (2020), Acuña (2021); Babu et al. (2015), pp. 1001–1008; Howden (2018); Nyqvist and Serres (2020), pp. 37–40.

## 4.2 VOD in Brazilian Underground Mines

According to Heider,[48] Brazil officially has 76 underground mines in operation and important projects that include underground mining. It is observed that in this count the author considers some conglomerates of mines as a single mine since many are connected underground or are very close, such as the mines of the Jacobina complex in the state of Bahia.

From the methodology used, based on extensive and current bibliographic research, and contact with Brazilian mines, it was possible to survey the underground mines that currently employ or have already carried out pilot tests for the implementation of a VOD system. Table 1 presents them (in alphabetical order) and summarizes the results obtained. Afterwards, the application of VOD technology in the ventilation systems of each mine is discussed in more detail.

### 4.2.1 Ipueira Mine

In the Ipueira mine of the Brazilian company Cia de Ferro Ligas da Bahia (Ferbasa) chromium ore (the main mineral, chromite) is exploited using the mining methods Sublevel open stoping and Sublevel caving.[49] It is basically formed by six interconnected development and mining panels arranged longitudinally in the North-South direction (Ipueira II, III, IV, VA, VB, and VIA). The mine has two main air inlets (main ramps), which are in Ipueira II and V, in addition to some raises for air entrance. The main ventilation (MV) system is by exhaustion and consists of 10 axial fans with an installed power of 1655 kW, located on the surface, arranged 2 by 2 (in parallel) in the raises of the mining panels Ipueira II, III, VA, VB, and VIA.[50]

According to Costa,[51] in 2013 Ferbasa implemented an automation project for the main ventilation of the mine, developed over approximately three years, with which it was possible to observe a considerable reduction in the energy consumption. As reported by Oliveira et al.,[52] the VOD system generated energy savings of around 71% compared to the classic system, equivalent to 8,290,648 kWh/year. The average specific consumption before and after the automation of the main ventilation system was 10.90 kWh/t ROM and 3.13 kWh/t ROM. In this way, all the main fans at the Ipueira mine became automated, being controlled from a command center on the surface. The fans are driven by variable frequency drives (VFDs), making it possible to adjust the rotation speed and consequently, the amount of fresh air supplied consistent with the calculated required flow, thus avoiding unnecessary

---

[48] Heider (2017), pp. 13–16.

[49] Fernandes and Ayres da Silva (2020), p. 1.

[50] de Costa (2017); de Oliveira et al. (2018), pp. 12–13.

[51] de Costa (2017).

[52] de Oliveira et al. (2018), pp. 12–13.

Table 1 List of Brazilian mines with VOD

| Mine/ Corporation | Location | Ore | Mining method | Depth (m) | VOD implementation method | Control forms and automation elements | References |
|---|---|---|---|---|---|---|---|
| Córrego do Sítio I/AngloGold Ashanti | Santa Bárbara (Minas Gerais) | Gold | Sublevel open stoping | 800 | Up to level 2 in MV & AV | VFDs | de Costa and da Silva (2020), pp. 555–560; de Vilhena Costa and Margarida da Silva (2020), pp. 1–8; de Costa (2019); Fernandes and Ayres da Silva (2020), p. 1 |
| Cuiabá/ AngloGold Ashanti | Sabará (Minas Gerais) | Gold | Sublevel open stoping | 1350 | – Up to level 2 in MV<br>– Up to level 4 in AV | – VFDs<br>– Sensors at some points in the mine | de Costa and da Silva (2020), pp. 555–560; de Vilhena Costa and Margarida da Silva (2020), pp. 1–8; da Costa (2019); Machado (2011); Oliveira (2021); de Vieira et al. (2021), p. 2; da Silva et al. (2021), pp. 282–294 |
| Extremo Norte/ Nexa Resources | Vazante (Minas Gerais) | Zinc (main), lead & silver | Vertical retreat mining | 300 | Up to level 2 in MV | VFDs | Fernandes and Ayres da Silva (2020), p. 1; de Vieira et al. (2021), p. 2; Pereira (2018) |
| Fazenda Brasileiro/ Equinox Gold | Barrocas (Bahia) | Gold | Sublevel open stoping | 1070 | Up to level 2 in MV | VFDs | Fernandes and Ayres da Silva (2020), p. 1; de Vieira et al. (2021), p. 2; Danfoss (2012); Equinox Gold Corp. (2021) |
| Ipueira/Ferbasa | Andorinha (Bahia) | Chrome | Sublevel open stoping/ Sublevel caving | 500 | – Up to level 2 in MV<br>– Levels 1, 2 and 5 in AV | – VFDs<br>– Sensors for gas monitoring at some points in the mine<br>– Optical fiber over 18 km | de Vilhena Costa and Margarida da Silva (2020), pp. 1–8; Fernandes and Ayres da Silva (2020), p. 1; de Costa (2017); de Oliveira et al. (2018), pp. 12–13 |
| Jacobina/ Yamana | Jacobina (Bahia) | Gold | Sublevel open stoping | 685 | – Up to level 2 in MV<br>– Up to level 5 in AV | – VFDs<br>– Sensors for gas monitoring at some points in the mine | Fernandes and Ayres da Silva (2020), p. 1; Yamana Gold Inc. (2020); BMS (2021) |

| Mine/ Corporation | Location | Ore | Mining method | Depth (m) | VOD implementation method | Control forms and automation elements | References |
|---|---|---|---|---|---|---|---|
| Lamego/ AngloGold Ashanti | Sabará (Minas Gerais) | Gold | Sublevel open stoping | 600 | Up to level 2 in MV & AV | VFDs | de Costa and da Silva (2020), pp. 555–560; Fernandes and Ayres da Silva (2020), p. 1; Oliveira (2021); Filho (2021) |
| Pilar/Jaguar Mining | Santa Bárbara (Minas Gerais) | Gold | Sublevel open stoping/ Cut and fill | 265 | Up to level 2 in MV | VFDs | Fernandes and Ayres da Silva (2020), p. 1; Tavares et al. (2021), p. 12; Brasil Mineral (2021); Oliveira et al. (2021b), pp. 19–21; RPA (2020) |
| Serra Grande (Mina III)/ AngloGold Ashanti | Crixás (Goiás) | Gold | Sublevel open stoping | 1100 | – Up to level 2 in MV & AV<br>– Tests (pilot scale) of level 5 in AV | – VFDs<br>– Sensors for temperature and gas monitoring at some points in the mine | Fernandes and Ayres da Silva (2020), p. 1; de Vieira et al. (2021), p. 2; Hansson and Baldeón (2019), p. 8; Oliveira et al. (2021a), pp. 16–18 |
| Vazante/Nexa Resources | Vazante (Minas Gerais) | Zinc (main), lead & silver | Vertical retreat mining | 550 | – Up to level 2 in MV<br>– Tests (on pilot scale) of levels 4 & 5 in AV | VFDs | de Costa and da Silva (2020), pp. 555–560; Fernandes and Ayres da Silva (2020), p. 1; de Vieira et al. (2021), p. 2; Pereira (2018); Maciel (2015) |
| Project Aripuanã/Nexa Resources | Aripuanã (Mato Grosso) | Zinc (main), lead & copper | Sublevel bench stope/ Vertical retreat mining | 110 | – Up to level 2 in MV<br>– Levels 1, 2 and 5 in AV | – VFDs<br>– Sensors for gas monitoring at some points of the mine | de Vieira et al. (2021), p. 2; Nexa. Annual Report (2020); Oliveira et al. (2020), pp. 14–16 |

Abbreviations: MV—Main Ventilation, AV—Auxiliary (or secondary) ventilation, VFD—Variable frequency drive

electrical energy consumption.[53] During shift intervals and in the period corresponding to "peak hours", the fans are programmed to work with lower rotations.[54] A ventilation raise is also equipped with ultrasonic sensors to measure the airflow in real-time, showing the information in the supervisory software (FactoryTalk View SE da Rockwell Automation) of the mine control center.[55] In emergency cases, there is the possibility of automatic reversal of the mine airflow, helping to remove polluting gases. The VOD control strategies adopted in the MV can therefore be classified as levels 1 and 2.

The auxiliary ventilation system (AV) is a blower one, basically composed of 15 axial fans with an installed power of 1122 kW associated with 700 to 1200 mm diameter vinylone ducts.[56] According to Oliveira et al.,[57] secondary fans used to cause a huge waste of energy, as they were activated by a soft starter and due to the absence of a supervisory system for monitoring and remote control. The same authors mention that in 2016 there was an expansion of the automation project of underground stationary equipment for the secondary ventilation, compressed air, and pumping system (in pilot phase), making the mine more integrated. Among the preliminary studies carried out to improve, at that time, the mine secondary ventilation system in terms of energy, productivity, and occupational safety/health, those by Costa[58] stand out. According to his research, the reduction in energy consumption was estimated at approximately 26%, based on the development and production schedule of the mining panels in the four-year period 2014–2017. His studies also include the implementation of a level 5 VOD control strategy on a pilot scale. Tests were then carried out with carbon monoxide (CO) detection sensors installed in specific locations on the main ramp of the mine with the primary objective of regulating secondary ventilation according to the amount of CO detected. As reported by Oliveira et al.,[59] the implementation of automation (levels 1, 2 and 5) to the secondary ventilation system actually generated a relevant energy saving, of 28%, compared to the classic system, equivalent to 1,021,815 kWh/year. The average specific consumption before and after this automation was 3.78 kWh/t ROM and 2.35 kWh/t ROM transported, respectively. As support for carrying out underground automation, the mine has a fiber optic communication infrastructure of more than 18 km underground, with ring redundancy, following certification standards. The automation system currently integrates all critical functions such as main and secondary ventilation systems, compressed air system, pumping system and autonomous/teleremote operation of certain mobile equipment such as LHDs, fan drill and

[53] de Oliveira et al. (2018), pp. 12–13.

[54] de Costa (2017).

[55] de Oliveira et al. (2018), pp. 12–13.

[56] de Costa (2017); de Oliveira et al. (2018), pp. 12–13.

[57] de Oliveira et al. (2018), pp. 12–13.

[58] de Costa (2017).

[59] de Oliveira et al. (2018), pp. 12–13.

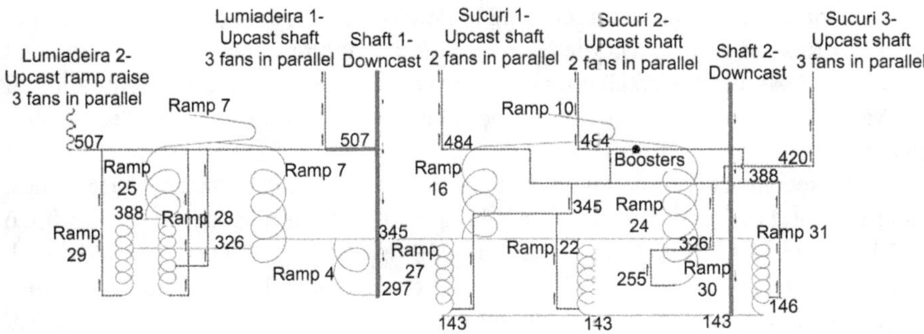


**Fig. 2** Simplified flowchart of part of the Vazante Mine

drilling jumbo (carried out by Ferbasa's technical team in partnership with Epiroc and ABB companies).

### 4.2.2 Vazante & Extremo Norte Mines

In the Vazante mining complex of Nexa Resources located in the municipality of Vazante (Minas Gerais), VOD technology is also used. It contains polymetallic mines with zinc as the main product and lead and silver as by-products. The mining method is Vertical retreat mining (VRM).[60]

The main mine is Vazante, with a length of more than 5 km, and two regions: Sucuri, to the north, and Lumiadeira, to the south.[61] The intake airways are the main access ramps and two shafts. The main ventilation system works on exhaustion and has 17 axial fans. Most of them (13) is located on the surface and associated in parallel in the following exhaust shafts (Fig. 2): Lumiadeira 01 (3250 hp fans), Lumiadeira 02 (3350 hp fans), Sucuri 01 (2250 hp fans), Sucuri 02 (2200 hp fans) and Sucuri 03 (3700 hp fans). In the underground, there are 2 Boosters Rp-9 of 250 hp and 2 Boosters N-484 of 250 hp.[62]

At the Extremo Norte mine, the intake airway is the main ramp. Its main ventilation system is also made up of fans that work on exhaustion. It is a parallel association of 3 fans of 600 hp located on the surface in Shaft 3.[63]

In both mines, the VOD control strategies employed are levels 1 and 2. There is a modulation of the rotation frequency of the main fans (from VFDs) in the shift change periods when the staff and equipment are considerably reduced underground.[64] Airflow variations can also be programmed automatically according to the blast and aeration period. The RockWell's PlantPAx® DCS is used, which was

[60] Fernandes and Ayres da Silva (2020), p. 1.

[61] Pereira (2018).

[62] de Vieira et al. (2021), p. 2.

[63] de Vieira et al. (2021), p. 2; Pereira (2018).

[64] de Vieira et al. (2021), p. 2.

adapted for mine ventilation. According to Pereira[65] in 2018, a study was carried out by Nexa Resources that recorded a reduction in monthly energy consumption of around 15.3% with the installation of frequency inverters in the main ventilation of the Vazante mine and a reduction in the rotation of 8 main mine fans during shift changes.

In the secondary ventilation system of the Vazante and Extremo Norte mines, there is a total of 1 centrifugal fan of 250 hp and 70 axial fans, being 5 of 75 hp, 6 of 100 hp, 43 of 125 hp, 3 of 200 hp, 9 of 250 hp and 4 of 350 hp, associated with cylindrical and twin flexible PVC ducts with diameters of 1200 m and 1000 mm.[66] Regarding automation, a level 5 VOD pilot (complete solution) on a working face at the Vazante mine was performed for approximately one year by a consulting firm specializing in VOD systems for mining.[67] Moreover, Maciel[68] reports positive preliminary tests with secondary and primary fans of the mine in 2015, aiming at their automatic shutdown or reduction of their speeds during certain periods that generated a reduction in monthly energy consumption of about 15.6%.

### 4.2.3 Córrego do Sítio I Mine

AngloGold Ashanti's Córrego do Sítio I gold mine is in Santa Bárbara (Minas Gerais). The mining method is Sublevel open stoping.[69] The main ventilation system airflow is 440 $m^3/s$.[70] The system consists of 6 exhaust fans: 2500 hp axial fans in parallel, 2500 hp centrifugal fans in parallel, and 2125 hp axial fans in parallel.[71] Their motors are coupled to drive frequency inverters that allow the adjustment of flow and pressure to meet the mine's operational needs. The main ventilation fans presently operate with an average rotation of 70% during shifts and 50% in shift changes, providing an economy of 53% in relation to the available operating power.[72]

As described by Costa and Silva,[73] the auxiliary ventilation of the mine was, in 2019, composed of dozens of secondary axial fans programmed to be switched off during the front shift changes without explosive blasting, and are reset once activities resume, providing an annual savings of approximately 21%.

65 Pereira (2018).

66 de Vieira et al. (2021), p. 2.

67 Pereira (2018).

68 Maciel (2015).

69 Fernandes and Ayres da Silva (2020), p. 1.

70 de Vilhena Costa and Margarida da Silva (2020), pp. 1–8.

71 da Costa (2019).

72 de Vilhena Costa and Margarida da Silva (2020), pp. 1–8.

73 de Costa and da Silva (2020), pp. 555–560.

### 4.2.4 Cuiabá Mine

AngloGold Ashanti's Cuiabá mine extracts gold ore in Sabará (Minas Gerais) using the Sublevel open stoping. It is one of the deepest Brazilian mines today (with a depth of 1350 m) and features a Champer spray cooling system.[74] Its ventilation system accounts for 44% of the electricity consumed in the mine.[75] The main ventilation system has a nominal capacity of 935 m$^3$/s. Basically, the intake airways are the main shaft, the cooled air shaft, an auxiliary raise, and the access ramp located on level 03 of the mine, while the air exhaust is basically carried out by three main fan systems arranged on the surface (SV02, SV03 e SV04), installed in the 5 raises of the mine. Each exhaust system is composed of centrifugal fans associated in parallel, featuring the SV02 system 3 fans system with a power of 1000 hp, the SV03 3 fans of 1600 hp, and the SV04 2 fans of 1600 hp. Attached to the cooling plant are three fans with a nominal flow of 340 m$^3$/s.[76] In addition to the centrifugal fans, there are also boosters in the main ventilation system of the mine.[77] The mine has a wireless communication network (Wi-Fi). In 2011, it was already possible to monitor and control the main fans located on the surface from the supervisory system in the central control room of the mine.[78] In addition to receiving information about their flows, static pressure, vibration and temperature levels, electric current, voltage, and motor rotation, it is possible to control the rotation speed of the exhaust fans at certain times, in agreement with the mine's operation, from VFDs coupled to their engines. According to Machado,[79] the automation system is composed of programmable controllers from SIEMENS, model SIMATIC S7-1200, and an HMI, for local operation, also from SIEMENS, model SIMATIC Multi Panel Touch.

In the secondary ventilation system, there are currently more than 100 axial fans of varying powers (40, 60, 75, 100, 125, 150, 200, and 250 hp) associated with 1.2 m and 1.0 m canvas ducts.[80] Many of them can be turned on and off automatically or have their rotation modified from the control room, according to the use or presence of people/equipment in the drifts. To this end, sensors are strategically installed on the ramp and in certain galleries of the mine.[81] There are also in the mine wireless sensors to control the required flow rate and cameras to assist in monitoring the ventilation doors, so that they are not opened unnecessarily, and damage can be observed.[82] In general, the VOD control strategies implemented in

[74] da Costa (2019); de Vieira et al. (2021), p. 2.

[75] de Vilhena Costa and Margarida da Silva (2020), pp. 1–8.

[76] da Costa (2019).

[77] Oliveira (2021).

[78] Machado (2011).

[79] Machado (2011).

[80] de Vieira et al. (2021), p. 2.

[81] Oliveira (2021); da Silva et al. (2021), pp. 282–294.

[82] de Costa and da Silva (2020), pp. 555–560.

the mine can be classified as up to level 4. According to Oliveira,[83] currently, such implementations guarantee a total reduction in energy consumption per year of 6,000,000 kW, approximately 21.2%.

### 4.2.5 Lamego Mine

At AngloGold Ashanti's Lamego mine, also located in Sabará (Minas Gerais), gold ore is extracted using the Sublevel open stoping mining method.[84] The nominal capacity of the main ventilation system is 500 m$^3$/s. It is exhaust and basically composed of 7 axial fans on the surface (4 in parallel with 250 hp of power and 3 in parallel with 100 hp) and boosters to optimize it.[85] Secondary ventilation is a blowing system, consisting of 16 axial fans: 8 of 40 hp, 3 of 250 hp, 3 of 42 hp, and 2 of 210 hp.[86]

The implementation of VOD technology in the main ventilation allows the remote start/stop of the 23 fans in the mine. During shift changes and when there is no mine operation, some of the fans are turned off.[87] As reported by Filho,[88] VOD system implementation was facilitated by the existence in the mine of the Smart Mine Underground software, developed in partnership with the Devex company, from 2011, for the monitoring and control of the entire underground operation from the control panel on the surface.

### 4.2.6 Serra Grande Mine

The Serra Grande unit of AngloGold Ashanti, located in Crixás (Goiás) comprises the following gold mines: Palmeiras, Mina Nova, Ingá, Pequizão and Mina III.[89]

Mine III is the deepest mine in the complex (1100 m). Its main ventilation system has a capacity of 900 m$^3$/s and is currently composed of 18 fans associated in parallel with powers ranging from 250 to 500 hp. The auxiliary ventilation system, on the other hand, is a blowing system, consisting of 30 fans with powers from 30 to 125 hp, associated with 800 mm to 1200 mm diameter ducts in PVC vinyl canvas reinforced with high tenacity polyester canvas.[90] Each installation is provided with a VFD controlled and monitored from the surface. The mine dampers for airflow

[83] Oliveira (2021).

[84] Fernandes and Ayres da Silva (2020), p. 1.

[85] de Costa and da Silva (2020), pp. 555–560.

[86] Oliveira (2021).

[87] de Costa and da Silva (2020), pp. 555–560; Oliveira (2021).

[88] Filho (2021).

[89] Oliveira et al. (2021a), pp. 16–18.

[90] de Vieira et al. (2021), p. 2.

and temperature adjustment are also controlled from the surface.[91] In addition to these VOD control strategies (levels 1 and 2), in 2019, the company Epiroc performed a six-month trial focusing on energy saving (maintaining air quality in accordance with the regulations) to control the ventilation at a mine development drift heading and on one single fan. The energy costs on that occasion represented 7% of the total mine production costs.[92] The drift was in the Inga Ramp (Orebody IX) at a depth of around 750 m from the surface. In the tests, the VOD system Serpent Automatic was used with the Serpent AVH125.90 fan (1.25 m diameter and 90 kW engine), Danfoss VLT Aqua Drive frequency inverter, and two sensors placed on a trolley located close to the drift heading: one sensor for carbon oxide (CO) and another for nitrogen dioxide ($NO_2$) emissions. Although the VOD system installed had no temperature measurement, separate temperature sensors were used to correlate with the installed system. Hansson and Baldeón[93] report a 47% reduction in power consumption when 98% of the operation mode was automatic.

### 4.2.7 Jacobina Mine

The Jacobina mining complex is located 10 km from the town of Jacobina (Bahia) and is comprised of five underground gold mines: Canavieiras (with Canavieiras South—CAS, Canavieiras Central—CAC and Canavieiras North—CAN sectors), João Belo (JBN), Morro do Cuscuz (MCZ), Morro do Vento (with Morro do Vento North—MVN and Morro do Vento South—MVS sectors) and Serra do Córrego (SCO). In all of them, the Sublevel open stoping is used. However, studies are currently also being carried out to review alternative mining methods with the use of backfill, especially the appropriateness of the Jacobina tailings for or hydraulic fill or paste fill applications and the use of cemented rock fill.[94]

Primary ventilation of the underground mines is provided using the main exhaust fans, ramps, and ventilation raises.

The auxiliary ventilation system to provide air to the working faces comprises auxiliary fans with rigid and flexible ventilation ducting. Table 2 resumes the underground ventilation fans (with air flowrates ranging from 120 $m^3/s$ to 200 $m^3/s$) employed in 2020 at the various underground mining sectors.

A program to expand the complex is currently being implemented, aiming at a significant increase in mine production. Its Phase 1, completed in 2019, included the installation of an advanced process control system to increase the mines' level of automation and an automation project for the underground mine ventilation system, which generated savings of over 25 million kilowatts.[95] Before automation, mine

[91] Hansson and Baldeón (2019), p. 8.

[92] Hansson and Baldeón (2019), p. 8.

[93] Hansson and Baldeón (2019), p. 8.

[94] Yamana Gold Inc. (2020).

[95] BMS (2021).

**Table 2** Ventilation fans at Jacobina mines[a]

| | Auxiliary fans (unit) | | | | | Main exhaust fans (unit) | | |
|---|---|---|---|---|---|---|---|---|
| Mine | 75 hp | 100 hp | 125 hp | 150 hp | Total | 250 hp | 550 hp | Total |
| JBN | 4 | 12 | – | 3 | 19 | 1 | 4 | 5 |
| MVS | 5 | 5 | 1 | – | 11 | 1 | 1 | 2 |
| MVN | – | 2 | – | – | 2 | 1 | – | 1 |
| SCO | – | – | – | – | – | – | – | – |
| CAS | 3 | 15 | – | 6 | 24 | – | 3 | 3 |
| CAC | – | 11 | – | 2 | 13 | – | 2 | 2 |
| SPARE | – | 5 | – | 1 | 6 | – | 4 | 4 |
| Total | 12 | 50 | 1 | 12 | 75 | 3 | 14 | 17 |

[a] Yamana Gold Inc. (2020)

fans had to be turned on or off manually. The underground communications are carried out through a Leaky Feeder system, which allows continuous contact between mine supervisors and operating- and service crews in various locations throughout the mine complex. There are also additional systems to improve the communication between the operators and the information center on the surface. In 2018 and 2019 the Newtrax system (from Sandvik) was implemented for controlling secondary fans, tracking underground personnel, and monitoring gas emissions.[96] Therefore, it was possible to reduce energy consumption by stopping secondary fans during shift changes from the Control Center on the surface and improve logistics in times of emergency by tracking the location of personnel in the mines; and monitoring gas levels in the underground working areas, allowing supervision to anticipate ventilation difficulties and minimize re-entry times after blasting. 100% of the main fan fleet and 78% of the secondary fans have automation.[97] The current supervisor and control system installed at Jacobina mines uses both programmable logic controllers (PLCs) and remote input/outputs (I/Os) from the Siemens Simatic line. The Phase 2 Expansion project of the Jacobina complex foresees the acquisition of new automation hardware compatible with the existing system and an update of the existing central processing units (CPUs).[98]

### 4.2.8 Pilar Mine

The Pilar mine and the Roça Grande mine are in the municipalities of Santa Bárbara and Caeté (Minas Gerais), respectively, and are part of Jaguar Mining Inc's Caeté Mining complex. Currently, the Roça Grande mine is on care and maintenance. At the Pilar mine, gold ore is extracted using two mining methods in use: Sublevel

[96] Yamana Gold Inc. (2020).

[97] BMS (2021).

[98] Yamana Gold Inc. (2020).

open stoping (, being the majority of the mined stopes filled using development waste) and Cut and fill, when ore geometry does not favor open mining.[99]

Ventilation for the Pilar Mine is an exhaust system. Summarizing, air is drawn down through the main mine ramp and up an exhaust raise near the ramp. Ventilation on the levels utilizes auxiliary fans and vent ducting.[100] It was not possible in the research to obtain the exact number of fans that currently make up each ventilation system, however, there are several records in the literature about improvements recently carried out aiming at its automation and better conditions of ventilation and temperature in underground.[101] One of them is the "Main Exhaust Project", which was implemented in 2021 with the installation in the mine of two larger centrifugal fans, each of 600 hp (generating a 70% increase in the main airflow), fully automated, monitored, and operated remotely, from the surface mine dispatch room (employment of level 1 and 2 VOD control strategies). To this end, a new data network structure (backbone) was installed with more robust and stable technology, allowing the expansion of the telephone system, with the use of extensions, Wi-Fi, computer, tablets, cameras, automation, monitoring of people at the mine, in addition to other features. In the project's next phase, the same technology will be available in another gold mine in operation by the company, the Turmalina Mine.[102]

### 4.2.9 Fazenda Brasileiro Mine

The Fazenda Brasileiro mine from Equinox Gold is in Barrocas (Bahia) and extracts gold ore using the Sublevel open stoping method.[103] The mine is mainly ventilated by a series of raises that exhaust air to the surface, and the declines and shaft, which offer fresh-air intake. The main ventilation system consists of 6 axial fans working on exhaust. The raises are connected to the main ramp in the active mining areas, and secondary ventilation fans (34 axial fans of 50, 75, 100, and 125 hp power) and tubing (vinyl canvas ducts with 0.8, 1, and 1.2 m in diameter) are used to carry fresh air into individual stoping areas.[104]

The main exhaust fans can be turned on/off automatically, as well as slowed down, from the control room on the surface. The VFD model used for underground control main ventilation and pumping systems is the VACON® X-series AC drives from Danfoss. Making it possible to adjust ventilation and pumping levels according to demand, they can cut energy consumption by around 30%.[105]

[99] Fernandes and Ayres da Silva (2020), p. 1; RPA (2020).

[100] RPA (2020).

[101] Brasil Mineral (2021); Oliveira et al. (2021b), pp. 19–21.

[102] Tavares et al. (2021), p. 12.

[103] Fernandes and Ayres da Silva (2020), p. 1.

[104] de Vieira et al. (2021), p. 2; Equinox Gold Corp. (2021).

[105] Danfoss (2012).

### 4.2.10 Aripuanã Project

The Aripuanã Project is a new underground mining project by Nexa Resources for the extraction and processing of zinc (main product), copper, and lead that has been implemented since 2019 and which has recently started production, in Serra do Expedito, 25 km from the city of Aripuanã (Mato Grosso). It is considered one of the largest zinc projects in the world with an expected production of 2.2 million tons of crude ore per year (containing zinc, lead, and copper) and estimated annual zinc production of 72.7 thousand tons.[106] The mining methods used are the Sublevel Bench Stope and the VRM.[107]

The main ventilation system was designed for a capacity of 400 $m^3/s$ of fresh air, basically counting on 4 axial fans of 400 hp, arranged in parallel. The auxiliary ventilation is a blowing system and consists of 28 axial fans associated with 1.40 m diameter vinylone ducts. Regarding automation, phase 1 of the project foresees the implementation of a VOD system with control strategies 1 and 2, aiming at changing the rotations of the main fans. In phases 2 and 3, the expansion of this automation is planned and the implementation of the VOD system with gas sensors in the auxiliary ventilation (levels 1, 2 and 5).[108]

### 4.2.11 Other Mines

In addition to the mines presented in the items above, the survey with mining companies on their ventilation systems evidenced the use or recent development of pilot-scale studies for the implementation of VOD systems in other mines in the country. In part of them, these studies are being carried out by companies specializing in VOD systems for mining. However, for reasons of confidentiality, the names of the mines and companies involved in these projects cannot be disclosed.

These are 3 medium-depth underground mines (less than 1000 m) that extract metallic minerals (2 in the southeast region and 1 in the country's central region) and 1 coal mine located in the country's southern region with shallow depth (less than 100 m). In all cases, the initial VOD control strategy can be classified up to level 2 and basically refers to the main ventilation system, allowing the activation/switching off and automatic control of the rotation of their axial exhaust fans.

## 5 Conclusions

Based on the results presented, it can be concluded that currently, approximately 20% of Brazilian underground mines have started automating their ventilation systems. Cuiabá and Ipueira Mines were the first mines in the country to develop

[106] Nexa (2020); Oliveira et al. (2020), pp. 14–16.

[107] de Vieira et al. (2021), p. 2.

[108] de Vieira et al. (2021), p. 2.

projects aimed at implementing a VOD system. Investment in this technology has achieved very satisfactory results in Brazilian mines, especially in terms of energy savings, with some of them reaching values above 50%, such as the Ipueira mine which reports an energy savings of approximately 71% when VOD is applied to its main ventilation system. Such positive results confirm what has been verified by other studies worldwide over the years.

It is observed that the automation takes place mainly on the fans. In a few mines, such as Mine III, there are other components of the ventilation system, such as dampers, which are automatically controlled from the surface.

Considering the classic classification of VOD control strategies (item 4.1.1), all mines employ (or in the case of the Aripuanã Project decided to use) at least level 1 ("Manual control") and level 2 ("Time of day scheduling") strategies in the automation of the fans that make up their main ventilation systems. Approximately 9% of mines have already extended (or in the case of the Aripuanã Project foresees to extend) such levels of automation to their secondary ventilation systems, making it possible to automatically turn on/off and adjust the flow rate of auxiliary fans when necessary. Regarding the level 4 control strategy ("Tagging"), it is currently implemented in about 3% of mines, allowing the distribution of the airflow consistent with real-time knowledge of the location of the equipment and/or manpower. Additionally, trials have already been carried out in 1% of mines. The level 5 strategy is used (or in the case of the Aripuanã Project will be employed) together with other control strategies as a fail-safe system in approximately 4% of mines. Moreover, pilot-scale tests have already been performed to consider environmental variables in the automation of the working faces ventilation (such as the real-time concentration of contaminant gases) in another 3% of mines.

It can also be determined that in most Brazilian underground mines this ventilation automation process up to level 2 can be classified more as an "In-house development", than as solutions acquired from traditional VOD software and hardware packages available on the market. However, for the implementation of more complex levels of automation, it is important to highlight the fundamental role of companies specialized in VOD technology in the country's mines, assisting or performing most pilot-scale tests in the mines.

Despite the observed benefits, most underground mines in the country (80%) do not use this technology. It is observed that an initial obstacle to the adoption of the automation process of ventilation systems in many of them is the lack of robust communication networks that allow automatic operation or remote control of ventilation assets. It is paramount to mention that this percentage does not include illegal mines, without registration in the Brazilian Ministry of Mines and Energy and excluded from official statistics. In these mines, focused on subsistence, rudimentary and artisanal mining techniques are usually used without a proper ventilation system.[109]

The use of VOD in the country primarily at the most basic levels (despite the success of many pilot-scale tests) can also be largely justified/explained by the growing need for higher investments in infrastructure, systems programming, multidisciplinary teams, maintenance (including calibration of components) as the level

[109] de Vieira et al. (2021), p. 2.

of system implementation increases, which can sometimes become unfeasible for the reality of the mine and not compensate for the benefits generated. On the other hand, some Brazilian mines also do not present very favorable characteristics for VOD use, as they require continuous airflow supply,[110] such as mines with radioactivity problems and hot ones.

Among the challenges faced for the implementation of VOD in the country, there are also cultural issues that sometimes interfere in the process of adapting the workforce to the new technology. In these cases, it is essential, before the implementation phase, adequate management aimed at preparing everyone for such changes, including, for example, efficient and transparent dissemination, information and communication process to explain how the new system will work, the expected benefits for the mine and its employees/operators regarding occupational health and safety and what to do if the system does not work as expected.[111] Pacheco,[112] and Brokering, Loring and Rutter[113] also mention the promotion of activities by the company, such as training, monitoring, and coaching for an easier adaptation to the new situation. In new projects with underground extraction, however, this issue is considerably mitigated, as the Aripuanã Project in the Brazilian state of Mato Grosso, which already contemplates the use of VOD since the beginning of the mine's lifetime.

The perspective is, therefore, that the use of technology will increasingly grow, in the country, in the coming years, especially in medium and large-scale underground metal mines, and that new projects of the same nature will already consider the automation of the ventilation system from the beginning. However, the trend is that the VOD control levels implementation follows a time-staggered approach, with the initial adoption of simpler levels, such as "Manual control" or the "Time of day scheduling", rather than going directly to more complex control strategies. In this way, the learning curve experienced by everyone with the new system implementation can be considered and a careful analysis of the benefits and costs involved at each stage can be carried out to know if it will effectively be worth the implementation of more control levels than those already in place.

## References

Acuña E (2021) Workshop Ventilation On Demand (VOD)

Acuña E, Allen C (2017) Ventilation control system implementation and energy consumption reduction at Totten Mine with Level 4 Tagging and future plans. In: Proc. First Int. Conf. Undergr. Min. Technol. Australian Centre for Geomechanics, Perth, pp 89–95

Acuña E, Feliú A (2014) Considering ventilation on demand for the developments of the New Level Mine Project, El Teniente, pp 813–821. https://doi.org/10.36487/ACG_rep/1410_59_Acuna. Accessed 22 May 2023

[110] Hardcastle and Kocsis (2004), pp. 51–57.

[111] Acuña and Feliú (2014), pp. 813–821; Gutiérrez et al. (2021), pp. 262–271; Acuña (2021).

[112] Pacheco (2021).

[113] Brokering et al. (2017), pp. 15–22.

Acuña E, Levesque M (2021) Auxiliary fan energy consumption profile under ventilation control strategies. In: 2nd Int. Symp. Mine Vent. South Am. (II SIVM). Vent. Undergr. Mines Increases Saf. Product. Universidad Nacional de Ingeniería, Lima, pp 442–457

Acunã E, Alvarez RA, Hardcastle SG (2014) A theoretical comparison of ventilation on demand strategies for auxiliary mine ventilation systems. In: 10th Int. Mine Vent. Congr. IMVC2014. The Mine Ventilation Society of South Africa, Sun City, pp 1–7

Acuña EI, Alvarez RA, Hurtado JP (2016) Updated Ventilation On Demand review: implementation and savings achieved. In: 1st Int. Conf. Undergr. Mining. UMining2016, Santiago, p 12

Allen CL (2008) Ventilation on demand (VOD) projects: Vale Inco Ltd. In: CIM (ed) CIM Maint. Eng. Oper. Conf. Comptes rendus du Colloq. ICM sur l'ingenierie Maint. l'exploitation Min., Val d'Or, pp 1–8

Ayres da Silva ALM, De Eston SM (2021) Brazilian ventilation standards: challenges and updates. In: 2nd Int. Symp. Mine Vent. South Am. (II SIVM). Universidad Nacional de Ingeniería, Lima, pp 348–365

Babu VR, Maity T, Prasad H (2015) Energy saving techniques for ventilation fans used in underground coal mines—a survey. J Min Sci 51:1001–1008. https://doi.org/10.1134/S1062739115050198. Accessed 22 May 2023

Baiden G, Bissiri Y, Hardcastle S, Kocsis C (2005) Justification of "on-demand" ventilation systems by means of simulation techniques. In: Appl. Comput. Oper. Res. Miner. Ind. Taylor & Francis, pp 457–462. https://doi.org/10.1201/9781439833407.ch61. Accessed 22 May 2023

Bartsch E, Laine M, Andersen M (2010) The application and implementation of optimized mine ventilation on demand (OMVOD) at the Xstrata Nickel Rim South Mine, Sudbury, Ontario. In: Hardcastle, McKinnon (eds) 13th US/North Am. Mine Vent. Symp. Miraco Mining Innovation, Sudbury, p 567

BMS (2021) Jacobina: Yamana project generates energy savings of more than 25 million kW, 2

Brasil Mineral (2021) Improvements inside Pilar mine. Bras Miner, 1

Brokering DR, Loring DM, Rutter CJ (2017) Practical implementation of VOD at the Henderson mine. In: Brune JF (ed) 16th North Am. Mine Vent. Symp. Colorado School of Mines, Colorado, pp 15–22

Chatterjee A, Zhang L, Xia X (2015) Optimization of mine ventilation fan speeds according to ventilation on demand and time of use tariff. Appl Energy 146:65–73. https://doi.org/10.1016/j.apenergy.2015.01.134. Accessed 22 May 2023

da Costa CV (2019) Replacement of diesel equipment with electric ones as an alternative to reduce the flow of the ventilation system. Federal University of Ouro Preto

da Silva JM, de Santos RCP, Filho AOS (2021) Expansion of human resources and infrastructure in Mine Ventilation in Brazil. In: 2nd Int. Symp. Mine Vent. South Am. (II SIVM). Vent. Undergr. Mines Increases Saf. Product. Universidad Nacional de Ingeniería, Lima, pp 282–294

Danfoss (2012) Ventilating a gold mine in Brazil. VACON® AC drives at Mineração Fazenda Brasileiro

de Costa JCA (2017) Studies for the implementation of an on-demand ventilation model in the Ipueira – Ferbasa mining complex Andorinha municipality – Bahia. Federal University of Campina Grande

de Costa LV (2019) Analysis via simulation of ventilation in an underground mine - case study Córrego do Sítio I mine. Federal University of Ouro Preto

de Costa LV, da Silva JM (2020) Strategies used to control the costs of underground ventilation in some Brazilian mines. REM - Int Eng J 73:555–560. https://doi.org/10.1590/0370-44672019730057. Accessed 22 May 2023

de Oliveira WL, de Alcântara WWM, Sarmento HJL, Ribeiro GÁV, Dias AJA, Carneiro CAS et al (2018) Automation project for stationary and mobile underground equipment at the Ipueira Mine. Minérios Miner 400:12–13

De Souza E (2018) Cost-saving strategies in mine ventilation. CIM J 9:7. https://doi.org/10.15834/cimj.2018.9. Accessed 22 May 2023

de Vieira JMC, da Silva WT, Ayres da Silva ALM (2021) Ventilation systems in Brazilian underground mines. In: University of São Paulo (ed) 29th Int. Symp. Sci. Technol. Initiat. USP São Paulo, São Paulo, p 2

de Vilhena Costa L, Margarida da Silva J (2020) Cost-saving electrical energy consumption in underground ventilation by the use of ventilation on demand. Min Technol 129:1–8. https://doi.org/10.1080/25726668.2019.1651581. Accessed 22 May 2023

Dicks F, Clausen E (2017) Ventilation on demand. Min Rep 153:334–341

Equinox Gold Corp (2021) Technical Report on the Fazenda Brasileiro Gold Mine, Bahia State, Brazil. Vancouver

Fernandes PT, Ayres da Silva ALM (2020) Overview of variant underground mining methods in Brazilian mines. In: 28th Int. Symp. Sci. Technol. Initiat. USP São Paulo. University of São Paulo, São Paulo, p 1

Filho AOS (2021) Technologies and practices that lower the cost and increase the reliability of the ventilation system. Federal University of Ouro Preto

Ge Q, Zhu W, Zhang W, Chen Q, Yang Z (2019) Ventilation-on-Demand system in Pulang copper mine. In: Proc. 11th Int. Mine Vent. Congr. Springer Singapore, Singapore, pp 116–125. https://doi.org/10.1007/978-981-13-1420-9_11. Accessed 22 May 2023

Gonen A (2021) Energy savings in auxiliary ventilation systems of underground mines. Int J Eng Technol Manag Res 8:72–82. https://doi.org/10.29121/ijetmr.v8.i10.2021.1055. Accessed 22 May 2023

Griffith M, Pinedo J (2022) Dynamic activity tracks and ventsim control. In: Vent. User Conf. 2022. Howden, Brisbane

Gutiérrez C, Mejías R, Vásquez R, Hueichapan J, Hurtado JP, González D et al (2021) Study for the expansion of the VOD pilot in the el teniente mine Codelco Chile. In: 2nd Int. Symp. Mine Vent. South Am. (II SIVM). Vent. Undergr. Mines Increases Saf. Product. Universidad Nacional de Ingeniería, Lima, pp 262–271

Gyamfi S (2020) Considerations and development of a ventilation on demand system in Konsuln Mine. Luleå University of Technology

Hansson J, Baldeón D (2019) Impact on energy consumption and air quality control with ventilation on demand in Serra Grande mine, Brazil. Svenska Bergteknikföreningen, Bergteknikdagen, Stockholm, p 8

Hardcastle SG, Kocsis CA (2001) Ventilation design for an automated underground metal mine. In: Wasilewski S (ed) 7th Int. Mine Vent. Congr. EMAG, Krakow, Poland, pp 779–786

Hardcastle SG, Kocsis C (2004) The ventilation challenge. CIM Bull:51–57

Hardcastle SG, Gangal MK, Udd JE, Grenier MG, Klinowski GW (1995) Mine ventilation and optimization. In: Proc. 26th Int. Conf. Saf., vol II. Mines Res. Institutes, Katowice, pp 183–199

Hardcastle SG, Gangal E, Leung E (1998) Green and economic mine ventilation with an integrated air management system. In: Singhal R (ed) 7th Int. Symp. Mine Plan. Equip. Sel. Balkema, Rotterdam, pp 785–794

Hardcastle SG, Gangal MK, Schreer M, Gauthier P (1999) Ventilation on Demand - Quantity or Quality - A Pilot Trial at Barrick Gold's Bousquet Mine. In: 8th U.S. Mine Vent. Symp., pp 31–38

Hardcastle SG, Kocsis C, O'Connor D (2006) Justifying ventilation-on-demand in a Canadian mine and the need for process based simulations. In: Mutmansky, Ramani (eds) 11th US/North Am. Mine Vent. Symp. 2006. Taylor & Francis Group, London, pp 15–27

Hartman HL, Mutmansky JM, Ramani RV, Wang YJ (1997) Mine ventilation and air conditioning, 3rd edn. Wiley, New York

Heider M (2017) Overview of underground mining in Brazil. In-the-Mine, pp 13–16

Howden (2018) Mine site health and safety, 2

Jacinto C, Anacleto H, Quispe J, Baca A (2021) Intelligent ventilation through a cyber-physical system. In: 2nd Int. Symp. Mine Vent. South Am. (II SIVM). Vent. Undergr. Mines Increases Saf. Product. Universidad Nacional de Ingeniería, Lima, pp 71–87

Kocsis K-CP (2009) New ventilation design criteria for underground metal mines based upon the "life-cycle" airflow demand schedule. University of British Columbia

Kocsis CK, Hardcastle SG, Hall R (2004) The benefit of using mine process simulators to design a "life-cycle" mine ventilation system. In: SME Annu. Meet. SME, Denver, Colorado, p 7
Kocsis C, Hardcastle S, Keen B (2008) A heat study and the modelling of future climatic conditions at Vale Inco's Coleman Mc Creedy East Mine. In: 12th US/North Am. Mine Vent. Symp., pp 203–210
Kolobe K, Rutter C, Shea N (2021) Evolution of the Henderson mine ventilation system. In: Mine Vent. CRC Press, London, pp 538–547. https://doi.org/10.1201/9781003188476-55. Accessed 22 May 2023
Kumar Katta A, Davis M, Kumar A (2020) Assessment of greenhouse gas mitigation options for the iron, gold, and potash mining sectors. J Clean Prod 245:118718. https://doi.org/10.1016/j.jclepro.2019.118718. Accessed 22 May 2023
Lagowski K (2013) Air supply on demand. CIM Mag 8:58–60
Machado HG (2011) Risk management in underground mines - assessment of deep mine ventilation. Federal University of Ouro Preto
Maciel W (2015) Implementation of on-demand ventilation in an underground mine in the Northwest of Minas Gerais. Faculty of Northwest Minas Gerais
McPherson MJ (2018) Subsurface ventilation engineering. Mining Ventilation Services, Inc, Berkeley
Moreau K, Laamanen C, Bose R, Shang H, Scott JA (2021) Environmental impact improvements due to introducing automation into underground copper mines. Int J Min Sci Technol 31:1159–1167. https://doi.org/10.1016/j.ijmst.2021.11.009. Accessed 22 May 2023
Nel AJH, Arndt DC, Vosloo JC, Mathews MJ (2019) Achieving energy efficiency with medium voltage variable speed drives for ventilation-on-demand in South African mines. J Clean Prod 232:379–390. https://doi.org/10.1016/j.jclepro.2019.05.376. Accessed 22 May 2023
Nensen P, Lundkvist A (2005) From 167 GwH to 72 GwH – Ventilation on Demand in LKAB's Iron Ore Mine, Malmberget
Nexa (2020) Annual Report Nexa. Nexa
Nyqvist J, Serres M (2020) ABB discusses the benefits of ventilation on demand. Can Min J:37–40
Oliveira BM (2021) Mine Ventilation, referencing the application in the Cuiabá-Lamego Mine
Oliveira T, Rezende F, Santos M (2020) Aripuanã Project. In-the-Mine 84:14–16
Oliveira T, Rezende F, Santos M (2021a) Continuity with innovation. In-the-Mine 90:16–18
Oliveira T, Rezende F, Santos M (2021b) Production at a sustainable level. In-the-Mine 90:19–21
Pacheco JL (2021) Workshop Ventilation On Demand (VOD)
Pacheco JL, Cruz JP, Acunã E, Araya N (2018) Implementation system on demand in the Andina Division. In: Prim. Simp. Int. en Vent. Universidad de Santiago de Chile, Minas Sudamérica, Chile
Pandey JK, Mischo H, Drebenstedt C (2015) Ventilation on demand for improving viability of underground mines in India. J Mines Met Fuels 63:119–125
Pereira FC (2018) Study of energy savings with the implementation of the on-demand ventilation system at Vazante Mine. Federal University of Ouro Preto
Rönnblom W (1996) Ventilation-by-demand. In: 9th Can. Inst. Min. Oper. Conf. CIM, Val d'Or, Quebec, p 5
RPA (2020) Technical Report on the Caeté Mining Complex, Minas Gerais State, Brazil. Toronto
Rustan A (1979) Review of developments in monitoring and control of mine ventilation systems. In: 2nd Int. Mine Vent. Congr., Reno, pp 223–229
Semin MA, Levin LY, Maltsev SV (2020) Development of automated mine ventilation control systems for Belarusian potash mines. Arch Min Sci 65:803–820. https://doi.org/10.24425/ams.2020.135178. Accessed 22 May 2023
Shriwas M, Pritchard C (2020) Ventilation monitoring and control in mines. Mining Metall Explor 37:1015–1021. https://doi.org/10.1007/s42461-020-00231-8. Accessed 22 May 2023
Sjöström S, Klintenäs E, Johansson P, Nyqvist J (2020) Optimized model-based control of main mine ventilation air flows with minimized energy consumption. Int J Min Sci Technol 30:533–539. https://doi.org/10.1016/j.ijmst.2020.05.016. Accessed 22 May 2023

Tavares C, Alam C, Campos P, Gonzaga F (2021) Improvements in CCA/Pilar. In: Inc. JM (ed) Jaguar em foco, #78, Ag/Set 2021, Ano VII, Belo Horizonte, p 12

Tran-Valade T, Allen C (2013) Ventilation-On-Demand key consideration for the business case. In: Can. Inst. Min. Conv. Proc. Toronto 2013 CIM Conf. CIM, Canada

Tuck MA, Finch C, Holden J (2006) Ventilation on demand: a preliminary study for Ballarat Goldfields NL. In: Mutmansky JM, Ramani RV (eds) 11th U.S./North Am. Mine Vent. Symp. The Pennsylvania State University, Pennsylvania, pp 11–14

Udd J, Clarke RWD, Hardcastle SG (1996) Optimizing air management and energy costs in northern mines. In: 4th Int. Symp. Min. Sartic, Longyearbyen, Svalbard, Sweden

Wallace K, Prosser B, Stinnette JD (2015) The practice of mine ventilation engineering. Int J Min Sci Technol 25:165–169. https://doi.org/10.1016/j.ijmst.2015.02.001. Accessed 22 May 2023

Wu HW, Gillies ADS (2005) Real-time airflow monitoring and control within the mine production system. In: AusIMM (ed) 8th Int. Mine Vent. Congr., Brisbane, pp 383–389

Yamana Gold Inc. (2020) NI 43-101 Technical Report Jacobina Gold Mine, Brazil. Toronto

# Sustainable Activities Assessment Through Conceptual Sieving Method: Case Studies in Colombian Mining

**Gustavo Aristizabal-H and Oscar Jaime Restrepo Baena**

**Contents**




**Abstract** Despite the importance that industrial activities have to society, for some years certain corporative sectors, such as mining, have faced strong criticism from various circles of global public opinion. This has hampered important industrial projects, generating great economic losses for companies and regions. While some localized studies indicate that in specific areas negative impacts of mining have been greater than positive ones, other analyzes highlight the opportunities for regional development that mining projects executed under the principles of sustainability represents. This work is based on the premise—according to definition of



G. Aristizabal-H · O. J. R. Baena (✉)
Universidad Nacional de Colombia – School of Mines, Medellín, Colombia
e-mail: gaaristizabalh@unal.edu.co; ojrestre@unal.edu.co







W. Frenz, A. Preuße (eds.), *Yearbook of Sustainable Smart Mining and Energy 2022*, Yearbook of Sustainable Smart Mining and Energy - Technical, Economic and Legal Framework 2, https://doi.org/10.1007/978-3-031-41873-0_10

sustainability adopted here—that for a project to successfully contribute to the sustainable development of a region, it must ensure that its positive impacts last in the long term beyond project closure. Thus, it is crucial that companies can assess the potential of their sustainability activities to produce that kind of impacts. Through this work, Conceptual Sieving is presented as a tool for this type of analysis. Moreover, a pilot application is made to five mining projects located in Colombia. At the end, the results obtained from this application are discussed.

## 1 Introduction

> Those who love practice without theory are like the sailor who boards ship without a rudder and compass and never knows where may cast (Leonardo da Vinci).

The notion of sustainable development arose in the United Nations Conference on Human Environment held at Stockholm in 1972. Afterwards, it was conceptualized through the Our Common Future report as a "development that meets the needs of the present without compromising the ability of future generations to meet their own needs".[1]

This concept has been applied in multiple ways in society, as well as embraced by organizations, people, and governments for all over the world. However, its application at business sector has been especially significant. For instance, the model of Triple Bottom Line[2] has been used by many firms as a guide for the incorporation of sustainable development aspects in their processes. In this model, sustainability is the principle of ensuring that our actions today do not limit the range of social, environmental, and economic options open to future generations.

A more global consensus for the incorporation of sustainable development concept in many subjects of society is the 2030 Agenda and the Sustainable Development Goals (SDGs).[3] This Agenda is a plan of action that sees poverty eradication as an indispensable requirement for sustainable development. In that sense, the seventeen SDGs are an integrated set of tools to be adopted by governments, companies, and civil society to address the global challenges related to poverty, inequality, climate, environmental degradation, prosperity, peace, and justice.

In mining industry, the concept of sustainable development was firstly incorporated in 1993 through the Whitehorse Initiative, which was carried out by the Canadian Mining Association.[4] Subsequently, in 1998 an international initiative was achieved, which was called the Global Mining Initiative (GMI). This effort was leaded by the world nine biggest mining companies at that time, as a preparation for the World Summit on Sustainable Development that took place in Johannesburg – South Africa in 2002.

---

[1] ONU (1987).

[2] Elkington (1997).

[3] ONU (2015).

[4] Fitzpatrick et al. (2011), pp. 376–384.

The GMI derived to the two years project: Mining, Metals and Sustainable Development (MMSD), which results were published in the Breaking New Ground report. It would be considered the foundational document of sustainability in mining. In this document, the role of mining is seen from the perspective of contribution that this industry can make to environmental conservation, social concerns, and governance systems, seeking increase the social welfare of current generations without undermining the well-being of future generations.[5]

Mines can only be located where there are mineral deposits of interest. For this reason, mining projects are often located in rural areas far from cities. This often means that, especially in developing countries, the populations neighboring mining projects, before their arrival, have significant deficiencies in their quality-of-life indices. In this sense, sustainability in mining could be understood as a set of strategies and activities aimed to help this industry contribute to the development of the areas of influence of its operations.

Mining industry supplies essential materials that constitute many products that are used daily in society.[6] This condition implies the exploitation of big amounts of non-renewable natural resources (NRNR). Hence, many groups argue that extraction of NRNR should decrease over the time.[7] These claims are based on several facts such as about 72% of key drivers of greenhouse gas emissions are in the form of CO2 due to combustion of fossil fuels and cement clinker production.[8]

Nevertheless, while recycling and reversed mining—plus a significant decrease in consumption—could help to reduce raw material exploitation, nowadays does not seem to be entirely foreseeable when these practices will drive a significant decrease in minerals and metals demand, or even fully supply it. Consequently, extractive activities could continue as today's scale for some more decades. Especially, if it is considered that small, and artisanal scale mining (ASM) it is the job of at less 40 million people in the world as well as the livelihood for 150 million people in 80 countries of global south.[9]

On the other hand, it is also said that despite the extraction of NRNR entails negative impacts it can contribute to the social, environmental, and economic dimensions of sustainability, as well as foster the reduction of poverty.[10] This is especially important in so far as 2030 Agenda points out poverty eradication as an indispensable factor for sustainable development.[11] In that sense, there are experiences of harmonious mining projects with communities.[12]

There is a knowledge stream that advocates for a mining industry that not only can mitigate its impacts but also can promote sustainable development in its zones

---

[5] IIED (2002).

[6] Azapagic (2004), pp. 639–662.

[7] Young and Septoff (2002).

[8] Olivier and Peters (2020).

[9] IGF (2017).

[10] Shen et al. (2015), pp. 15–26; Jenkins and Yakovleva (2006), pp. 271–284.

[11] ONU (2015).

[12] Slack (2012), pp. 179–184.

of influence. This type of management is feasible through the generation of useful and longstanding capital for future and current generations,[13] as well as by the fostering economic development based on new infrastructure and productive projects at local level. [14] Moreover, the contribution to the reinforcement of local governance through the support of economic planning and innovative institutional frameworks can also enhance the human capital.[15] Por otra parte, la minería es fundamental para soportar la transición energética, en tanto la fabricación de equipos para la generación de energía renovable requiere diversidad de metales y minerales.[16]

However, social pressure is increasing on mining companies to minimize their impacts,[17] since it is considered that the physical basis of territorial development can be affected when natural resources are degraded.[18] Thus, despite claims that mining impacts can be upgraded through sustainability management, social conflict around extractive activity appear to have intensified since the turn of the century.[19]

Concerning to Colombia, Rudas and Espitia (2013)[20] carried out an analysis for the period 2000–2012. They found that in this country certain indicators of life quality show more favorable values in non-mining towns than in those where mining has been for years. Such is the case of: multidimensional poverty, index of unsatisfied basic needs, housing deficit, infant mortality rate, and illiteracy. Although, this does not establish that mining is the cause of this panorama; it does allow to affirm that extractive activities for more than a decade did not lead to a substantial improvement of these indicators in those populations. Likewise, Rudas (2012)[21] states that for contributions of mining to be reflected positively in regions, solid institutions are required, especially at local level.

Nowadays, the main social challenges of mining sector in Colombia still imply the reinforcement of sustainability practices, especially towards understanding the needs of the local communities and the creation of mechanisms of participation to compensate for the loss of livelihoods, health effects and other social consequences of mining activities.[22]

This paper is grounded on the premise that robust theoretical notions can drive effective practical implementations. In that sense, this effort seeks to offer a tool for the conceptual pertinence assessment of sustainability activities planned by firms in order to contribute to sustainable development. The information for this work was

---

[13] ICMM (2012); Shen et al. (2015), pp. 15–26.

[14] Shen et al. (2015), pp. 15–26; Jenkins and Yakovleva (2006), pp. 271–284; ICMM (2012); Hamann (2003), pp. 237–254.

[15] Di Boscio (2010).

[16] Jiskani et al. (2022), p. 132055; Watari et al. (2021), p. 127698.

[17] Shen et al. (2015), pp. 15–26.

[18] Almaguer (2007), pp. 1–8.

[19] Vargas (2019), pp. 873–880.

[20] Rudas and Espitia (2013).

[21] Rudas (2012).

[22] Betancur et al. (2018), pp. 538–553.

obtained from specialized literature, sustainability corporative reports, interviews with stakeholders and field direct observations in the zones of influence of five mining projects in Colombia. These cases were included to add a practical component to this work. These five mines are especially relevant for Colombian Government due to their dimension, term, location, and level of contribution to national economy.

In this chapter authors establishing first the theoretical bases of Conceptual Sieving framing it as a tool for the assessment of sustainability activities. Subsequently, the methodology is explained, both to build this tool as well as to make a pilot application of it. Then, the components and functioning of the Conceptual Sieving are described and the steps of the case studies selection process are presented. Afterwards, the results of pilot application and field work are discussed. Finally, the conclusions and limitations of this work, as well as further developments are also outlined at the end of this writing.

## 2 Theoretical Framework

In this work, corporate sustainability is defined as the application of the concept of sustainable development at the organizational level, which in turn refers to the organizational actions that address the dimensions: economic, environmental, and social in an interconnected and interdependent way.[23] On the other hand, sustainability in mining is a relatively recent–widely debated topic. To build this framework, indexed literature from web-databases was review. Hence, several recurrently referenced concepts were identified related to main aspects of sustainability management in mining projects. Below is a relation of some of these conceptual convergences.

### *2.1 Added Value to Society from Mining Projects*

Minerals and metals have played a crucial role in the development of humanity; however, their extraction has also caused significant negative impacts.[24] This must be kept in mind, since the opposition of communities, local governments or civil society to mining has not only affected projects in operation but also the start of new ones. This can happen regardless of their legal, technical, or financial feasibility.[25]

For this reason, the success of the extractive business relays not only on operational efficiency, but also on aligning its economic interest with other values of society through more sustainable business models.[25] This approach can not only

[23] Hahn et al. (2015), pp. 297–316.

[24] Giurco and Cooper (2012), pp. 3–12.

[25] Hamann (2003), pp. 237–254; Prno and Scott (2012), pp. 346–357; Slack (2012), pp. 179–184.

[26] Shen et al. (2015), pp. 15–26.

help make the industry's contribution to sustainable development more effective, but also boost business competitiveness.[27]

Traditionally, sustainability in mining industry tends to focus on managing short- and medium-term direct impacts. Above all if these are produced in the operational stage of the mine. That is: water management, energy efficiency, cleaner production, restoration of vegetation cover and actions philanthropic.[28] Nevertheless, despite advances in social and environmental management in mining, the public perception of this industry remains negative,[29] which produces an adverse effect on projects and therefore on the sector competitiveness.[30]

The competitiveness of mining industry requires, in addition to the full execution of existing projects, a continue emergence of new projects. Thus, social acceptance becomes a determining factor,[31] which implies demonstrating that mining contributes effectively to sustainable development.[32] In other words, exploiting minerals and metals in a clean and fair way is becoming an intangible value that can affect the viability of extractive projects. To achieve this, mining companies must see sustainability as an alternative to traditional business models.

According to Veiga et al. (2001), the benefits for a community thanks to a mine managed with the principles of sustainable development include new values for the zone of influence of the project, such as: economic diversification, direct employment, business development, drinking water, electricity, and infrastructure. As a result, thinking beyond traditional socio-environmental management implies that mining is an engine for the environmental and socioeconomic improvement of a region. In short, sustainability in mining implies going beyond legal compliance, approaching processes from a more holistic perspective.[33]

In this regard, Fuisz-Kehrbach (2015)[34] analyzed the sustainability actions carried out by a sample of 16 mining companies of various scales at the international level. It was found that 90% of these actions were due to conventional control and mitigation of their direct environmental impacts, added to philanthropic actions and traditional social management. Additionally, close to 70% of sustainability activities were developed only in the operational phase of the mines' life cycle.

Without mining projects there are no minerals or metals to sell. Therefore, the fact that mining companies can execute its projects is a crucial factor, not only for its competitiveness but also to supply society needs. In this sense, the socio-environmental aspects are proving to be variables with the capacity to stop an extractive project indefinitely.

---

[27] Senge et al. (2008) and Porter and Kramer (2011).

[28] Fuisz-Kehrbach (2015), pp. 101–115.

[29] Fonseca et al. (2014), pp. 70–83.

[30] ICMM (2012).

[31] Thomson and Boutilier (2011), pp. 1779–1796; Prno and Scott (2012), pp. 346–357.

[32] ICMM (2012); Maffini et al. (2014), pp. 84–93.

[33] Zuleta et al. (2013), pp. 79–94.

[34] Fuisz-Kehrbach (2015), pp. 101–115.

According to Porter (2002),[35] a strategy to increase the competitiveness of a company is differentiation. In this way, companies identify attributes in their products that potential customers can value. Companies then try to achieve those attributes. The company that manages to do this better than its competitors, then it will have managed to differentiate itself. This usually implies that the products exceed the minimum attributes that customers expect to obtain. The means for differentiation are specific to each sector, and could be product functionality, price, or distribution mode, among other aspects.

Additionally, Porter and Kramer (2011)[36] argue that well-being of communities and competitiveness of companies are strongly intertwined, due to successful firms need prosper communities able to offer crucial public assets as well as an environment that give support to businesses. Thus, considering that communities need successful companies that offer jobs and opportunities to create wealth, companies can generate economic value by creating social value from clusters that support business activity.

In summary, although the minimum that mining companies are expected to do is to meet the law, this has often not been enough to achieve social acceptance of the projects. Eventually, differentiating by going beyond what is mandatory, as well as carrying out voluntary actions that result in lasting-positive impacts in the area of influence of projects, can be a way towards social acceptance.

## 2.2 *Holistic Approaches of Sustainability in Mining*

If the industry wants to upgrade the effectiveness of its contribution to sustainable development and, in general, creatively face all the central issues of sustainability, articulation across sectors is crucial.[37] In this sense, a holistic vision of sustainability helps to identify business and innovation opportunities that underlie socioeconomic and environmental challenges.[38]

There are several reference frameworks aimed to facilitate the articulation of sustainable development dimensions with industry, including mining. The livelihood of people can be affected or improved by many factors. Therefore, tools to think about such complexity are needed. Hence, a more holistic perspective of sustainability could help to consider the aspects that constitute its environmental, social, and economic dimensions, as well as the interactions between them and with stakeholders in the short and long term.[39] This type of perspective should converge

[35] Porter (2002).

[36] Porter and Kramer (2011).

[37] Senge et al. (2008).

[38] Hart and Milstein (2003), pp. 56–69.

[39] Lozano (2015), pp. 32–44.

in specific sustainability activities with lasting benefits in the zones of influence of mining projects.

Through various initiatives, it has been sought to incorporate the SDGs into mining activity in such a way that it contributes to its fulfillment. Among the most representative initiatives is the Mapping Mining to the Sustainable Development Goal.[40] Here it is argued that the mining industry can mobilize considerable human, physical, technological, and financial resources to achieve the SDGs, since its projects are often located in fragile areas from the social and ecological point of view.

It is argued that by properly managing sustainable development through this industry, jobs can be generated, innovation stimulated, and investment and infrastructure achieved that lead to long-term structural changes. However, if sustainable development is poorly managed, mining can also lead to environmental degradation, population displacement, inequality, and increased conflict, among other problems. Specifically, this framework raises in relation to SDG 17 to focus on collaboration and leverage through actively dialog with Governments, civil society, and development partners, as well as joint with bottom-up grassroots movements and top-down leadership initiatives.

On the other hand, in 2010 the United Nations Development Program (UNDP) proposed a range of potential contributions to sustainable development by mining. This organization proposed a scheme of priorities in relation to the SDGs, where mining companies could promote the achievement of these objectives. Through its direct operations mining generates employment and economic growth in low-income countries, while, through articulations with governments and civil society it can make the benefits extend beyond the useful life of the mine. Hence, extractive industry can have a positive impact on the natural environment, climate change and social capital (PNUD (2016)).

UNDP framework distinguishes the mining impact degree on each target, ranging from indirect, moderately direct, to very direct. It also identifies the positive impacts of mining that can be enhanced as well as the negative ones that can be prevented or mitigated. In general, SDGs represent an opportunity for upgrading monitoring of the mining industry to promote inclusive socio-economic development, for which data disaggregation and sectoral reporting are crucial.[41]

ICMM proposes another approach by the articulation between the SDGs and its Ten Principles for the Mining Industry[42] (see Fig. 1). It is argued that mining, unlike other industries, has the potential to contribute to various objectives transversally. This is due to the positive impacts that companies and their operations can have on communities, ecosystems, and economies. This scheme provides a relationship

[40] WEF (2016).

[41] Cole and Broadhurst (2021), pp. 233–243.

[42] ICMM Mining Principles (ICMM 2020): (1) Ethical Business, (2) Decision-Making, (3) Human Rights, (4) Risk Management, (5) Health and safety, (6) Environmental Performance, (7) Conservation of Biodiversity, (8) Responsible Production, (9) Social Performance and (10) Stakeholder Engagement.

**Fig. 1** Mining contribution to SDGs (Source: ICMM • The Sustainable Development Goals Seen: 20/07/2021)

between the SDGs and these 10 principles. It can be noted that principle number nine: "Social Performance" is the only one that cuts across all the SDGs.

In addition to these general frameworks, there are some corporate standards that provide firms with highly applicable elements to incorporate sustainability aspects into industrial activities. These schemes provide indicators and monitoring tools that allow detailed monitoring of the goals set in terms of sustainability.

Reporting sustainability activities and results is an increasingly frequent activity in the industry. With this, the practice of measuring, disclosing and being responsible for organizational performance towards the goal of sustainable development before internal and external stakeholders can be defined. A generally accepted reporting model is the Global Reporting Initiative (GRI), which includes a special supplement for the mining and metals sector.[43]

[43] GRI (2010).

Finally, another important standard widely applied is the Towards Sustainable Mining™ (TSM), which is the Mining Association of Canada's award-winning commitment to responsible mining. This tool is described as a program established in 2004 which main objective is to enable mining companies to meet society's needs for minerals, metals, and energy products in the most socially, economically, and environmentally responsible way.

By these kinds of approaches is possible to achieve a wider vision of mining impact. This could drive sustainability management to more effective actions that not only minimize risks but give also major and lasting benefits for companies and stakeholders.

## 2.3 *Capabilities Strengthen*

Mining companies have gradually understood that sustainability is something more than percentage of profitability—after taxes—that has been invested in social and environmental issues, having more to do with the way in which that profitability has been obtained.[44] Additionally, since mining returns have been obtained from natural capital, part of it should be reinvested in the effective promotion of social and productive capabilities in the areas of influence of projects.[45] Above all, since the specific characteristics of the mining industry promote the development linkages and generate powerful spillovers in local economies.[46]

In that sense, the importance of capabilities accumulation and technological upgrading for economic growth in local communities has been widely acknowledged.[47]

Sustainability management should help strengthen community organizations and local institutions through programs that foster capabilities that promote empowerment of populations,[48] generating conditions that foster collaborative learning between stakeholders,[49] as well as capability building in local communities[50] through dynamics that last indefinitely after mine closure.

This is quite important for countries endowed with natural resources, where the recent technological developments open the way to new windows of opportunity, so that institutional and corporative efforts should be oriented to promote and strengthen local capabilities.[51]

44 Hamann (2003), pp. 237–254.

45 Almaguer (2007), pp. 1–8.

46 Anzolin and Pietrobelli (2021), pp. 395–399.

47 Chang and Andreoni (2020), pp. 324–351.

48 Thomson and Boutilier (2011), pp. 1779–1796; Hamann (2003), pp. 237–254.

49 Milanez and de Puppim (2013), pp. 427–434.

50 McLellan and Corder (2013), pp. 37–46.

51 Anzolin and Pietrobelli (2021), pp. 395–399.

Dawkins (2006)[52] states that an action is altruistic when it contributes to the well-being of someone at the expense of the person that carry out the action. Under this approach, corporate altruism occurs when companies provide a benefit to actors without their involvement or compensation. In this way, the activity is one way from the one who provides to the one who receives, and the actor is a passive subject receiving benefits. On the other hand, when processes engage communities and institutions as well as entail commitments from them, collective skills are fostered. Thus, populations strengthen capabilities aimed to enhance their livelihoods by acquiring greater access and more responsible management of resources available to them.[53]

## 2.4 *Innovations in Mining in the Context of Sustainability*

An activity is considered innovative when it involves the implementation of a new or significantly improved product or process.[54] On the other hand, if this is linked to the concept of sustainable development,[55] it could be said that an innovation is sustainable if its benefits tend to last indefinitely being available to future generations.

What justifies the effort involved in sustainability activities in mining activity is that its benefits last long after the project has ended.[56] Hence, innovation in mining can address reduction in environmental impact, increase production, and upgrade mineral recovery.[57] Overall, in the framework of sustainability companies must think beyond the projects and use their profitability to boost post-mining economies in the regions.[58]

In general, Endl et al. (2021)[59] found that all innovation concepts show repercussions on individual SDGs. Consequently, many innovation concepts have positive impacts on the environmental SDGs. For instance, R+D+I collaboration and multi-sector partnerships facilitate good innovation practices transfer and most positive impacts.

Sustainable innovation processes are needed that involve aspects such as the recycling of metals and minerals, technologies to enable extraction in deposits declared unprofitable,[60] integrate locally produced renewable energy supply on a

[52] Dawkins (2006).

[53] Adams (2008); Zimmerman (2000); Rappaport (1984), pp. 1–7.

[54] OECD (2018).

[55] UN (1987).

[56] Almaguer (2007), pp. 1–8.

[57] Kashan et al. (2022), p. 102575.

[58] Milanez and de Puppim (2013), pp. 427–434.

[59] Endl et al. (2021), p. 101440.

[60] Shen et al. (2015), pp. 15–26; Jenkins and Yakovleva (2006), pp. 271–284; ICMM (2012); Hamann (2003), pp. 237–254.

large scale,[61] among other aspects. In short, innovation is required to play a key role to effectively take advantage of the opportunities generated by rapid and high-income generation, especially in regions rich in mineral resources.[62]

However, innovation often has important obstacles in mining activity because changes use to affect each stage of mine life cycle as well as stakeholders' networks.[63] Thus, promoting innovations in companies that contribute to sustainable development demands certain types of capabilities, as well as significant changes in technology and management, in order to achieve cleaner and more productive organizational structures and innovative processes.[64] This is a complex long-term effort that implies overcoming organizational resistance.[65] If it is framed in sustainability remuneration can be important. As the promotion of these capacities has a significant impact on long-term corporate growth.[66]

## 3 Methodology

The theoretical convergences above described were used to define the four conceptual sieves of the tool. Despite their relevance, this work does not pretend that these convergences are the only ones in the literature, nor the most important ones related to sustainability management in mining. Its application is intended to serve as an example of how Conceptual Sieving works. On the other hand, the information for the pilot application was gathered from sustainability corporative reports. All of this was complemented by interviews, and on-site observations in the zones of influence of five mining projects.

In general, companies from all sectors periodically publish corporate sustainability reports. Although these types of reports contain valuable information, they are also subject to criticism. For example, it is said of its lack of veracity because what is reported is not always equivalent to what is executed.[67] It is also argued that these documents lack a holistic approach, since their indicators are usually analyzed in isolation and that their diffuse structure also allows unsustainable companies to camouflage themselves among those that are.[68]

On the other hand, Pucker (2021)[69] argues that despite such reporting has become widespread, for example environmental damage is still growing. This author state

[61] McLellan and Corder (2013), pp. 37–46.

[62] Kinnear and Ogden (2014), pp. 42–53.

[63] Gruenhagen and Parker (2020), p. 101540.

[64] Silvestre and Neto (2014), pp. 809–817.

[65] Börjesson et al. (2014), pp. 120–140.

[66] Yang (2012), pp. 34–46.

[67] Fuisz-Kehrbach (2015), pp. 101–115.

[68] Fonseca et al. (2014), pp. 70–83.

[69] Pucker (2021).

problems with sustainability reporting and even with sustainable investing is related to nonstandard metrics, insufficient auditing and unreliable ESG ratings.

However, at the same time it is recognized that these reports can serve to know the position of the companies regarding certain issues, as well as the priorities that they have defined in the framework of sustainability management.[70]

Conceptual Sieving is a tool for the assessment of sustainability activities reported by companies. This analysis has the purpose of infer the potential of these actions to achieve their objectives. This can also be useful to identify the sustainability vision of companies. In brief, in this work is proposed that analyzing the conceptual bases of the actions and strategies that make up the socioeconomic and environmental management of companies can help understand how they conceive of sustainability. Since the objective of this work is not to evaluate veracity, it is not verified if what is reported corresponds to what has been implemented. In a shorthand, the sort of actions that companies include in their reports is a reflex of their sustainability vision, regardless of whether these activities are implemented or not.

Although the source of information that feeds the Conceptual Sieving are the sustainability reports, in-depth interviews and onsite observations were also carried out to complement this assessment from a more qualitative approach. In the case of the pilot application carried out in this study, authorized visits were included in mining operations, as well as to neighboring populations. Dialogues with stakeholders at the community and institutional level were also held. These conversations involved conducting individual interviews with a small number of respondents to explore their views.

In this sense, places were visited where the sustainability activities of mining projects had been implemented for years. This allowed to observe results and listen to the actors involved. In summary, Conceptual Sieving is aimed to obtain a quantitative evaluation of the project's sustainability activities regarding to the potential of the durability of its positive impacts. On the other hand, the interviews and direct observations in the field serve to obtain indications that help to corroborate the results of the evaluation regarding to this inferred durability potential.

## 4 The Conceptual Sieving

Like the technique used for the classification of granular materials, in this proposal each sustainability activity reported by the company "go through" a series of sieves associated with certain concepts. Thus, all these actions are analyzed in the light of such concepts, one after another. This implies that each activity can "be retained" in any of the sieves or "pass through" them all (see Fig. 2). At the end, sets of activities are classified according to the sieves they have crossed and in which they have been retained. For the practical application presented in this paper, the concepts that

[70] Fuisz-Kehrbach (2015), pp. 101–115.

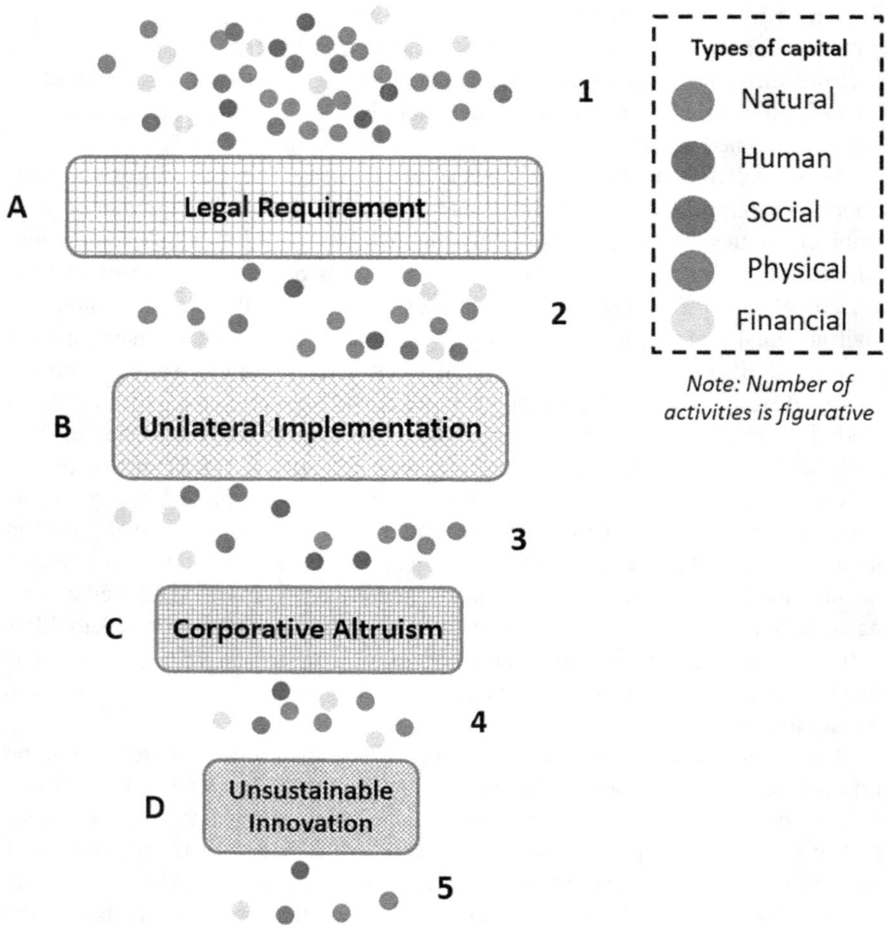


**Fig. 2** Conceptual Sieving components (Source: Self-Made)

determine the sieves are related to the sustainability of actions carried out by mining companies. As a result, an assessment of the potential of positive effects—benefits—of these actions to last after mining project closure is obtained.

The process begins with activities reported review in the latest corporate sustainability report published by the company. Given their nature, these actions are usually related to the socioeconomic and environmental dimensions of industrial activity. In order to facilitate the classification, before sieving as such, each activity is cataloged according to dimension of sustainability to which it mainly points.

For this, the Sustainable Livelihoods Framework (SLF) (DFID 1999)[71] was used, which is designed to rise the understanding of the factors that affect the livelihoods

[71] Department for international development (DFID) (1999).

**Fig. 3** Pentagon of Sustainable Livelihoods Capitals (Source: DFID 1999)

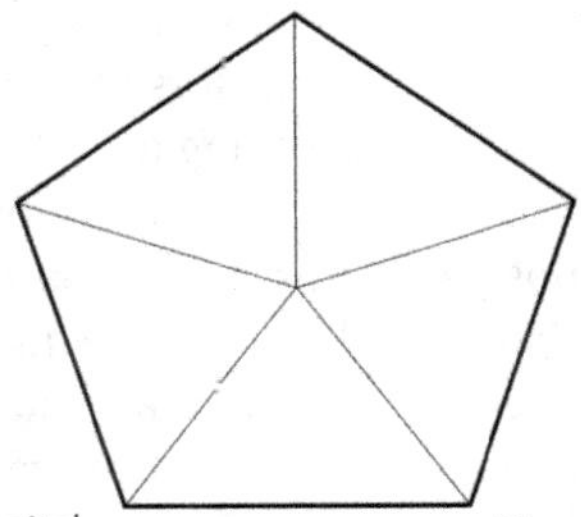


of people, particularly those who live below the line. of poverty. For years this has been useful to determine if the activities are more focused on contributing to the social, physical, economic, human, or natural capital of populations of influence of an industry or a productive project. In general, considering diversity of livelihoods can be a productive analytical tool when approaching aspects of social sustainability.[72]

A livelihood is sustainable when it has the capacity of endure and maintain or enhance its strengths indefinitely, without depleting the renewable natural resources that support it.[73] This approach seeks to gain an accurate and realistic understanding of communities' strengths and how they endeavor to convert these into positive livelihood outcomes. Consequently, SLF is primarily concerned with people.[74] The central element of this framework is the pentagon of capitals, which can be used to outline the level of access that people have to them (see Fig. 3).

The SLF framework identifies five core interconnected asset categories or types of capital upon which livelihoods are built. Accordingly, each category by its own is not sufficient to yield all the varied livelihood outcomes that people seek. Thus, human capital represents the skills, knowledge and good health that together enable people to labor. Social capital refers to mutual benefit networks and connectedness between individuals with shared interests, that increase people's trust, reciprocity, and ability to work together, reach more formalized agreements and expand their access to wider institutions, such as political or civic bodies. Natural capital is the term used for the resources and services provided by biotic and abiotic factors of nature, whereas physical capital comprises the basic infrastructure and producer goods needed to support livelihoods. Finally, financial capital denotes the economic resources that people use. Overall, capitals are better understood if they are thought more as building blocks of interconnected livelihoods, as each can induce the growth of the others. In that sense, the SLF can be used both to the planning of new sustainability activities or to assessing the results of implemented ones.

---

[72] Segerstedt and Abrahamsson (2019), pp. 610–619.

[73] Serrat (2017).

[74] Department for international development (DFID) (1999).

A primary idea of this type of framework is the sustainable development planning should be done participatorily with the communities of interest, based on a dialogue that addresses the factors that perpetuate poverty.[75]

Activities classified according to SLF (1 in Fig. 2) go to the first conceptual sieve (A in Fig. 2). According to first conceptual convergence posed in the theoretical framework of this paper, this stage seeks to identify sustainability activities reported by companies that exceed the basic socioeconomic and environmental performance. Therefore, their entail a differential factor that represents a value added to society. This first sieve is called "Legal Requirements". Accordingly, reported activities that coincide with law obligations of the country where the project is developed, as well as established licensing or prior consultation commitments, are retained here. Thus, mandatory activities—not differential ones—are retained, while those carried out voluntarily and that have an added value for society—differential ones—(2 in Fig. 2) continue to the second sieve of the series.

The following conceptual sieve (B in Fig. 2) is based on the second conceptual convergence of the theoretical framework of this writing. Here, the establishing alliances between the company and its stakeholders is highlighted. This, to give a more holistic approach to sustainability activities, as well as to enhance their relevance and social-institutional acceptance. In this sense, the second sieve called "Unilateral Implementation" retains the reported sustainability activities that do not show cooperation with one or more stakeholders. These can be national and local governments, the local population—including mining communities—and universities, among others.[76] To the extent that activities pass sieves, they accumulate categories. For instance, in point 3 of Fig. 2, the activities that have already passed-through first two sieves are called: sustainability activities with added value to society and articulated with stakeholders.

The third sieve in series (C in Fig. 2) concerns to capability building in communities and institutions related to company's activities. According to theoretical framework, this can be a way to improve appropriation and durability of benefits promoted by the industrial projects in their areas of influence. Therefore, activities where people appear as passive subjects receiving benefits are retained. This is because report does not show their contribution to the development of activity. This sieve allows the passing-through of activities that engage communities and institutions, promoting their capability development, while retaining what could be classified as corporate altruism. Thus, the actions that continue to the fourth sieve besides to have "added value to society" and are "articulated with stakeholders" also involve people and institutions in the development of the activity, promoting the development of capacities (4 in Fig. 2).

Finally, the fourth sieve (D in Fig. 2) aims to classify the activities according to their novel and enduring character according to the fourth conceptual convergence proposed in the theoretical framework. Accordingly, this sieve retains the activities reported by the companies that from this perspective could not be classified as sustainably innovative. Finally, the group of activities that pass-through all sieves are

[75] Heemskerk (2005), pp. 82, 87.

[76] Svobodova et al. (2020), p. 101724.

classified as: sustainability activities with added value to society, articulated with stakeholders, capability strengthener and sustainably innovative (5 in Fig. 2). As stated in the theoretical framework of this article, in the case of extractive industry, these characteristics are important so that mining projects can make an effective contribution to the sustainable development of their areas of influence.

## 5 Pilot Application: Cases Selection

In order to carry out a first application of the Conceptual Sieving, various mining projects in Colombia with high national relevance. It was ensured that these mines were in construction or operation phase, seeking that their sustainability management processes had a certain degree of maturity. To form a group with some of the most important mining projects of a country, which in turn they declare that sustainability management is a priority issue, can be a way to obtain a sample of more representative activities of sustainability management in mining that takes place in that country.

To select these projects, the following criteria were considered: (1) high national participation in royalties' generation, (2) these are considered a priority by national Government, (3) their information about sustainability is publicly accessible, (4) they arouse interest in the national public opinion and (5) it was possible to establish an agreement with the company to take part in this study. The latter aspect to enable complementary field work.

Regarding the first criterion and according to the National Mining Agency of Colombia (ANM) (2020), among the different minerals and metals exploited in the country, coal has the largest share in the generation of royalties with 83% of the total. In second place, gold generates 8% and then nickel 7.6%. In turn, although with a lower level of generation, but with a high social and economic relevance in some regions of the country, there are emeralds, limestone, ceramic clay, copper, and iron. Therefore, mining companies that operate on a large scale in these eight sectors tend to be the ones that generate the highest royalties in Colombia. To identify which large-scale projects in these sectors are considered a priority by the Colombian Government, those that were officially classified as Projects of National and Strategic Interest (PINE)[77] or as Projects of Regional and Strategic Interest (PIRE)[78] were verified. This means that the Government considers that they have a high impact on economic and social development, as well as the power to promote the objectives of the National Development Plan.

As a third criterion, it was verified which projects had a website with public access to information about its sustainability management, including periodic reports. Subsequently, the interest that these projects aroused in national public was measured. For this, through the Google Trend™ tool, the level of search that the

[77] EITI Colombia (2016).

[78] MINENERGY (2017).

**Table 1** Selection of mining projects for the pilot application (Source: Self-Made)

| Sector \ Criterion | Number of projects that meet each criterion | | | | Number of projects that meet at least 3 criteria | Number of projects participating in the study |
|---|---|---|---|---|---|---|
| | High royalties (Total number of projects) | Goberment priority (PINE / PIRE) | Public access to sustainability information | High public interest | | |
| Precious metals | 4 | 2 | 3 | 4 | 3 | 3 |
| Non precious metals | 3 | 2 | 1 | 2 | 1 | 1 |
| Coal | 3 | 3 | 2 | 2 | 2 | 0 |
| Limestone | 2 | 0 | 2 | 1 | 1 | 0 |
| Ceramic clays | 1 | 0 | 1 | 1 | 1 | 1 |
| Total | 13 | 7 | 9 | 10 | 8 | 5 |

name of project presented was checked. Projects that registered zero searches were considered not to arouse interest.

Finally, it was verified how many of projects analyzed met at least three of four previous criteria. Hence, a definitive list of eight projects was formed. Next, these firms were contacted to present them this study and to know their interest in participating on it. At the end, five mining projects participated (see Table 1).

# 6 Pilot Application: Findings and Analysis

Based on the proposed methodology, corporate sustainability reports of the same year were taken from the five mining projects participating in the study. All reported activities were identified and totaled 595. Subsequently, based on SLF model criteria,[79] the focus of these actions was determined. It was found that sustainability activities reported by these mining projects for the year of sample focused their management on human and natural capital with 39% and 31% respectively. Consequently, the reinforcing of individual skills at academic and professional level of populations of interest, as well as the impacts management on renewable natural resources determined the sustainability management in these companies for reported year (see Fig. 4).

On the other hand, activities oriented towards physical capital accounted for 14% of the total. This reflects a certain interest of the companies in the generation of infrastructure, while social capital (13%) and financial capital (3%) presented the lowest values. The lower volume of activities aimed at promoting social capital in the areas of influence of projects entails a smaller number of activities focused on strengthening the capabilities of collaboration, partnership, and reciprocal trust in the communities of interest. Finally, the low value of financial capital could be influenced by the scarce presence of actions such as donations, among others related to corporate altruism.

Thus, for the year of the sample, analyzed projects oriented their sustainability management mainly to prevention, compensation, and mitigation of the impacts of their operations on water, soil, air, and biodiversity. But, on a slightly higher scale,

[79] Department for international development (DFID) (1999).

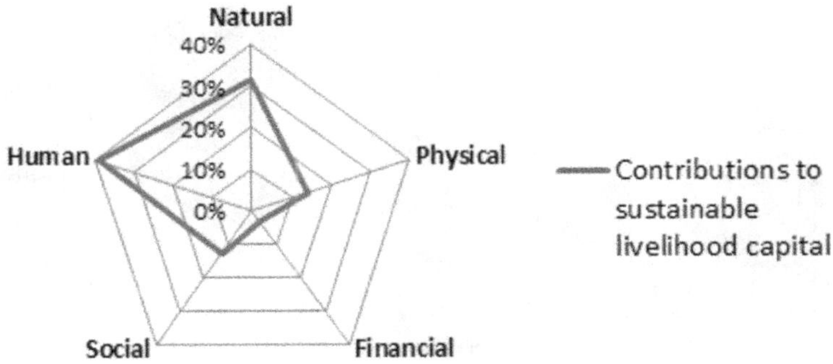


**Fig. 4** Focus of sustainability activities reported by five mining projects analyzed (Source: Self-Made)

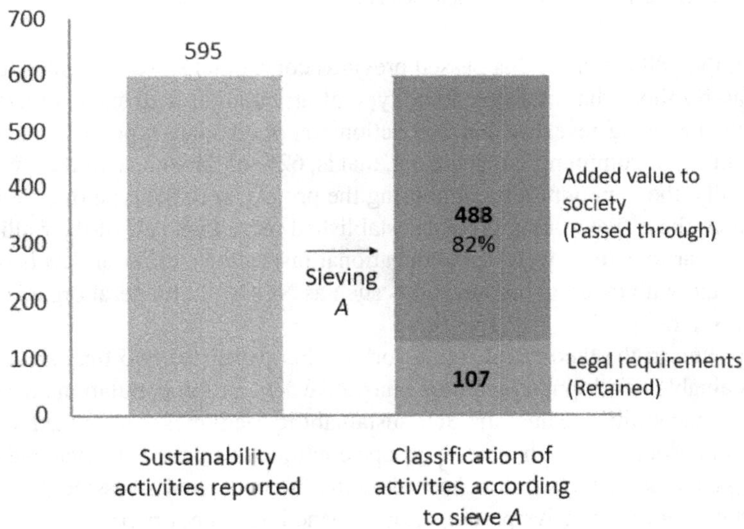


**Fig. 5** Sieve A results—legal requirements (Source: Self-Made)

they helped strengthen the job skills of the population in their area of influence. Nevertheless, the generation of basic infrastructure and capabilities promotion of communities for cooperation did not show similar levels. In other words, although these companies contributed to the development of people's skills to work, this apparently was not balanced with actions for them to learn to work collectively.

After this initial analysis, the 595 sustainability activities reported were sent to sieve A (Legal Requirements) in order to select which ones were mandatory for the companies. In this sense, it was verified that 107 of these activities coincided with social and environmental requirements of Colombian law. Therefore, 488 activities (82%) conceptually passed-thorough this sieve as they represented voluntary actions (see Fig. 5).

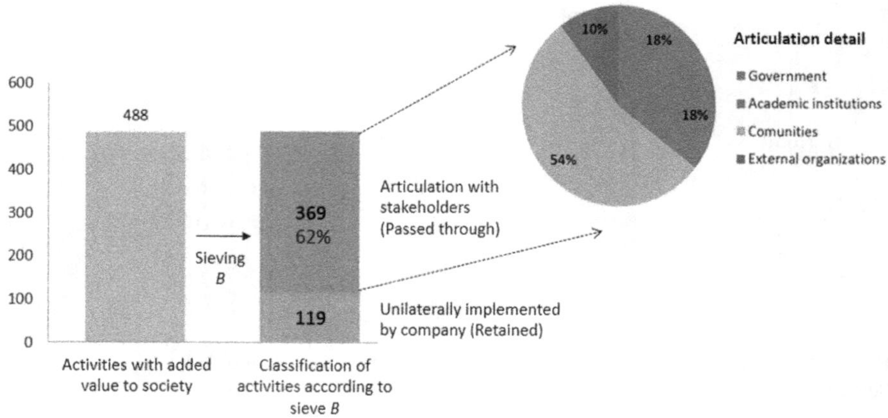


**Fig. 6** Sieve B results—unilaterally implementation (Source: Self-Made)

Next, the 488 activities that passed previous conceptual sieve continued to sieve B to classify those that included some type of articulation with stakeholders. This Conceptual Sieving revealed that 369 actions reported some type of alliance with organizations or communities of interest, that is, 62% of the total activities reported. Specifically, the communities neighboring the project were the type of stakeholder with which these five mining projects established more than half of their alliances. The rest of articulations were with educational institutions, different levels of government and with external organizations such as NGOs, multilateral organizations, and banks, among others (see Fig. 6).

According to the theoretical framework of this paper, the two final sieves state that sustainable development requires innovative actions and populations with solid collective capabilities. Thus, the 369 sustainability activities with added value to society and articulated with interest groups continued to sieve C to evaluate which ones engaged their beneficiaries. According to this conceptual assessment, 41 activities did not report an active participation of beneficiaries nor partner organizations in the action execution. As a result, they were retained in sieve C. This left as a result that 55% of the total activities reported (328) continued to the last sieving (see Fig. 7).

Finally, these 328 activities reached sieve D where their sustainably innovative character was analyzed. As a result, it was obtained that 61 activities could be classified as such (see Fig. 7).

Consequently, from the perspective of this tool 10% from 595 sustainability activities reported could be comprehensively classified as: contributors of value to society, articulated with interest groups, capability strengthener and sustainably innovative. The results of the Conceptual Sieving are summarized in Table 2 and some activities analyzed are specified.

In Table 2 is presented some of 595 sustainability activities reported by five mining companies participating in this study. It is presented how some activities meet criteria and pass certain sieves. Although then they can be filtered later due to the

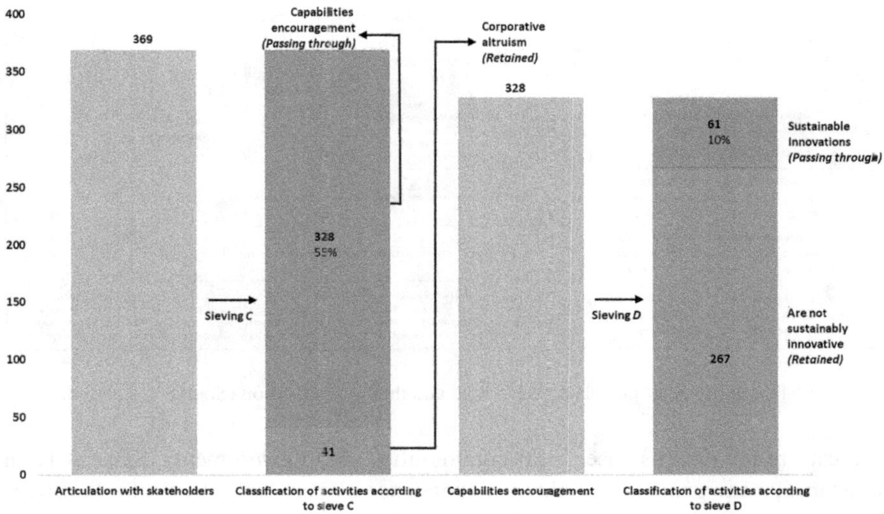


**Fig. 7** Sieves C & D results—corporative altruism & activities that are not sustainably innovative (Source: Self-Made)

**Table 2** Summary of sieving results (Source: Self-Made)

| *Initial amount of activities: 595* | | Number of retained activities | Examples of retained activities | Number of activities that pass through (Meet the criterion) | % | Passing through criterion | Examples of activities that pass through the sieve |
|---|---|---|---|---|---|---|---|
| Sieves | Retention factor | | | | | | |
| *A* | Coincidence with a legal requirement | 107 | - Sewage treatment<br>- Rehabilitation of tailings areas<br>- Royalties payment | 488 | 82% | Added value to society | - Support for local ventures<br>- Clean energy cooperation with environmental authority<br>- Forest conservation |
| *B* | Unilaterally implemented by the company | 119 | - Forest conservation<br>- Road safety education carried out by the company | 369 | 62% | Articulation with stakeholders | - Support for local ventures<br>- Clean energy cooperation with environmental authority<br>- Gas consumption subsidies |
| *C* | Corporative Altruism | 41 | - Gas consumption subsides<br>- Emergency Fire Brigade loan<br>- Laboratory equipment donation | 328 | 55% | Capabilities encouragement | - Support for local ventures<br>- Clean energy cooperation with environmental authority<br>- Accompaniment to local suppliers |
| *D* | Activity is not sustainably innovative | 267 | - Accompaniment to local suppliers<br>- Hiring local labor<br>- Academic scholarships | 61 | 10% | Sustainable innovation | - Support for local ventures<br>- Clean energy cooperation with environmental authority<br>- Training to local Government in projects presentation to national Government |

meeting of some retention factor. For example, "forest conservation" in most cases was an activity carried out by mining projects on their own, without the participation of stakeholders. Although it can be characterized as added value to society (it passes-through sieve A), it is in turn an activity disjointed from its socioeconomic context. Accordingly, it is retained in sieve B. This disarticulation can affect its long-term durability after the mining project closure.

Similarly, the "emergency brigade loan" passes-through sieves A and B. Therefore, acquires the character of value added to society and articulated with

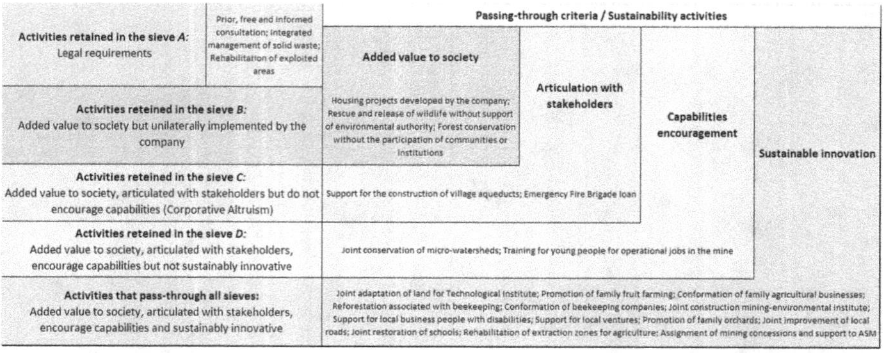


**Fig. 8** Sustainability activities observed in field and their classification (Source: Self-Made)

interest groups (in this case, local communities, and Government). Since it is an assistance, it can generate population dependence on company support for the attention of this type of events. Thus, it does not build capabilities and is retained in sieve C characterizing it as corporate altruism.

On the other hand, actions such as "support for local suppliers of mine" and "hiring local labor" pass-through first three sieves. Both are usually innovative processes. Nonetheless, in this case they could not to be sustainable as their benefits depend on the presence of company in the territory and mining is a temporary activity. For this reason, they are retained by sieve D, being characterized as: added value to society, articulated with interest groups, capability strengthener, but not sustainably innovative.

Finally, 61 from 595 sustainability activities reported by these five mining projects for the year of the sample pass-through all the sieves. In other words, 10% of the total activities acquired the four established sustainability criteria. From the point of view of this proposal this means that more comprehensive activities have more potential for their positive effects to last in the territory after mining project closure.

## 7 On Site Observations and Interviews Developing

Activities were randomly chosen among those that passed-through the four sieves and that it was also possible to observe aspects of their implementation in the field. Likewise, with another group of activities that did not pass all the sieves, the same exercise was carried out. In total, thirteen activities that passed-through all sieves were observed in the field, as well as ten that were retained. In Fig. 8 are listed these 23 activities characterized according to the sieves they passed-through and which they were retained. For example, "Support for the construction of village aqueducts" passed-through sieves A and B. Thus, it is classified as added value to society and articulated with stakeholders. However, in this case the activity did not include

**Fig. 9** Forest conservation on company land without stakeholder involvement Conformation of beekeeping companies (Source: Own file)

processes of community appropriation of given infrastructure, nor operation-maintenance training. Therefore, it is very possible that benefits of this facilities will not be maintained after the closure of the mining project. Hence, this activity is retained in sieve D and is classified as an activity that is not sustainably innovative. Figure 9(a) shows an image of one of the activities retained in sieve B while Fig. 9(b) shows an activity that pass-through all sieves.

Likewise, actors involved in these activities were interviewed to know their opinion and participation in their development. These interviews were open conversations focused on exploring two topics:

Type of articulation and participation with stakeholders for the development of sustainability activity.

Community capabilities promoted through the activity so that its positive impacts are maintained in the long term after the mining project.

In total, 57 people were interviewed, including members of institutions and communities of interest, as well as external participants involved in the development of activities. People interviewed had some interest in reported sustainability activities and therefore were part of one of following project stakeholders:

- Local, regional, and national Governments
- Communities of interest
- Local educational institutions
- Universities
- Multilateral banking

- ASM in the area of influence of project
- NGOs

Observations and inquiries focused on verifying the degree of practical implementation of conceptual aspects related to the sustainability of the activities. In this sense, concerning to the ten activities that did not pass all the sieves, it was verified the low level of knowledge and appropriation of communities and local governments with regard these actions, which were led mainly by the mining companies. Thus, it is likely that the dynamics promoted, and benefits generated by these activities last only as long as mining company is present in the territory. Hence, its benefits will not be available to future generations.

On the other hand, regarding actions that met the four sieving criteria, a greater degree of knowledge and appropriation of activities by beneficiaries was noted. To such an extent that in some of them company only carried out monitoring activities.

From a more anecdotal but noteworthy perspective, some people stated that scholarship-type programs can be contradictory. With this, the most qualified young people go to the cities in search of opportunities and often do not return. These people suggest that scholarship programs should be aligned with entrepreneurship and infrastructure generation programs to encourage these "formed brains" to remain in the area and help in its development.

Finally, people from local communities were asked about who they consider more important: local government or mining company. All people consulted stated that company is more important than municipality. This statement indicates the leading role that mining companies have in the areas of influence of their projects, which may have adverse effects on the already marked institutional weakness of these small towns.

## 8 Limitations and Further Research

Given that the sample of sustainability activities comes from specific year, this does not catalog the vision or management of companies in the medium or long term.

The year of sustainability activities sample is not specified. That is because the objective of this work was mainly to have material for the application and exemplification of the tool.

Due to confidentiality, the names of the companies that participated in the pilot application are not mentioned.

The theoretical consensuses presented are not the only ones, nor the most recurrent in the literature. A new application of this tool, based on a review that identifies the main conceptual agreements on sustainability in mining industry, would be useful for the sector. This could yield a broader analysis of the sample.

Conceptual Sieving can also be useful to assess the pertinence of company's sustainability actions with the vision and goals that it has set itself. Thus, it is possible to quantify and identify that both the current management and implementation are on the path of what companies seek to achieve.

These activities assessment have a conceptual character. Therefore, there could be subjectivity in the analyzes when these are carried out by an external agent to the evaluated organization. The sieving products will be valuable as long as the implementation is carried out on the basis of agreed criteria and aligned with the vision and goals of organization.

New applications of this tool in other theoretical contexts, different than sustainability, as well as in practices other than mining, could be a matter of inquiry.

## 9 Conclusion

Through the theoretical review on which this work was based, recurring concepts were identified in indexed literature regarding to sustainability in mining. These are considered relevant so that the positive impacts of mining projects last in the territory in the long term after the mine closure. In this sense, the identified concepts refer to attributes that sustainability activities should have, which were: added value to society, articulation with stakeholders, strengthening of local capabilities and sustainable innovation. Through the pilot application, the pertinence of sustainability activities carried out by five mining projects in Colombia was analyzed from the perspective of these concepts. The outcomes of sieving showed that for a specific year of management, 82% of the actions implemented by these projects added value to society, 62% were articulated with stakeholders, 55% strengthened local capacities and 10% were sustainably innovative.

**Acknowledgments** This work would not have been possible without the support of the Ministry of Science, Technology and Innovation of Colombia and the School of Mines of the Universidad Nacional de Colombia—Medellin Campus and their joint Graduate Program. Likewise, the practical application was possible thanks to the openness and investigative will of mining companies from the gold and ceramic clay sectors in the department of Antioquia as well as the metal sector in the department of Córdoba in Colombia. We would also want to thank academic colleagues for their valuable criticism and feedback: Carlos Adrian Saldarriaga and Antonio Romero (RIP) (Universidad Nacional de Colombia), Marcello Veiga (University of British Columbia), Juan Lucena, Jessica Smith, and Nicole Smith (Colorado School of Mines), Jürgen Kretschmann and Peter Goerke-Mallet (Technische Hochschule Georg Agricola), Peter Knights and Deanna Kemp (University of Queensland—UQ) as well as Anthony Hodge (former president of ICMM). Finally, we would want to thank the Australian Government Endeavor Program for its financial support for a research internship at the Sustainable Minerals Institute—UQ as well as its students and researchers for their valuable comments and contributions to the development of this effort.

## References

Adams R (2008) Empowerment, participation and social work, 4th edn

Agencia Nacional de Minería (ANM) (2020) Transferencias de regalías y compensación por mineral. https://www.anm.gov.co/?q=regalias-contraprestaciones-economicas&field_tipo_de_regal_a_y_o_contra_value=TransferenciasSistemaGeneralRegalias. Accessed 22 May 2023

Almaguer RC (2007) Desarrollo local sostenible y comunidad minera: un enfoque sociocultural. Revista Futuros 17(V):1–8

Anzolin G, Pietrobelli C (2021) Local content policies: why mining need consistent policy packages to support capabilities development. Extr Ind Soc 8:395–399

Azapagic A (2004) Developing a framework for sustainable development indicators for the mining and minerals industry. J Clean Prod 12:639–662

Betancur CB, Loaiza UJC, Denich M, Borgemeister C (2018) Gold mining as a potential driver of development in Colombia: challenges and opportunities. J Clean Prod 199:538–553

Börjesson S, Elmquist M, Hooge H (2014) The challenges of innovation capability building: learning from longitudinal studies of innovation efforts at Renault and Volvo Cars. J Eng Technol Manag 31:120–140

Chang HJ, Andreoni A (2020) Industrial policy in the 21st century. Dev Change 51:324–351

Cole MJ, Broadhurst JL (2021) Measuring the sustainable development goals (SDGs) in mining host communities: a South African case study. Extr Ind Soc 8:233–243

Dawkins R (2006) The selfish gene, 3rd edn. Oxford University Press

Department for international development (DFID) (1999) Sustainable livelihoods guidance sheets

Di Boscio N (2010) Mining enterprises and regional economic development: an exploratory analysis of sustainable development model. Ph.D Thesis, The London School of economics and Political Science, London, UK

EITI Colombia (2016) PINES. https://www.eiticolombia.gov.co/es/informes-eiti/informe-2016/marco-institucional/pines/. Accessed 22 May 2023

Elkington J (1997) Cannibals with forks: the triple bottom line of 21st century business. Capstone Publishing Limited - Oxford Centre for Innovation, Oxford, United Kingdom

Endl A, Tost M, Hitch M, Moser P, Feiel S (2021) Europe's mining innovation trends and their contribution to the sustainable development goals: blind spots and strong points. Resour Policy 74:101440

Fitzpatrick P, Fonseca A, McAllister ML (2011) From the Whitehorse mining initiative toward sustainable mining: lessons learned. J Clean Prod 19:376–384

Fonseca A, McAllister ML, Fitzpatrick P (2014) Sustainability reporting among mining corporations: a constructive critique of the GRI approach. J Clean Prod 84:70–83

Fuisz-Kehrbach SK (2015) A three-dimensional framework to explore corporate sustainability activities in the mining industry: current status and challenges ahead. Resour Policy 46:101–115

Giurco D, Cooper C (2012) Mining and sustainability: asking the right questions. Miner Eng 29:3–12

Global Reporting Initiative (GRI) (2010) Sustainability reporting guidelines & mining and metals sector supplement. GRI, Amsterdam, The Netherlands

Gruenhagen JH, Parker R (2020) Factors driving or impeding the diffusion and adoption of innovation in mining: a systematic review of the literature. Resour Policy 65:101540

Hahn T, Pinkse J, Preuss L, Figge F (2015) Tensions in corporate sustainability: towards an integrative framework. J Bus Ethics 127:297–316

Hamann R (2003) Mining companies' role in sustainable development: the 'why' and 'how' of corporate social responsibility from a business perspective. Dev South Afr 20(2):237–254

Hart SL, Milstein MB (2003) Creating sustainable value. Acad Manag Exec 17(2):56–69

Heemskerk M (2005) Collecting data in artisanal and small-scale mining communities: measuring progress towards more sustainable livelihoods. Nat Resour Forum 29:82, 87

ICMM (2012) In brief: mining's contribution to sustainable development – an overview. ICMM, London, UK

ICMM (2020) Mining principles: performance expectations. ICMM, London, UK

Intergovernmental Forum on Mining, Minerals, Metals and Sustainable Development (IGF) (2017) Global trends in Artisanal and Small-Scale Mining (ASM): a review of key numbers and issues. IISD, Winnipeg

International Institute for Environmental Development (IIED) (2002) Breaking new ground. IIED, London, UK

Jenkins H, Yakovleva N (2006) Corporate social responsibility in the mining industry: exploring trendesarrollo sostenible in social and environmental disclosure. J Clean Prod 14:271–284

Jiskani IM, Moreno-C BM, Rehman AU, Fernandez CJM, Uddin S (2022) Implications to secure mineral supply for clean energy technologies for developing countries: a fuzzy based risk analysis for mining projects. J Clean Prod 358:132055

Kashan AJ, Lay J, Wiewiora A, Bradley L (2022) The innovation process in mining: integrating insights from innovation and change management. Resour Policy 76:102575

Kinnear S, Ogden I (2014) Planning the innovation agenda for sustainable development in resource regions: a central Queensland case study. Resour Policy 39:42–53

Lozano R (2015) A holistic perspective on corporate sustainability drivers. Corp Soc Responsib Environ Manag 22:32–44

Maffini GC, Marques KJ, Kruglianskas I, da Barbieri RLA, Schoproni BR (2014) Management for sustainability in companies of the mining sector: an analysis of the main factors related with the business performance. J Clean Prod 84:84–93

McLellan BC, Corder GD (2013) Risk reduction through early assessment and integration of sustainability in design in the minerals industry. J Clean Prod 53:37–46

Milanez B, de Puppim OJA (2013) Innovation for sustainable development in artisanal mining: advances in a cluster of opal mining in Brazil. Resour Policy 38:427–434

MINENERGÍA (2017) Proyectos de interés estratégico PIREs. https://www.minenergia.gov.co/proyectos-de-interes-regional-estrategico-pire/. Accessed 22 May 2023

OECD/Eurostat (2018) Oslo Manual 2018: guidelines for collecting, reporting and using data on innovation, 4th edition, The Measurement of Scientific, Technological and Innovation Activities, OECD Publishing, Paris/Eurostat, Luxembourg. https://doi.org/10.1787/9789264304604-en. Accessed 22 May 2023

Olivier JGJ, Peters JAHW (2020) Trends in global CO2 and total greenhouse gas emissions: 2020 Report, (PBL publication number: 4331). PBL Netherlands Environmental Assessment Agency, The Hague

PNUD (2016) How can mining contribute to the sustainable development goals? https://www.un.org/africarenewal/news/how-can-mining-contribute-sustainable-development-goals. Accessed 22 May 2023

Porter M (2002) Competitive advantage: creating and sustaining superior performance. The Free Press, NY

Porter ME, Kramer MR (2011) Creating shared value. Harvard Business Review, January–February

Prno J, Scott SD (2012) Exploring de origins of 'social license to operate' in the mining sector: perspectives from governance and sustainability theories. Resour Policy 37:346–357

Pucker KP (2021) Overselling sustainability reporting. Harvard Business Review, Sustainable Business Practices, May–June. https://hbr.org/2021/05/overselling-sustainability-reporting. Accessed 22 May 2023

Rappaport J (1984) Studies in empowerment: introduction to the issue. Prev Hum Serv 3(2-3):1–7

Rudas LG (2012) Minería e hidrocarburos en Colombia: ¿Sostenibles par a las finanzas públicas y el desarrollo? En: Minería, territorio y conflicto en Colombia, Universidad Nacional de Colombia

Rudas G, Espitia J (2013) Minería en Colombia, Institucionalidad y territorio, paradojas y conflictos, Contraloría General de la Republica. Contraloría General de la República, Bogotá

Segerstedt E, Abrahamsson L (2019) Diversity of livelihoods and social sustainability in established mining communities. Extr Ind Soc 6:610–619

Senge P, Laur J, Scheley S, Smith B, Kruschwitz N (2008) The necessary revolution: how individuals and organizations are working together to create a sustainable world. Brodway Books, New York

Serrat O (2017) The sustainable livelihoods approach. In: Knowledge solutions. Springer, Singapore. https://doi.org/10.1007/978-981-10-0983-9_5. Accessed 22 May 2023

Shen L, Muduli K, Barve A (2015) Developing a sustainable development framework in the context of mining industries: AHP approach. Resour Policy 46:15–26

Silvestre BS, Neto RS (2014) Are cleaner production innovations the solution for small mining operations in poor regions? The case of Padua in Brazil. J Clean Prod 84(1):809–817
Slack K (2012) Mission impossible? Adopting a CSR-based business model for extractive industries in developing countries. Resour Policy 37:179–184
Svobodova K, Vojar J, Yellishetty M, Molnarova KJ (2020) A multi-component approach to conceptualizing the reputation of the mining industry from a stakeholder perspective. Resour Policy 68:101724
Thomson I, Boutilier R (2011) The social license to operate. In: Darling P (ed) SME mining engineering handbook, Ch. 17.2. Society of Mining Metallurgy and Exploration, Littleton, Colorado, pp 1779–1796
Towards Sustainable Mining (TSM). https://mining.ca//wp-content/uploads/dlm_uploads/2021/07/TSM-fact-sheet-English-2021.pdf. Accessed 22 May 2023
United Nations (1987) Report of the World Commission on Environment and Development: our common future. United Nations Publications, Switzerland
United Nations (2015) Transforming our world: the 2030 agenda for sustainable development. https://sustainabledevelopment.un.org/content/documents/21252030%20Agenda%20for%20Sustainable%20Development%20web.pdf. Accessed 22 May 2023
Vargas G (2019) Social mobilisation in Colombia's extractive industries, 2000–2015. Extr Ind Soc 6:873–880
Veiga M, Scoble M, McAllister ML (2001) Mining with communities. Nat Resour Forum 25:191–202
Watari T, Nansai K, Nakajima K, Giurko D (2021) Sustainable energy transitions require enhanced resource governance. J Clean Prod 312:127698
World Economic Forum (WEF) (2016) Mapping mining to the sustainable empowered lives. Resilient nations. Development Goals: An Atlas
Yang J (2012) Innovation capability and corporate growth: an empirical investigation in China. J Eng Technol Manag 29:34–46
Young J, Septoff A (2002) Digging for change: Towardesarrollo sostenible a responsible minerals future, an NGO and community perspective. The Mineral Policy Center. http://bankwatch.org/documents/eir_mineralsfuture.pdf. Accessed 22 May 2023
Zimmerman M (2000) Empowerment theory: psychological, organizational, and community levels of analysis. In: Handbook of community psychology, vol 1. New York
Zuleta TA, Asencio GJ, Leyva CD, Montero PJM (2013) Sustentabilidad empresarial de proyectos mineros: el análisis multicriterio como perspectiva acertada para su evaluación. Minería y Geología 29(4):79–94

# The Mongolian Raw Materials Industry

**Thomas Hollenberg and Bayanmunkh Myagmarsuren**

## Contents



**Abstract** Mongolia, which is only very sparsely populated with a total of around 3.4 million inhabitants, is one of the ten most resource-rich countries in the world. In the southern part of the country, the Gobi Desert, there are gigantic deposits of copper, silver, gold, coal, fluorspar, and other raw materials. Coal, copper, and gold are among the most important deposits and account for about 89% of Mongolia's export economy and over 20% of the country's gross domestic product. Accordingly, mining is now the most important industrial sector in the country from an economic point of view.

T. Hollenberg (✉) · B. Myagmarsuren
German-Mongolian Institute for Resources and Technology GMIT, Raw Materials and Process Engineering, Ulaanbaatar, Mongolia
e-mail: th@dr-hollenberg.com; bayanmunkh@gmit.edu.mn


W. Frenz, A. Preuße (eds.), *Yearbook of Sustainable Smart Mining and Energy 2022*, Yearbook of Sustainable Smart Mining and Energy - Technical, Economic and Legal Framework 2, https://doi.org/10.1007/978-3-031-41873-0_11

But Mongolia also has molybdenum, tungsten, zinc, uranium, and rare earths elements, in the extraction of which China has so far had a monopoly. Without a secure supply of raw materials, especially in the field of rare earths, Germany threatens to lose competitiveness in important future technologies such as electromobility, digitization and the energy transition.

With appropriate foresight, the government of the FRG had already passed its first raw materials strategy in 2010. Following this, Chancellor Dr. Angela Merkel signed a raw materials agreement during the first state visit to Mongolia by a German head of government on 13th October 2011. This was intended to secure access to rare earths elements for German industry. At the time this raw materials agreement was signed, the exploitation of the raw materials had only just begun and international mining groups were competing for the mining licenses and for the extraction of the raw materials.

The present report provides an overview of mining activities in Mongolia ten years after the conclusion of the German-Mongolian Raw Materials Agreement.

## 1 History

Mongolia is one of the ten most resource-rich countries in the world. Coal, copper, and gold are among the most important deposits and account for about 89% of Mongolia's export economy and over 20% of the country's gross domestic product. Accordingly, mining is now the most important industrial sector in the country from an economic point of view.

Since 1922, when the then people's government decided to take over the coal mining in Nalaikh, near the capital Ulaanbaatar, mining activities have been progressing steadily. However, systematic, and scientific geological prospecting and exploration of Mongolia only started in 1939.

The beginnings of modern mining go back to the 1940s and 1950s. In 1943–1948, the Yugzur, Tumentsogt and Burentsogt tungsten mines started their production and fluorspar mining also started. In 1950, the Zuunbayan oil field was developed, but due to limited reserves, it was already exhausted in the mid-1960s (Fig. 1).

The important phase of mining development began in the 1970s and 1980s as the Modot Placer gold mine, the Khajuu-Ulaan, Khar-airag, Urgen, Chuluut tsagaan del, Borundur, Berkh fluorspar mines, the Erdenet copper-molybdenum mine and the brown coal mines -Mine Baganuur started operations. Here, the copper and fluorspar mines were developed on the basis of joint venture agreements with the former Soviet Union.

After the collapse of the Soviet Union, Mongolia made a peaceful transition to a democratic parliamentary system of government in 1990. Without significant financial and technical assistance from the Soviet Union, Mongolia needed fresh foreign investment to develop the economically vital mining sector.

At the beginning of the 1990s, geological exploration on oil and other raw materials as well as rare earth elements was started again in joint ventures with companies from western industrialized countries and Russia. Today more than 150

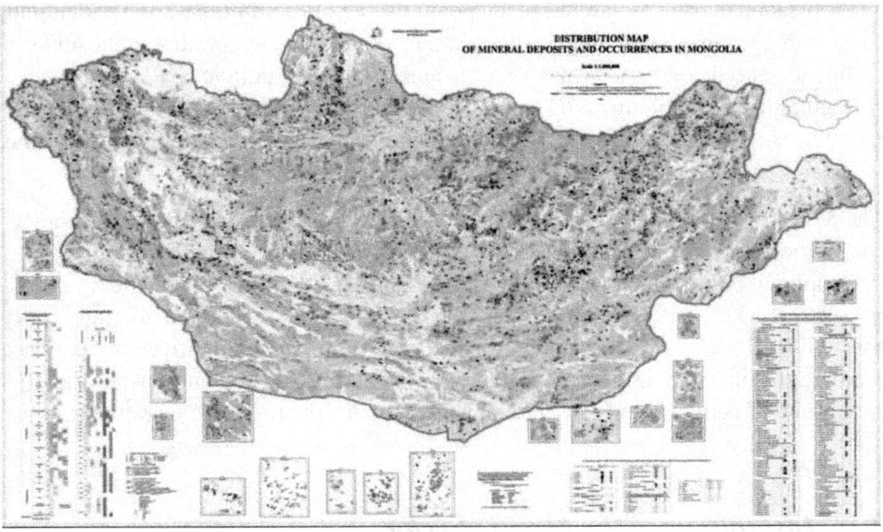


**Fig. 1** Mineral deposits in Mongolia

companies mine coal, copper, molybdenum, gold (primary and secondary deposits), tin (primary and secondary deposits), silver, tungsten, fluorspar, iron, and uranium.

## 2 Mining Policy

On January 16, 2014, the Parliament of Mongolia passed the Law Approving the "State Policy on Natural Resources 2014–2025" by Resolution No.18. The law aims to "secure primary national interests based on the development of excellent and responsible mining operations through a reliable private sector". The focus is on "developing a multisectoral and balanced economic structure in the short and medium term".

The main principles of the policy are "to provide the population with sufficient social and economic benefits from the mining industry while taking into account the development of the resource sector and Mongolian national development policy based on the Millennium Development Goals".

In this regard, the Ministry of Mines and Heavy Industry intends to implement the following strategic objectives, taking into account policy guidance, compliance with policies, medium and long-term strategies, and ensuring program and project developments:

- Ensuring political analyzes taking geological laws into account
- Mining Laws and Policies
- Fuel and oil laws and regulations
- Laws and guidelines for heavy industry

- Assistance in providing managerial positions in public administration and human resources management, legal advice and in expanding cooperation in the areas of finance and development of economic and investment policies;
- regulating the implementation of laws, policies, programs and projects;
- Internal audit assessments of legal, financial, program, project and investment as a basis for risk assessment.

The development of the resource sector up to 2050 was outlined as part of the objective of the Ministry of Mines and Heavy Industry (MMHI) to establish an integrated development policy planning system for the geological, mining, oil and heavy industry sectors. The basis of this policy is, through the implementation of strategically and politically relevant measures on deposits, to improve competitiveness, increase productivity in mineral extraction, increase labor productivity and ensure the efficiency and effectiveness of resource management and the national austerity program.

## 2.1 *Institutions of the Mining Sector*

### 2.1.1 Important Government Institutions

Ministry of Mining and Heavy Industry (MMHI)

The main goal of the Ministry of Mines and Heavy Industry is to build a balanced, multi-pillar economy and increase the utilization of natural resources through the development of transparent and responsible mining and heavy industry.

The main tasks are the development and formulation of laws, regulations, guidelines, medium and long-term strategies, programs and project developments related to geology and mining, the fuel and oil sector and heavy industry. Furthermore, the ministry has to carry out political analyzes as well as carry out internal controls and monitoring in the implementation of the tasks. In addition, the Ministry is responsible for the financial review of budget financing, programs, projects and capital expenditures, conducting internal audit and risk management.

Mineral Resources and Petroleum Authority (MRPAM)

The Mineral Resources and Petroleum Authority assists the Ministry in drafting state development policies in the field of geology, mining, and petroleum, to provide consumers and investors with prompt, objective, and factual assistance, so that the resource and petroleum industry strengthens the development of the national economy and state policy in this sector is implemented.

MRPAM Petroleum Exploration and Petroleum Products Departments are organized to ensure, monitor and support the product sharing agreements in accordance with the Petroleum Law of Mongolia and relevant government decisions.

### Ministry of Nature, Environment and Tourism (MNET)

One of the main tasks of the Ministry of Nature, Environment and Tourism is to implement national policies and legislation related to environmental protection, wise use, and restoration of natural resources. In addition, the MNET is also responsible for the decision-making and approval of regulations and administrative guidelines to be followed by the administrative authorities in matters of environmental protection.

Mining or exploration license holders must submit their environmental protection plan to the MNET within 30 days of receiving their permit. The Ministry is responsible for receiving and reviewing this plan. The Ministry is also responsible for collecting 50% of the companies' environmental protection budgets each year. These funds can be used by the MNET for their intended purpose if a company has not fully implemented its environmental protection plans.

### Ministry of Labor and Social Protection (MLSP)

The Ministry of Labor and Social Affairs is to develop and implement employment policies to promote employment, create decent working conditions and ensure human resources development by improving vocational training.

In addition, the Ministry is responsible for job creation, poverty alleviation, lowering the cost of living, as well as regulating issues related to the employment of foreign nationals on the territory of Mongolia and vice versa.

### Mongolian Tax Administration (MTA)

The Mongolian Tax Administration consists of the State Tax Administration (MTA), the provincial (aimags), capital and district tax offices and the state tax inspectors in the administrative units (soums). The MTA is under the direct supervision of the Treasury Department.

The main functions of the MTA are to organize the implementation of tax legislation, to provide information and advice to taxpayers and to conduct training on the implementation of tax legislation and the collection of taxes.

### Mongolian Customs Office (MCO)

The Customs Board is a government administrative agency responsible for the nationwide implementation of customs legislation and is under the supervision of the Ministry of Finance.

Mongolian National Audit Office (MNAO)

The National Audit Office is Mongolia's supreme audit authority. It is not permitted to monitor the actions of the National Audit Office and the Auditor General of Mongolia or assign them any duties. The National Audit Office has full authority to audit the functions of governmental organizations regardless of their funding sources, with the exception of the Mongolian parliament.

Nuclear Energy Commission (NEC)

The Nuclear Energy Commission is responsible for the peaceful use of radioactive minerals and nuclear energy on the territory of Mongolia, research into nuclear technologies, radiation protection and training of related human resources.

The Nuclear Energy Commission issues and revokes licenses to handle radioactive materials or minerals, and coordinates and monitors scientific research in the field of nuclear energy.

The NEC consists of four departments, the Administration Department, the Nuclear Engineering Regulatory Department, the Nuclear Safety Regulatory Department and the External Relations Department.

State Agency for Policy and Coordination on State Property (SAPCSP)

The mission of the State Agency for State Property (SAPCSP) is to formulate, implement and improve guidelines for the management of state property, to increase returns through good management of state property or state holdings, and to ensure transparent and fair competition in procurement activities at a level that corresponds to international standards.

The Government Procurement Regulations and the "Regulation on Keeping Records of Persons Whose Right to Participate in Tenders is Restricted" were adopted by Government Decision No. 274 of July 3, 2019. The government agency responsible for implementation was established under the responsibility of the Prime Minister of Mongolia on the basis of Resolution No. 438 of December 18, 2019.

*Other institutions*

- Mongolian National Mining Association
- Geological Society of Mongolia
- Ministry of Finance
- Mongolian National Chamber of Commerce and Industry
- National Human Rights Commission of Mongolia
- Regulatory Agency of the Government of Mongolia
- General Agency for Specialized Inspection
- Provincial Governments: Citizen Representatives (Khural), Deputy Governors, Department of Policy Development, Environmental Inspectorate, NGO's

## 2.2 Legal Framework

The legal framework for exploration and mining in Mongolia is based on the following laws, guidelines, and regulations:

- Article 38 of the 1992 Constitution of Mongolia (amended 2001) requires the government to "take measures to protect the environment and to use and restore natural resources in a sustainable manner".
- The Minerals Law of Mongolia, revised in 2014, regulates "the prospecting, exploration and exploitation of minerals within the territory of Mongolia" and establishes environmental protection and regulations in exploration areas and areas surrounding open pit mines and mining operations.
- The Environmental Protection Law of Mongolia of 1995 regulates "relationships between the state, citizens, businesses and organizations in order to protect the right of human beings to live in a healthy and safe environment, an ecologically balanced social and economic environment, on the protection of the environment for present and future generations, the appropriate use of natural resources and the restoration of available resources".
- The 2011 Law on Information Transparency and Right to Information regulates "relationships to ensure the transparency of the state and the right of citizens and legal entities to seek and obtain information".
- The Law of Mongolia on Environmental Impact Assessment of 2012 aims to protect the environment, "prevent ecological imbalances, ensure minimal negative impact on the environment through the use of natural resources, and regulate relationships, that may arise in connection with the assessment of environmental impacts".
- The 2002 Law of Mongolia on Land regulates "the ownership and use of state-owned land and other related issues".
- The Law of Mongolia on State Supervision and Inspection (2003) regulates the state supervision and inspection structure, the legal basis and the conduct of state supervision and inspection.
- The Law of Mongolia on Subsoil of 1989 regulates "issues related to the use and protection of the subsoil in the interests of present and future generations".
- The Law on Prohibition of Mineral Exploration and Mining Activities in areas in the Headwaters of Rivers, Protected Water Reservoir Zones and Forested Areas "Long Named Law of 2009 prohibits "mining in headwaters of rivers, protected areas of water reservoirs and forest areas" and regulates the remediation activities carried out in these areas.
- The Model Agreement on Environmental, Mining and Infrastructure Development Issues Related to Mine Site Development and Job Creation (2016) outlines the basis for an agreement governing the relationship between a mining license holder and the provincial local government (Aimag) and of the administrative district (Soum) and issues of environmental protection, mineral extraction, infrastructure development and job creation.
- Mongolia committed to the implementation of the EITI Mining Transparency Initiative in December 2005 and was accepted as a candidate on September 27,

2007. Mongolia is currently an EITI compliant country making significant progress in EITI implementation. With around 1000 companies, Mongolia's EITI reports are among the most comprehensive in the world (EITI, 2015).

On June 12, 1992, Mongolia signed the United Nations Framework Convention on Climate Change (UNFCCC). The UNFCCC was ratified on September 20, 1993 and came into force in 1994.

- Mongolia ratified the Kyoto Protocol on December 15, 1999.
- Mongolia also signed the Paris Agreement on April 22, 2016.
- In 2011, Mongolia adopted a national action program on climate change, which includes concrete measures to respond to climate change for all important economic sectors.
- Mongolia started the green development policy in 2014, which includes key indicators to measure progress in implementing sector-specific action plans (UNFCC Intended Nationally Defined Contribution [INDC] 2014). In its INDC submission, Mongolia outlined the expected impact of its policies and plans to reduce total national greenhouse gas emissions excluding land use, land use change and forestry by 14 percent by 2030 (UNFCC INDC, 2014). Mongolia also intends to contribute to global efforts to reduce greenhouse gas emissions by increasing energy production from renewable sources from 7.62 percent in 2014 to 30 percent by 2030. On September 25, 2015, Mongolia adopted the 2030 Agenda for Sustainable Development along with the United Nations Sustainable Development Goals (SDGs).

Given the importance of the extractive economy, Mongolia is making ongoing efforts to change its legal environment and align it more closely with global standards. In November 2017, the Mongolian Parliament amended some articles of the Minerals Law of Mongolia, which came into force on July 8, 2006. The changes changed the process for obtaining new exploration licenses to an open bidding system. The procedures for applying for and obtaining a license have also been affected by the changes.

The Corporate Income Tax Law (CIT) of Mongolia, which came into force in June 2006, was amended to include the term beneficial owner, which was changed to beneficial owner in 2018 (as in defined in the Law of Mongolia on Combating Money Laundering and Terrorist Financing of May 2013).

In March 2019 Parliament revised tax laws including the General Taxation Law (GTL) and the Corporate Income Tax Law (CIT Law), both of which came into force on 1 January 2020. The definition of beneficial owner has been introduced separately in the GTL as a person (i.e. natural person, legal entity or entity) that

- holds more than 30% of the shares or participation rights or exercises more than 30% of the voting rights.
- is entitled to dividends from any entity that owns (either itself or indirectly through one or more levels of an ownership chain) exploration or mining licenses for minerals, petroleum, radioactive minerals, or land tenure and use rights.

Considered are transactions involving a sale, in whole or in part, by a beneficial owner (whether onshore or offshore) to a Mongolian legal entity of shares, voting rights and participation rights, exploration or mining licenses for minerals, radioactive minerals or petroleum, and land and usage rights owns. As a "sale of rights" it is subject to income tax at 10% of the value of the mining license instead of the previous 30%. The income earned by the beneficial owner from the sale of shares is considered to be the taxable income of the Mongolian legal entity from the sale of rights and not subject to withholding tax (contrary to the CIT Law of 2006).

The sale of Shares by non-beneficial owners is subject to income tax at 10% based on the value of the selling price of the Shares. Tax evasion, misrepresentation of income from the sale of rights, failure to provide documentation to be used in the valuation, or failure to report a change of beneficial owner may be grounds for authorities to terminate a mining license.

In March 2019, Parliament enacted an amendment to raise the minimum gold royalty from 2.5% to 5% of sales value.

In October 2019, the Constitutional Court of Mongolia issued a final ruling on Article 47 of the Commodity Law regarding concerns about double charging of royalties. In November 2019, Parliament amended Article 47 of the Commodities Act (latest amendment) to remove the contradictory wording in the March 2019 amendment regarding possible double levying of royalties first from mining license holders and then from non-mining license holders own but buy minerals from the mining license holders and resell them directly or after processing to third parties. The latest change came into effect in November 2019. Royalty payment rates and procedures have not been affected by the latest change.

The Raw Materials Act does not include:

- Water
- Mineral oil and natural gas
- Radioactive raw materials
- Common raw materials such as sand, gravel and granite deposits that can be used as building materials

Exploration licenses are initially granted for a period of three years and can be renewed up to three times for a further three years each. The holder of an exploration license may hold it for a maximum of 12 years from the date of initial issuance. Exploration license holders must apply for a renewal at least one month prior to the expiration of the license term. Exploration license holders are required to pay annual fees for each hectare of the licensed area as follows:

- In the first year, the fee is 145 MNT per hectare.
- In the second year, the fee is 290 MNT per hectare.
- In the third year, the fee is 435 MNT per hectare.
- From the fourth to the sixth year, the annual fee is 1450 MNT per hectare.
- From the seventh to the ninth year, the annual fee is 2175 MNT per hectare.
- From the tenth to the twelfth year, the annual fee is 7250 MNT per hectare.

Under the Minerals Act, holders of exploration licenses are required to prepare and submit an environmental protection plan for the licensed area. If an environmental protection plan is not submitted to the governor of the local government agency, the permit may be suspended and possibly revoked. Exploration permit holders are also required to submit annual environmental protection activity reports. If no annual environmental protection activity report is prepared on the basis of the environmental protection plan, the permit can also be suspended and possibly revoked. Exploration permit holders are also required to contribute to an environmental compensation fund based on the permit environmental protection plan. The contributions make up 50% of the environmental protection budget for the respective year and must be paid into a special-purpose account designated by the local administrative authority. The local government can use funds from the fund to take action to implement the environmental protection plan if the permit holder fails to meet its environmental obligations.

Mining licenses are issued for an initial term of 30 years and are renewable twice for additional 20-year terms. The Mining Resources and Petroleum Agency (MRPAM) may decide whether to grant an extension depending on resource reserves. Licenses for radioactive raw materials are subject to a different regulation.

Mining license holders must apply for their license renewal at least two years prior to expiration and pay 21,750 MNT per hectare annually to the beneficiary for the mining license. The annual royalty per hectare for deposits of limestone, coal and minerals destined for domestic production is 7250 MNT. These fees are payable annually in advance. (Article 33.1, Raw Materials Act.)

Mining license holders and their contractors are encouraged to hire Mongolian nationals. The quota for foreign staff is currently limited to 10% of the total number of employees. Mining license holders are also required to prepare an Environmental Management Plan based on the Environmental Impact Assessment along with an annual Environmental Protection Plan or Management Plan. The annual plan must be submitted to and approved by the Ministry of Environment and Tourism. An annual report on the implementation of the environmental management plan must be prepared and approved by a state commission. Mining license holders are required to pay 50% of the budget required to implement the environmental management plan into a special bank account designated by the Ministry of Environment and Tourism for environmental remediation every year. These monies will be returned to the mining license holder during the mine's closure based on its compliance with the obligations set out in the Environmental Impact Assessment Report.

The Mongolian legal system is liberal when it comes to foreign ownership of raw materials. For example, there is no general prohibition on a foreign entity or person indirectly holding a mineral concession. Subject to compliance with minimum capital investment requirements under the Investment Law, foreign legal entities or individuals can easily apply for and obtain an exploration or mining license through ownership of a legal entity in Mongolia (which may be newly incorporated or acquired). According to the Investment Law, each foreign investor must invest at least USD 100,000 if the company is a foreign-invested enterprise (Article 3.1.5, Investment Law). The Investment Act defines a foreign-invested company as a company in which 25% or more of the shares are held by foreign investors.

However, if a state-owned foreign legal entity owns or wishes to acquire more than 33% of the shares of a mining company, its investment must be approved by the Mongolian government agency responsible for foreign investment (currently the National Development Agency) (Tables 1, 2, 3 and 4).

**Table 1** Number of mining licenses by commodity type—12/31/2020

| | Mining license | | Area size | |
|---|---|---|---|---|
| | Count | Percent | Hectare | Percent |
| **Total** | **1696** | **100%** | **1,653,800.5** | **100%** |
| Gold (placer) | 440 | 25.9% | 155,901.5 | 9.4% |
| Coal | 303 | 17.9% | 769,825.9 | 46.5% |
| Fluorite | 198 | 11.7% | 35,350.3 | 2.1% |
| Gold (placer) | 91 | 5.4% | 77,192.5 | 4.7% |
| Iron | 76 | 4.5% | 48,442.2 | 2.9% |
| limestone | 73 | 4.3% | 41,066.3 | 2.5% |
| Common mineral, gravel, sand | 70 | 4.1% | 3684.6 | 0.2% |
| Common mineral, mud | 46 | 2.7% | 2632.4 | 0.2% |
| Common mineral, aggregate stone | 34 | 2.0% | 9240.3 | 0.6% |
| Tungsten | 29 | 1.7% | 9062.1 | 0.5% |
| Common mineral, sand | 24 | 1.4% | 1333.4 | 0.1% |
| Concentrated metal | 23 | 1.4% | 26,976.6 | 1.6% |
| Gypsum | 23 | 1.4% | 7975.6 | 0.5% |
| Copper | 18 | 1.1% | 84,900.5 | 5.1% |
| Gold (placer), copper | 14 | 0.8% | 86,764.3 | 5.2% |
| Tin | 12 | 0.7% | 5386.5 | 0.3% |
| Earth rare metal | 9 | 0.5% | 7.435.4 | 0.4% |
| Graphite | 8 | 0.5% | 18,106.4 | 1.1% |
| Copper, molybdenum | 8 | 0.5% | 14,328.3 | 0.9% |
| Common mineral, sand stone | 8 | 0.5% | 1553.5 | 0.1% |
| Uranium | 7 | 0.4% | 76,866.1 | 4.6% |
| Molybdenum | 7 | 0.4% | 8922.5 | 0.5% |
| Marble | 7 | 0.4% | 2020.0 | 0.1% |
| Copper, gold | 5 | 0.3% | 8807.4 | 0.5% |
| Molybdenum, tungsten | 5 | 0.3% | 5047.1 | 0.3% |
| Aggregate stone | 5 | 0.3% | 1698.9 | 0.1% |
| Iron, zinc | 5 | 0.3% | 1126.4 | 0.1% |
| Manganese | 4 | 0.2% | 7507.0 | 0.5% |
| Tungsten, Tin | 4 | 0.2% | 2833.9 | 0.2% |
| Мөсөн шүү | 4 | 0.2% | 1449.0 | 0.1% |
| Silver | 4 | 0.2% | 1050.4 | 0.1% |
| Gold (placer), silver | 3 | 0.2% | 9587.7 | 0.6% |
| Aluminium | 3 | 0.2% | 1370.9 | 0.1% |
| Other | 126 | 7.0% | 118,354.7 | 7.2% |

*Source*: Cadastre Division, MRPAM

In order to promote environmentally and socially responsible mining activities, the government is in the process of drafting a mining law in addition to the raw materials law. Currently, both mining and exploration activities are regulated by the Commodities Act. The draft mining law aims to regulate mining separately, from the determination of reserves to the closure of the mine. If the draft mining law is passed, exploration and mining activities will be governed by separate laws (Figs. 2, 3 and 4).

**Table 2** Foreign license holders by country—12/31/2020

| | Holder | License | Area size | |
|---|---|---|---|---|
| | Count | Count | Hectare | Percent |
| **Total** | **243** | **393** | **1,231,408.3** | **100%** |
| China | 130 | 163 | 325,801.5 | 26.46% |
| Virgin Islands | 30 | 56 | 107,954.8 | 8.77% |
| Hong Kong | 14 | 27 | 66,897.5 | 5.43% |
| Singapore | 14 | 54 | 313,228.4 | 25.44% |
| Canada | 12 | 23 | 220,308.9 | 17.89% |
| South Korea | 10 | 12 | 10,489.2 | 0.85% |
| Australia | 5 | 5 | 26,641.3 | 2.16% |
| Russia | 4 | 5 | 2490.4 | 0.20% |
| Great Britain | 3 | 4 | 6038.7 | 0.49% |
| Holland | 3 | 15 | 106,872.3 | 8.68% |
| Germany | 2 | 2 | 603.9 | 0.05% |
| Bermud | 2 | 3 | 7314.3 | 0.59% |
| United States of America | 2 | 4 | 2442.3 | 0.20% |
| Luxemburg | 2 | 6 | 13,528.2 | 1.10% |
| Japan | 2 | 3 | 3015.1 | 0.24% |
| Kazakhstan | 1 | 1 | 136.5 | 0.01% |
| Malaysia | 1 | 2 | 2416.5 | 0.20% |
| Ukraine | 1 | 1 | 245.0 | 0.02% |
| Bulgaria | 1 | 1 | 3700.9 | 0.30% |
| Russia, China | 1 | 1 | 47.5 | 0.00% |
| Russia, South Korea | 1 | 1 | 118.2 | 0.01% |
| Barbados | 1 | 3 | 9183.0 | 0.75% |
| Virgin Islands, Singapore | 1 | 1 | 1934.0 | 0.16% |

*Source*: Cadastre Division, MRPAM

## 2.3 Deposits

For clear understanding, it is important to briefly explain the term "raw material deposit" in the sense of the Mongolian mining legislation and then to explain the classification system of reserves and resources (Fig. 5).

The most important terms of the Raw Materials Act are explained below:

- A "mineral deposit" as defined by the Natural Resources Law of Mongolia is a deposit where the quality and reserves are proven and economically mineable. Reserves are registered by the Minerals Counsel (also called Professional Reserves Counsel) in the National Registry of Reserves, which is maintained at MRPAM (Mineral Resources and Petroleum Authority).

Mongolia's Natural Resources Law distinguishes between:

- Deposits of strategic importance

**Table 3** Origin of foreign license holders in cooperation's—12/31/2020

| | Holder | License | Area size | |
|---|---|---|---|---|
| | Count | Count | Hectare | Percent |
| **Total** | **96** | **141** | **309,034.3** | **100%** |
| Mongolia, China | 46 | 64 | 91,278.5 | 29.5% |
| Mongolia, South Korea | 9 | 14 | 24,162.5 | 7.8% |
| Mongolia, Russia | 8 | 10 | 1347.8 | 0.4% |
| Mongolia, Virgin Islands | 6 | 10 | 35,233.8 | 11.4% |
| Mongolia, Singapore | 4 | 6 | 28,137.9 | 9.1% |
| Mongolia, Japan | 4 | 7 | 13,386.0 | 4.3% |
| Mongolia, Germany | 3 | 4 | 5911.9 | 1.9% |
| Mongolia, Great Britain | 2 | 2 | 5135.2 | 1.7% |
| Mongolia, Czech Republic | 2 | 5 | 28,576.6 | 9.2% |
| Mongolia, Singapore, Hong Kong | 1 | 1 | 1469.5 | 0.5% |
| Mongolia, Kazakhstan | 1 | 1 | 1789.4 | 0.6% |
| Mongolia, United States of America, Virgin Islands, South Korea | 1 | 1 | 3766.7 | 1.2% |
| Mongolia, Bulgaria | 1 | 3 | 1083.7 | 0.4% |
| Mongolia, Canada | 1 | 1 | 3957.7 | 1.3% |
| Mongolia, Canada, China | 1 | 1 | 6352.8 | 2.1% |
| Mongolia, Luxemburg | 1 | 1 | 3273.9 | 1.1% |
| Mongolia, Russia, Ukraine | 1 | 1 | 46.9 | 0.0% |
| Mongolia, Hong Kong, Japan | 1 | 3 | 43,809.3 | 14.2% |
| Mongolia, Poland | 1 | 1 | 1561.3 | 0.5% |
| Mongolia, Hungary | 1 | 1 | 87.4 | 0.0% |
| Mongolia, Hong Kong | 1 | 4 | 8665.5 | 2.8% |

*Source*: Cadastre Division, MRPAM

Deposit whose size may have potential impact on the national security, economic and social development of the country at the national and regional levels, or which has the potential to contribute more than five percent annually to the total gross domestic product.

- Deposits of common raw materials

Deposits containing sediments and rock concentrations in sufficient quantities that could be used as construction material.

- Deposits of conventional raw materials

Deposits that do not meet the criteria of one and two.

### 2.3.1 Deposits of Strategic Importance

Two annexes to 2007 Decree No. 27 of the Parliament of Mongolia deal with the issue of strategically important land deposits. The first list contains 15 deposits that have already been recognized as strategically important, while the second refers to

**Table 4** List of companies operating in Mongolia that are registered on international stock exchanges—12/31/2020

| | Stock exchange | Symbol | Currency | Average price | | | | | | Market value /2021.I.06/ |
|---|---|---|---|---|---|---|---|---|---|---|
| | | | | 2019 | 2019.XII | 2020.XI | 2020.XII | 2020.XII / 2019.XII | 2020.XII / 2020.XI | |
| Turquoise Hill Resources Ltd. | NYQ | TRQ | USD | 1.036 | 0.639 | 9.23 | 11.91 | 1864% | 129.0% | 2.71B |
| Peabody Energy Corporation | NYQ | BTU | USD | 22.07 | 9.476 | 1.21 | 1.86 | 19.6% | 153.7% | 335.45M |
| RioTinto pic | LSE | RIO.L | GBp | 4342.7 | 4372.2 | 4736.8 | 5493.8 | 125.7% | 116.0% | 71.92B |
| Mongolia Energy Corporation Limited | HKG | 0276.HK | HKD | 0.121 | 0.078 | 0.86 | 1.01 | 1295% | 117.4% | 208.82M |
| Mongolian Mining Corporation | HKG | 0975.HK | HKD | 0.412 | 0.695 | 0.49 | 1.14 | 164.0% | 232.7% | 1.30B |
| South Gobi Resources Ltd. | HKG | 1878.HK | HKD | 0.949 | 0.858 | – | – | – | – | 136.35M |
| Centerra Gold Inc. | TOR | CG.TO | CAD | 9.131 | 10.086 | 12.37 | 14.160 | 140.4% | 114.5% | 4.45B |
| Erdene Resource Development Corporation | TOR | ERD.TO | CAD | 0.207 | 0.175 | 0.43 | 0.39 | 222.9% | 90.7% | 115.04M |
| Entrée Resources Ltd. | TOR | ETG.TO | CAD | 0.407 | 0.39 | 0.47 | 0.52 | 133.3% | 110.6% | 115.27M |
| Silver Elephant Mining Corp. | TOR | ELEF.TO | CAD | – | – | 0.43 | 0.42 | – | 97.7% | 87.95M |
| Steppe Gold Ltd | TOR | STGO.TO | CAD | 0.849 | 0.915 | 2.19 | 2.23 | 243.7% | 101.8% | 174.84M |
| Kincora Copper Limited | VAN | KCC.V | CAD | 0.097 | 0.039 | 0.08 | 0.08 | 205.1% | 100.0% | 24.98M |
| Altan Rio Minerals Limited | VAN | AMO.V | CAD | 0.01 | – | 0.12 | 0.15 | – | 125.0% | 12.74M |
| Xanadu Mines Limited | ASX | XAM.AX | AUD | 0.071 | 0.033 | 0.04 | 0.04 | 121.2% | 100.0% | 33.01M |
| Aspire Mining Limited | ASX | AKM.AX | AUD | 0.028 | 0.13 | 0.07 | 0.09 | 69.2% | 128.6% | 50.76M |
| Haranga Resources Limited | ASX | HAR.AX | AUD | – | – | – | – | – | – | – |
| Tian Poh Resources Limited | ASX | TPO.AX | AUD | 0.09 | – | – | – | – | – | 19.33M |
| TerraCom Limited | ASX | TER.AX | AUD | 0.512 | – | 0.15 | 0.16 | – | 106.7% | 120.33M |
| Viking Mines Limited | ASX | VKA.AX | AUD | 0.011 | – | 0.02 | 0.03 | – | 150.0% | 19.32M |
| Banpu Public Company Limited | SET | BANPU.BK | THB | 13.987 | 11.695 | 7.62 | 10.09 | – | 132.4% | 57.85B |
| Petro Matad Limited | LSE | MATD.L | GBp | 5.301 | 3.488 | 2.48 | 2.91 | 83.4% | 117.3% | 26.24M |
| Wolf Petroleum Limited | ASX | WOF.AX | AUD | – | – | – | – | – | – | 2.71B |
| Elixir Energy Limited | ASX | EXR.AX | AUD | – | – | 0.13 | 0.11 | – | 84.6% | 335.45M |

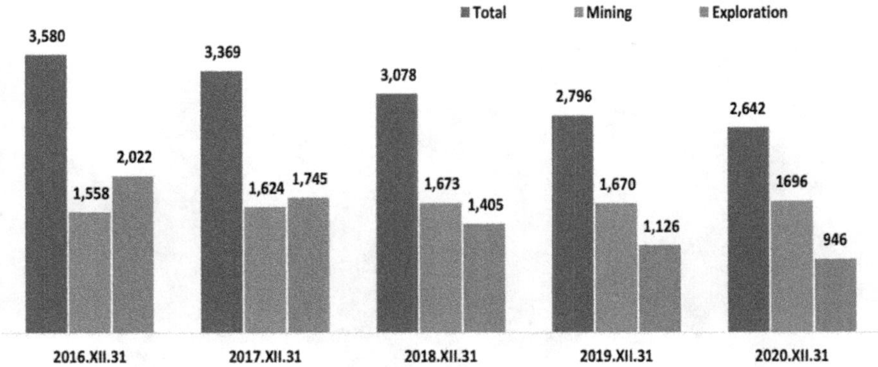


**Fig. 2** Number of valid licenses from 2016 to 2020 (MRPAM 2023)

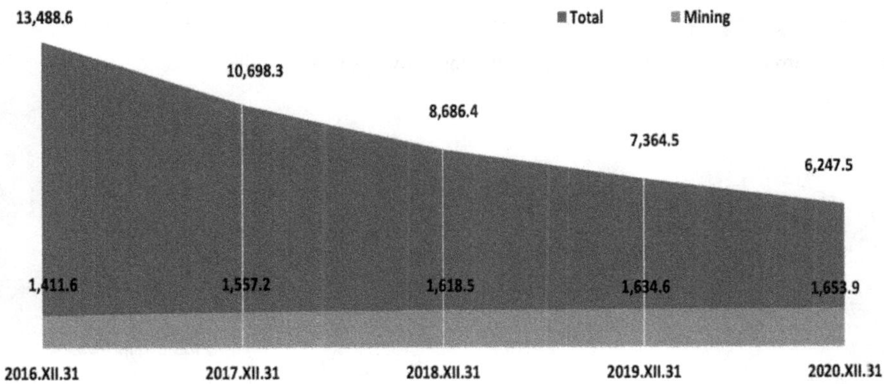


**Fig. 3** Licensed area in 1000 ha from 2016 to 2020 (MRPAM 2023)

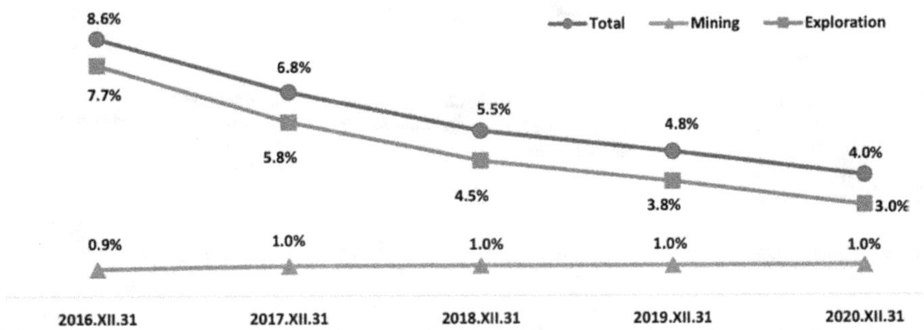


**Fig. 4** Percentage of the total licensed area on the territory of Mongolia from 2016 to 2020 (MRPAM 2023)

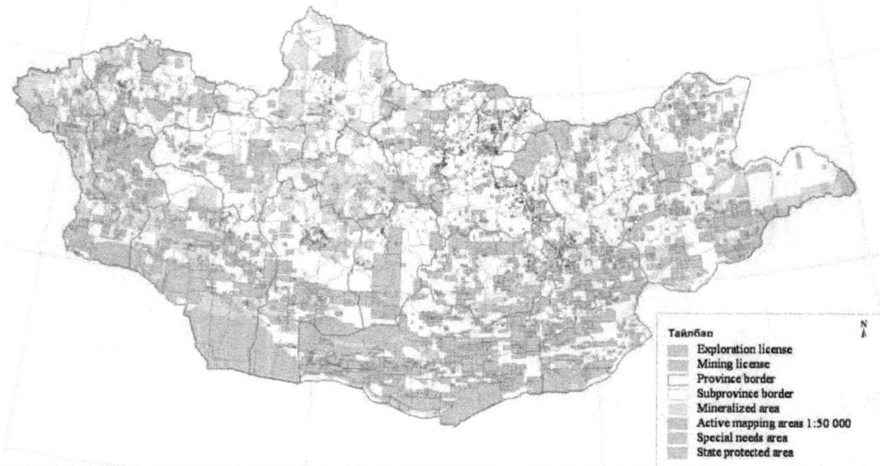


**Fig. 5** Current mapping, exploration, and mining activities in Mongolia

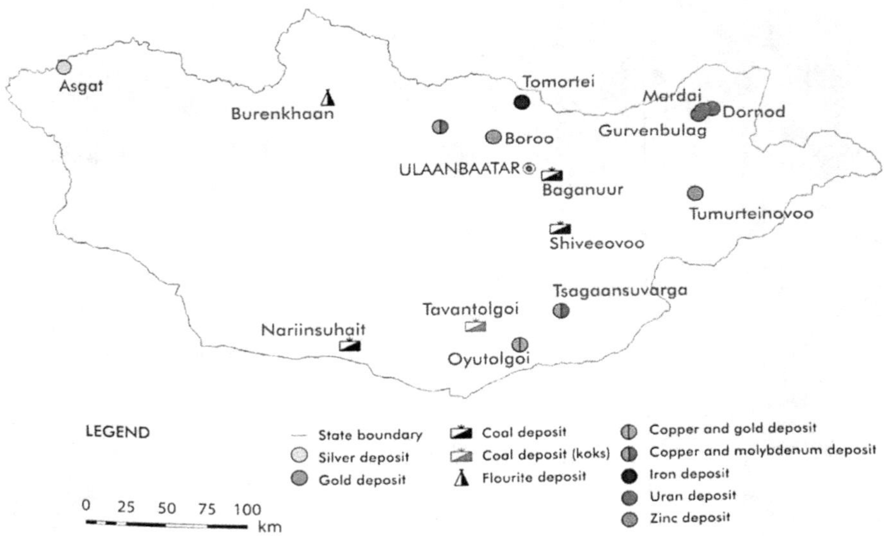


**Fig. 6** The 15 strategic resource deposits

a further 39 deposits from which further deposits of strategic importance can be selected after further evaluation. In January 2015, the Gatsuurt Gold Deposit was added to the Strategic Deposit List. Gatsuurt is located in a forested river basin, an area where mining is prohibited. In order to facilitate the exploitation of the mine, it has been declared a deposit of strategic importance. The 15 strategic resource deposits are shown in Fig. 6 and briefly described in Table 5.

**Table 5** Deposits of strategic importance. Reserves estimated at start of production

| Lagerstätte | Rohstoff | Größe, Gehalt | Bemerkungen |
|---|---|---|---|
| Tavan Tolgoi | Bituminous Coal | 6.420 Mio. t Bituminous Coal, thereof 1.400 Mio. t Coking Coal | Operator SOE, Erdenes TT |
| Nariin Sukhait | Bituminous Coal | Reserve estimation 125 to 380 mio. t | 3 independent mining companies located approximately 55 km from the Chinese border |
| Baganuur | Lignite | 600 mio. t | In operation since 1978. Currently 40% of local consumption is produced |
| Shivee Ovoo | Lignite | 646.2 mio. t | In operation since 1990. Planning of a power plant for energy export to China. |
| Mardai | Uranium | 1.104 t, mineral content 0.119% $U_3O_8$ | None of these deposits have been active in recent years. Licensing disputes and complicated regulatory and legal frameworks create complications |
| Dornod | Uranium | 28.868 t, mineral content 0.175% $U_3O_8$ | |
| Gurvan Bulag | Uranium | 16.073 t, mineral content 0.152% $U_3O_8$ | |
| Tumurtei | Iron ore | 229.3 Mio. t, mineral content 51.15% Fe | One of the largest iron ore deposits in Mongolia. State owned |
| Oyu Tolgoi | Copper, Gold | Measured and Inferred Resources: 2.2 Mrd. t Ore, 21.2 Mio. t Cu, 772.9 t Au | In production since 2013. Owned by Turquoise Hill Resources (66%) and Mongolia represented by EOT |
| Tsagaan Suvarga | Copper, Molybdenum | Reserves: 16 Mio. t oxidic ore, 250 mio. t primary sulfidic ore mit 1.6 mio. t Cu and 66.000 t Mo | Licensed to Mongolyn Alt Corp.; promising project. |
| Erdenet | Copper, Molybdenum | 1.2 Mio. t, mineral content 0.51% Cu and 0.012% Mo | In operation since 1978. Russian-Mongolian joint operation. Annual production approx. 539,000 t of Cu concentrate, 2000 t of Mo concentrate |
| Burenkhaan | Phosphorite | 192.24 mio. t ore; mineral content 21.1% $P_2O_5$ | Environmental restrictions can become problematic. |
| Boroo | Gold | 24.5 Mio. t ore; mineral content 1.6 g/t Au (ca. 39 t Au) | Deposit exhausted; from 2004 to 2012 about 56 t Au were produced. |
| Tomortein Ovoo | Zinc | 7.7 Mio. t ore, mineral content 11.5% Zn | Mined by Tsairt Minerals. |
| Asgat | Polymetallic, Silver | 6.4 Mio. t ore, mineral content 351.08 g/t Ag | Infrastructure problems. |
| Gastuurt | Gold | Measured and Estimated Resources: 398,000 ounces (12.4 t) Au | Mining license held by Centerra Gold. Negotiations between owners and the state are making little progress. |

The 2014 amendments to the Raw Materials Act address an important issue regarding the exploitation of deposits of strategic importance. The changes give the government the option of either exercising its right to take a 34–50% interest or charging a special royalty in lieu of that interest. The Natural Resources Law empowers the Mongolian parliament to designate mineral deposits of "strategic importance".

If the reservoir has been explored and developed with government funds, the government can own up to a 50% interest in a company that holds a license for a strategically important reservoir. If the deposit has been explored and developed by private companies, the state can hold up to 34% of the shares. License holders of a strategically important deposit must also publicly trade at least 10% of their shares on the Mongolian Stock Exchange.

### 2.3.2 Raw Material Reserves of Mongolia

MRPAM (Mineral Resources and Petroleum Authority) is responsible for registering and reporting Mongolia's mineral reserves. Table 6 summarizes the details of mineral reserves registered with the Mongolian State Mineral Reserve Fund over the past two years.

#### Classification of Ore Reserves and Resources

The abundance of mineral resources is usually stated as a reserve or resource and is further subdivided according to the degree of precision with which the size and quality of the deposit has been investigated and the modifying factors that may hinder or affect the economic extraction of the ore.

Mongolia has its own system of reporting on mineral reserves and resources. For this reason, published figures on resources and reserves from Mongolia are usually misinterpreted or misunderstood. The same figures are called "reserve" by some authors, experts or analysts, "resource" by others, and again "resources plus reserves" (Since democratization in 1991, the classification of mineral reserves, once focused on Russia, has changed several times. In October 2014, Mongolia became a member of CRIRSCO (Committee For Mineral Reserves International Reporting Standards).

Deposit evaluations and comparisons between national mineral resources require comparable systems, so the Mongolian classification system and its differences from internationally applied systems are outlined below.

In many western mining jurisdictions, deposit reserves are the assets of a mining company. Within the general reserve classification used in western mining jurisdictions, there are two subgroups:

- proved
- probable

**Table 6** Raw material production 2018–2020

| | Unit of measurement | 2018 | | 2019 | | 2020 |
|---|---|---|---|---|---|---|
| | | Plan | Report | Plan | Report | Plan |
| Gold (Total) | kg | 20,022.9 | 16,174.7 | 18,013.2 | 14,416.5 | 15,888.9 |
| Gold (Amount of gold that accompanies the concentrate) | kg | 8921.1 | 9110.0 | 6474.3 | 7615.5 | 4809.8 |
| Gold (Total amount of gold to be mined from hard rock and placer deposits) | kg | 11,101.7 | 7064.7 | 11,538.9 | 6935.0 | 11,079.1 |
| Gold (Amount of gold to be extracted from the hard rock | kg | 3338.6 | 1641.2 | 3863.8 | 2045.0 | 4698.1 |
| Gold (Amount of gold to be mined from placer deposits) | kg | 7763.1 | 5423.6 | 7675.1 | 4890.1 | 6381.0 |
| Building materials | | | | | | |
| Lime | ths.tons | 304.9 | 47.6 | 114.7 | 55.6 | 245.6 |
| Limestone | ths.tons | 1724.7 | 505.5 | 1761.8 | 412.4 | 1187.7 |
| Clinker | ths.tons | 0.0 | 0.0 | 14.0 | 0.0 | 814.0 |
| Cement | ths.tons | 2051.8 | 1077.9 | 1706.7 | 1706.7 | 869.2 |
| Mineral powder | ths.tons | 0.0 | 0.0 | 31.0 | 0.0 | 0.0 |
| Crystal | ths.tons | 55.8 | | 61.8 | | 61.8 |
| Marble | ths.tons | 5.0 | 3.2 | 231.6 | 3080.4 | 252.9 |
| Gypsum | ths.tons | 392.0 | 47.7 | 243.2 | 107.1 | 166.6 |
| Tungsten | tons | 1186.7 | 385.7 | 1544.7 | 165.5 | 621.7 |
| Copper | ths.tons | 1219.3 | 1321.1 | 1161.1 | 1259.5 | 1275.2 |
| Copper (Cathode) | tons | 10.8 | 12.1 | 10.0 | 9.7 | 10.0 |
| Manganese | ths.tons | 230.6 | 0.0 | 230.6 | 0.0 | 129.1 |
| Molybdenum | ths.tons | 4.9 | 5.5 | 4.8 | 5.3 | 4.7 |
| Silver | kg | 7988.9 | 2885.5 | 4229.3 | 2972.6 | 12,761.9 |
| Silver concentrate | ths.tons | 13.0 | 0.0 | 0.0 | 0.0 | 194.6 |
| Silver (Amount of silver accompanying concentrate | kg | 56,292.9 | 66,234.5 | 68,472.5 | 76,891.6 | 65,000.1 |
| Iron | ths.tons | 15,127.8 | 8134.6 | 14,193.0 | 8242.0 | 15,220.5 |
| Whole iron concentrate | ths.tons | 4800.3 | 2990.7 | 5176.6 | 1372.8 | 5087.5 |
| Iron powder concentrate | ths.tons | 7562.9 | 3924.5 | 7201.8 | 1168.0 | 6769.2 |
| Iron ore | ths.tons | 2764.6 | 1185.2 | 1814.6 | 2226.9 | 3260.9 |
| Powdered iron ore | ths.tons | 0.0 | 34.3 | 0.0 | 3474.3 | 103.0 |
| Uranium | tons | 5.7 | 0.0 | 16.5 | 0.0 | 22.3 |
| Fluorite | ths.tons | 1107.2 | 642.4 | 1805.9 | 358.7 | 2207.4 |
| Fluorite concentrate AG-97 | ths.tons | 134.1 | 65.3 | 219.9 | 58.3 | 129.0 |

(*continued*)

**Table 6** (continued)

| | Unit of measurement | 2018 | | 2019 | | 2020 |
|---|---|---|---|---|---|---|
| | | Plan | Report | Plan | Report | Plan |
| Fluorite concentrate AG-95 | ths.tons | 61.8 | 56.5 | 90.2 | 41.0 | 120.9 |
| Gravitational fluorite МГФ 55-85 | ths.tons | 0.0 | 0.0 | 6.3 | | 0.0 |
| Whole fluorite MG-92 | ths.tons | 39.5 | 7.8 | 28.0 | 2.8 | 28.2 |
| Whole fluorite MG-85 | ths.tons | 58.6 | 7.0 | 65.0 | 3.5 | 30.0 |
| Whole fluorite MG-80 | ths.tons | 10.0 | 2.3 | 16.3 | 1.1 | 12.7 |
| Whole fluorite MG-75 | ths.tons | 176.4 | 51.0 | 171.6 | 32.7 | 199.9 |
| Whole fluorite MG-65 | ths.tons | 0.0 | 9.7 | 1.6 | 15.1 | 6.2 |
| Fluorite ore ФР | ths.tons | 626.8 | 442.7 | 1207.0 | 204.3 | 1680.4 |
| Blue lead | ths.tons | 23.6 | 22.5 | 19.5 | 29.6 | 16.4 |
| Polishing stone | ths.tons | 26.3 | 5.1 | 26.3 | 0.0 | 0.0 |
| Tin | tons | 98.6 | 17.1 | 71.7 | 26.4 | 0.0 |
| Zinc | ths.tons | 112.9 | 118.0 | 136.3 | 123.7 | 129.6 |
| Zeolite | ths.tons | 39.7 | 0.0 | 39.7 | 0.0 | 105.2 |

*Source*: Mining Production and Technology Division; MRPAM

A proven raw material reserve is assumed to include what mining operations will actually produce based on current science and technology. (Since reserve calculations include factors such as commodity price, average ore grade, cut-off ore grade, and concentrate grade, it is easy to understand that an ore body may become larger or smaller over time depending on these and other variables.)

The classification currently prescribed by the Ministry of Mines divides mineral resources and reserves into groups (resources, geological reserves and economic reserves) according to the precision of their study (pre-feasibility or feasibility study, geology, prospecting, exploration) and on the basis of mining, processing, economic, market, legal, environmental, social, governmental and other factors that may affect economic extraction.

The terms resources and reserves as used in Mongolia cannot be compared to reserves or resource classes estimated using JORC or NI43-101 methods.

## 2.4 *Primary Mining Products and Mining Companies*

Mining and quarrying make up the largest part of Mongolian industry. At the end of 2019, the total production of mining reached 12,467.1 billion MNT, increasing by 11.0% compared to the same period last year.

The production of the most important raw materials in mining is divided into the following four sub-areas:

- Ore mining
- Coal mining

- Crude oil production
- Other raw materials

Total foreign trade turnover reached $932.3 million in January 2020, down 16.1% or $178.9 million from the same period last year. Imports exceeded exports, with a difference of $14.3 million. Mining products such as iron ore, coal, copper concentrate, zinc concentrate, fluorspar, molybdenum concentrate and crude oil accounted for 98.3% and 92.5% of total commodity exports, respectively.

In 2019, fluorspar production increased by 54,900 t or 54.3% and iron ore production increased by 2400 t or 37.7%. However, molybdenum concentrate production decreased by 200 t, or 3.3%, and fluorspar concentrate production decreased by 19,100 t, or 0.6%, year-on-year.

In addition, the production of copper metal concentrate in 2019 was 1.2624 million t. Compared to the previous year, this represented a decrease of 48,400 t or 3.7%, but an increase of 182 kt or 16.8% compared to 2014, and 740.4 kt or 2.4 times compared to 2010 (Fig. 7).

### 2.4.1 Coal

Mongolia is among the ten countries in the world with the largest coal reserves, with 173.3 billion tons of coal reserves from currently over 300 known deposits in fifteen basins. Currently, there are seven state-owned and local companies in eight open-pit mines, 28 domestic companies and joint ventures in 33 mining operations, and seven foreign investment companies in another eight open-pit mines (source: MPRAM).

In 2019, coal production reached 55.8 million tons. Compared to the previous year, this was an increase in production of 4.4 million tons or 8.5%. Exports accounted for 36.6 million tons of coal, or 65.6% of total coal production. Compared with the previous year, exports increased slightly by 0.3 million t or 0.9%.

Coal has become the leading export product. In 2019, hard coal accounted for 48.8 million tons or 87.5% of total coal production and lignite for 7.0 million tons or 12.5%.

Average coal production per capita was around 17.5 t in 2019, up 1.2 t or 7.3% year-on-year. Among them, hard coal production per capita was about 15.3 t in 2019, increasing by 1.2 t or 8.9% compared with 2018, while average lignite production per capita increased by 2.2 t. The average coal export per capita was around 11.5 t in 2019. It increased by 1.3 t or 13.1%.

In 2019, revenue from coal mining amounted to MNT 6.8 trillion. This was MNT 1.5 trillion or 28.6% more than the previous year. The share of coal mining in total industrial output reached 22.7% in 2019, with growth of 2.7%. Coal mining accounted for 39.5% of mining output in 2019, up 4.9% from 2018 (Figs. 8 and 9).

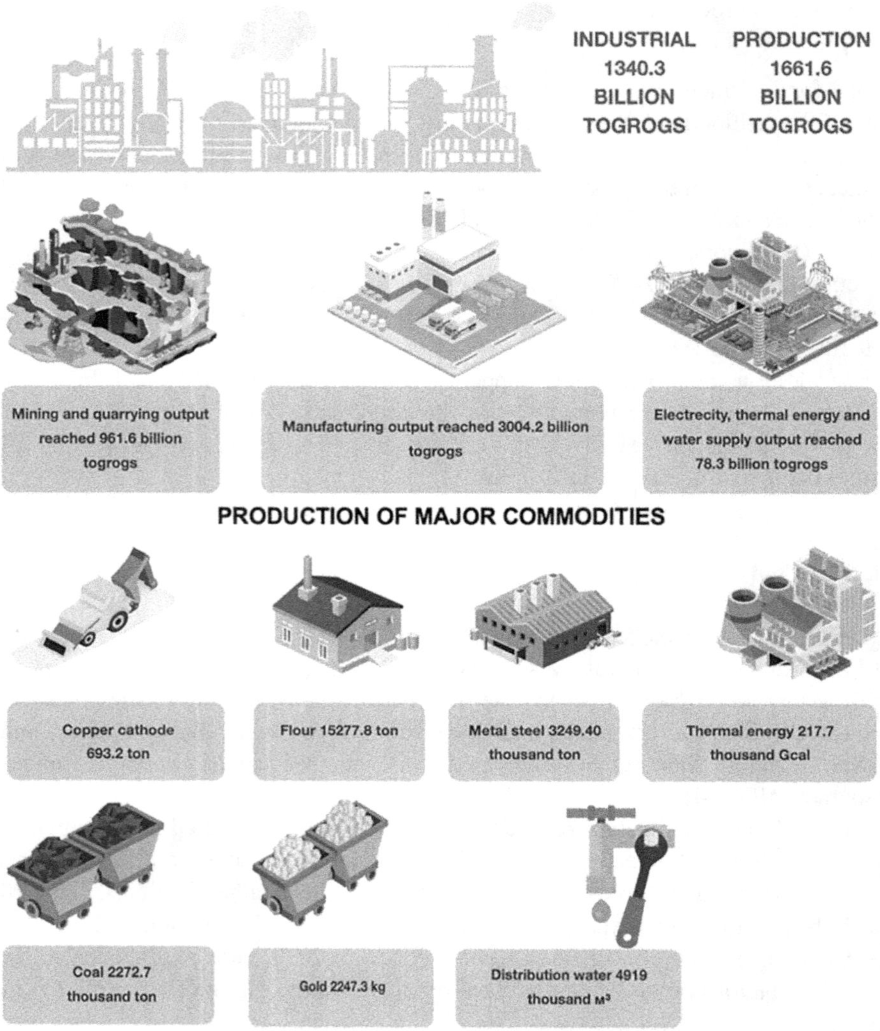


**Fig. 7** Raw material production and industrial production of Mongolia

### 2.4.2 Copper

On the territory of Mongolia, dozens of large and small copper and copper-molybdenum deposits have been identified, as well as more than 400 occurrences and mineralization zones of the various ore formations. These are located in twelve zones of Zich province. A total of 36.3 million tons of copper reserves were explored and identified.

Concentrate production in Erdenet accounts for more than 20 percent of the country's economy. The exploited Erdenet Ovoo deposit has ore reserves of around

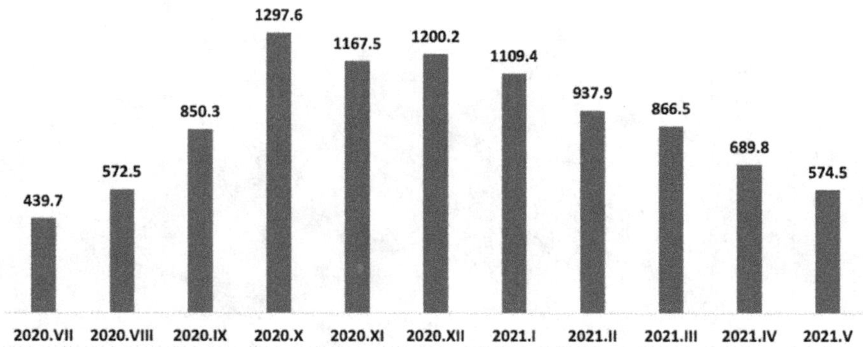


**Fig. 8** Mongolia's coal export (MRPAM 2023)

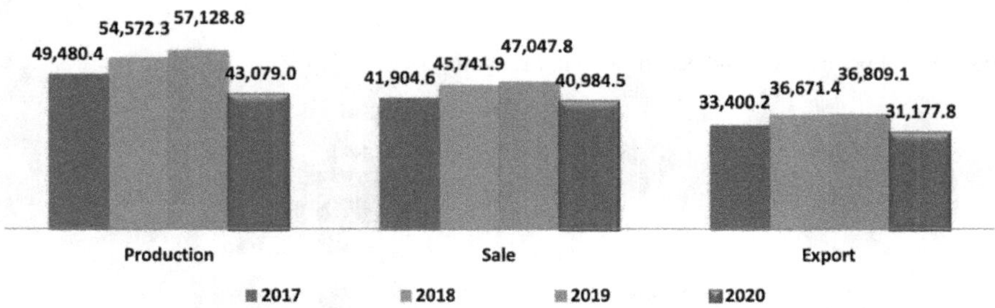


**Fig. 9** Mongolia's coal production, sales, and exports 2017–2020 in kt (MRPAM 2023)

1 trillion t. copper reserves amount to 4632 million t and 125,414 t of molybdenum reserves.

It is estimated that these reserves will last for another 25 years, increased by the fact that the Ministry of Mines has taken over five exploration licenses for the Erdenet Ovoo. The reserves grew by three million tons of copper and 55,000 tons of molybdenum. The value of these reserves is currently estimated at $24 billion.

**Oyu Tolgoi** is among the top 5 in the world in terms of reserves and mining capacities. According to current knowledge, the ore reserves of the deposit amount to 3380 million t. It is estimated that 31.1 million t of copper, 1328 t of gold and 7601 t of silver are mined.

Es ist drei Jahre her, dass die Regierung der Mongolei mit Rio Tinto und Ivanhoe Mines eine Investitionsvereinbarung zur Erschließung der Lagerstätte unterzeichnet hat. In dieser Zeit wurde die Aufbereitungsanlage zur Konzentratherstellung gebaut, die sich momentan im Probebetrieb befindet. Die Anlage hat eine Produktionskapazität von 100.000 t Erz pro Tag.

**Tsagaan Suvarga** copper and molybdenum deposit is located in Dornogovi Aimag, 220 km southwest of the Aimag center, 550 km south of Ulaanbaatar, 170 km west of Zuunbayan railroad station and 75 km from the center of Mandakh Soum.

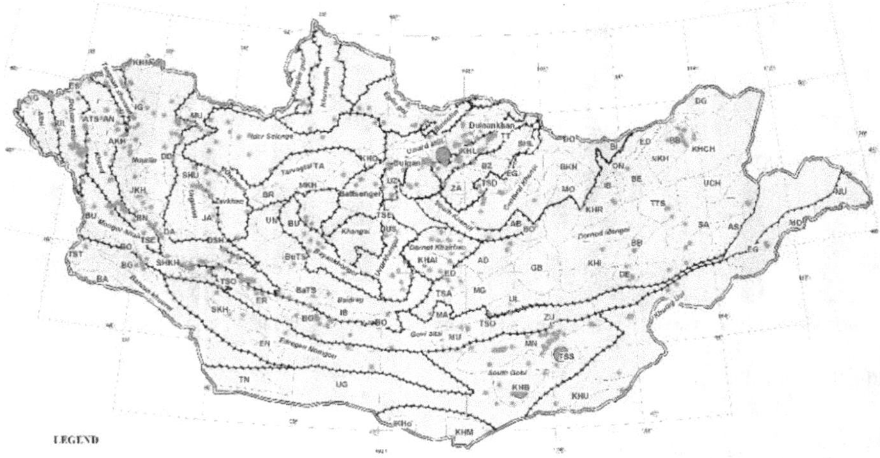

**Fig. 10** Copper deposits and occurrences in Mongolia

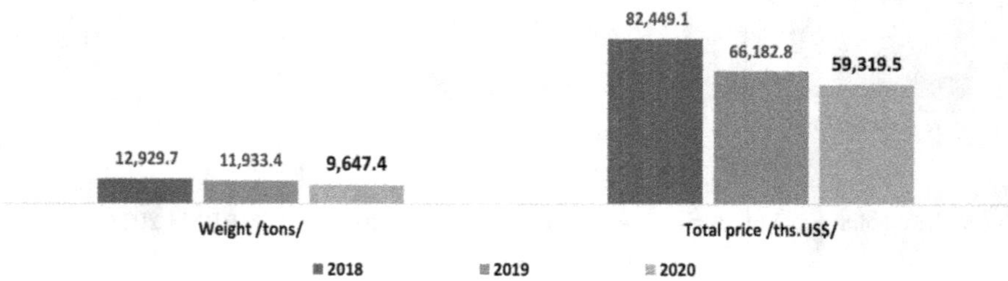


**Fig. 11** Copper cathode export in tons and price in thousand USD 2018–2020

The license for the Tsagaan Suvarga deposit is held by Mongolyn Alt LLC (MAK). The deposit has 236.2 million tons of ore reserves and may contain 568.6 thousand tons of copper and about 66,000 tons of Mo (Figs. 10 and 11).

In addition, metallogenic zones have been identified in the Mongolian Altai, Gobi-Altai, Tsagaan-Olom, Dundgovi and South Kherlen with copper, copper-zinc-pyrite, copper-skarn mineralization and copper-porphyry mineralization.

### 2.4.3 Gold

In connection with the adoption and implementation of the Law on Prohibition of Exploration and Mining in the Area of Headwaters of Rivers, Water Protected Areas and Forest Areas in 2009, gold production was reduced.

When the legal and fiscal environment of gold mining was improved in 2013 and gold production was encouraged by the government, there was an immediate increase in production. Consideration was given to further increasing gold production. To this end, a working group was jointly formed in 2016 by the Ministry of Finance, the Ministry of Mines and Heavy Industry, and the Bank of Mongolia to develop the Gold II program and began implementation in 2017 (Mineral Resources and Petroleum Authority).

In 2019, gold production was 16.3 t, which means that it decreased by 4.4 t or 21.1% compared to the previous year. 9.1 t or 55.8% of the total mined gold was exported, which means that gold export decreased by 5.7 t or 2.6 times compared to the same period the year before.

The average gold production per capita in 2019 was about 5.1 g and decreased by 1.4 g or 22.1% compared to last year. In addition, under the national program "Gold 2" prospecting was carried out on an area of 521,200 km$^2$, which is 33.3% of the total area of Mongolia, so that additional areas and deposits with gold deposits could be identified in detail.

### 2.4.4 Fluorite

Mongolia is one of the leading fluorite producers in the world and has extensive information on all aspects of fluorite ores in Mongolia. Fluorite mining in Mongolia began in 1946, and in the first decades only high-grade fluorite ore was mined and exported after hand sorting. In 1986, Mongolsovtsvetmet, renamed Mongolrostsvetmet in 1991/92—a Mongolian-Russian joint venture, started mineral processing of low-grade fluorite ores, i.e., production of fluorspar concentrates, in Bor Ondor.

Today, the fluorite ore processed at the Bor Ondor plant contains on average 30% CaF2 with an applied cut-off content of 26–28% CaF2 (oral communication, 2008). Both "metallurgical grades" >75% CaF2 and "flotation grades = acid grades" >97% CaF2 are produced and exported by rail mainly to Russia, USA, Ukraine and other CIS countries.

In 2008, another large fluorspar processing plant was opened near Dalanjargalan Sum by a Chinese American joint venture (Kevin Invest LLC). It focuses on the purchase of fluorite ore from Ninjas and from the recently discovered Burkhant fluorite deposit.

Most fluorite deposits in Mongolia are hydrothermal or epithermal in origin. Based on the major minerals, there are five types of epithermal fluorite deposits in Mongolia:

- Quartz Fluorite,
- Calcite-quartz-fluorite,
- Barite-quartz-fluorite,
- Aduralia (orthoclase) quartz-fluorite,
- Sulfide quartz fluorite (sulfides: e.g. pyrite, pyrrhotite, galena, sphalerite).

### 2.4.5 Iron Ore

Mongolia has been exporting iron ore and concentrate for more than ten years. According to statistics, there are a total of 250 iron deposits in Mongolia. Detailed geological explorations have been carried out on 25 of these.

The reserves of seventeen of them are estimated at 430 million tons of iron ore. The total reserves of iron ore deposits and deposits are estimated at 5.5 billion tons. In fact, Mongolia lags far behind the rest of the world in metallurgy. Therefore, it is necessary to process ores and concentrates based on large iron deposits such as Tumurtei and Tumurtolgoi in order to carry out value addition in the country.

### 2.4.6 Rare Earth Elements (REE)

In order to intensify the prospecting and exploration for rare earth deposits in Mongolia, the Ministry of Mines and Heavy Industry has developed the "Rare Earths" program. With the help of this program, mining on REE is to be advanced in order to achieve a positive effect on the national economy.

Currently there are six main deposits, about 80 occurrences and more than 280 mineralizations on the territory of Mongolia. Four deposits and several occurrences have very low grades or are very small, so they are currently economically insignificant. These include the Tsagaan Chuluut placer deposit in Omnodelger Soum, Khentii Aimag, which contains only 758 tons of monazite. Between 2009 and 2010, three of the four REE deposits were re-examined and re-sampled. The newly collected and analyzed samples confirm the old analytical data and allow a final evaluation of the deposits in international comparison.

The Khalzan Burgedei REE deposit is located in Khovd Aimag in western Mongolia. The deposit was discovered in 1983 and explored in detail in the following years. It is part of the Western Mongolia Lakes Zone, which contains several Middle Devonian REE deposits.

The Lugiin Gol REE deposit is associated with the Lugiin Gol complex. The Lugiin Gol carbonatites are not as abundant or volumetrically significant as the alkaline silicate rocks (syenites), but represent unique REE mineralization accessible for mining and metal extraction among the currently known occurrences and deposits of REE ores in Mongolia.

These ores occur predominantly as veins and dykes ranging from a few centimeters to 2 m wide and up to 1000 m long. The calcite is medium- to coarse-grained (0.1–15 mm) and forms a mosaic of crystals in the negative V position (70–98%), associated with additional amounts of primary automorphic apatite or REE carbonates (Kynicky et al. 2018).

The Mushgai Khudag REE deposit is located in Mandal-Ovoo, Umnugovi Aimag, southern Mongolia and was discovered in 1976, prospected in 1983/84 and explored in more detail between 1989 and 1994. Four types of mineralization can be distinguished:

1. Mineralized breccia with carbonite cement
2. Mineralized carbonatite
3. Magnetite-apatite ore
4. Apatite ore

The initial calculated resources are:

- 21,885,097 tonnes of ore (>0.5% REO) with an average grade of 0.88% REO ore, equivalent to 192,448 net REO in the main ore body
- 1,752,162 t of apatite ore grading 3.11% REO, averaging 54,492 t net REO, both in category in C2
- 1.6 million t of ore grading 0.63–0.65% REO or 24,700 t net REO in category P1.

This results in 23,637,259 tons of ore and 246,940 tons of net category C2.

## 2.5 *Mining Companies*

### 2.5.1 Erdenes Mongol LLC

Erdenes Mongol LLC was established on February 22, 2007 with the aim of taking over the state representation of Mongolia's strategic deposits and implementing their exploitation.

Erdenes Mongol LLC owns various companies responsible for the strategic deposits of mineral deposits and related infrastructure development, projects and programs to enhance their value and operations. In addition, Erdenes Mongoll LLC holds the state-owned interests in the mining companies.

Its functions are to implement strategic, government policies to increase profitability, attract foreign and domestic investors, raise capital, implement value-added industrial projects, and increase production capacity (Fig. 12).

The operating subsidiaries of Erdenes Mongoll LLC are

- Erdenes Tavan Tolgoi JSC,
- Erdenes Oyu Tolgoi LLC, holds 34% of Oyu Tolgoi LLC
- Baganuur JSC,
- Shivee Ovoo JSC,
- Mon-Atom LLC,
- Erdenes Shive Energy LLC,
- Erdenes Methane LLC which in turn holds 50% or more interest in Erdenes Asset Management LLC and
- Gashuun Sukhait Road LLC, which operates and maintains the Tavan Tolgoi-Gashuun Sukhait road under a public-private partnership.

### 2.5.2 Erdenes Tavan Tolgoi JSC

Erdenes Tavan Tolgoi JSC was established in 2010 to economically exploit the strategically important Tavan Tolgoi coking coal deposit. Geographically, the deposit is located 98 km east of Dalanzadgad, a center of South Gobi Province, and 15 km southwest of the city of Tsogtsetsii. The deposit is 550 km away from the national capital Ulaanbaatar.

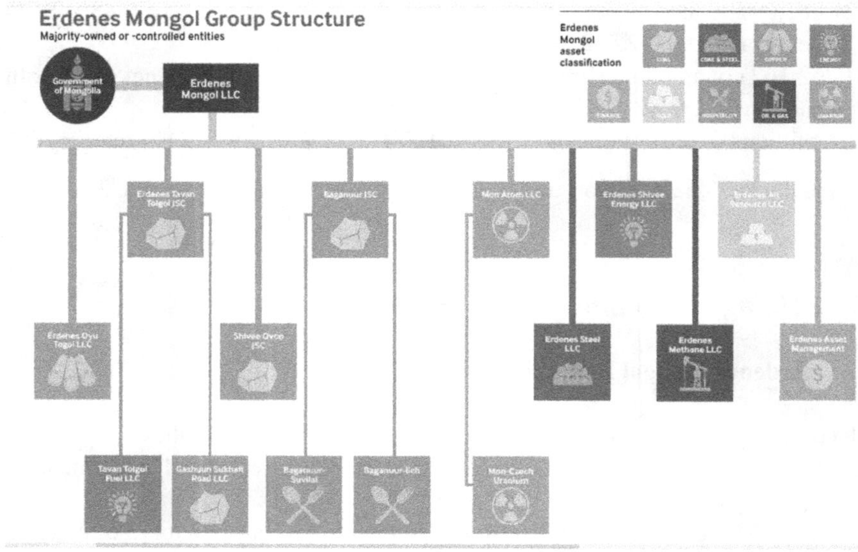


**Fig. 12** Corporate structure of Erdenes Mongol LLC

The deposit is located in a hilly region of the Gobi at an altitude of 1515–1560 m NN. The region is characterized by a harsh environment with 154 mm of precipitation, most of which falls in summer and evaporates completely.

### 2.5.3 Baganuur JSC

Baganuur JSC has been responsible for ensuring the supply of coal-fired power plants within the centralized power supply system since its establishment in 1978. In 1995, the company was privatized and transformed into and joint stock company. 75% of the shares are owned by the state or Erdenes Mongol LLC as trustee and 25% are owned by Mongolian citizens.

The Baganuur lignite deposit is located 130 km northeast of Ulaanbaatar, 15 km east of the center of Bayandelger Soum, Tuv Aimag, in the Tavansuvai steppe in the Kherlen River basin. It is connected to Maanit station by an 80 km railroad line. A paved road leads from Ulaanbaatar to Baganuur.

The maximum annual capacity is 4.0 million t. Depending on the needs of domestic consumers, 3.6 to 3.8 million t of coal are mined and 16.0 to 18.0 million $m^3$ of overburden are moved. Open pit mining supplies 60% of Mongolia's coal demand, which is more than 70% of the coal demand of the central region.

According to studies, Baganuur coal can be used not only for power generation, but also as a raw material for gas, fuel and briquette production, as well as for the production of chemical products. The total reserves of the deposit are 812.1 million tons.

### 2.5.4 Shivee Ovoo JSC

The Shivee Ovoo open pit mine is located 265 km south of Ulaanbaatar in the Shiveegobi Soum of the Gobisumber Aimag at 1200 m above sea level. In 1990, the Ministry of Mines decided to establish the Shivee Ovoo open pit mine as part of the mining operation in Nalaikh district. Preliminary and detailed explorations were carried out by the state in 1986–1988. The geological reserve for the Shivee Ovoo was approved by the National Resource Committee through Decree No. 33 by December 22, 1989. The Shivee Ovoo property covers an area of 29,500 hectares of land, the deposit comprises three main parts and contains reserves of 2708.7 million tons of coal.

Shivee Ovoo coal is classified as B2 lignite. The deposit contains a total of eight seams, of which Seams I and II are currently being mined. The total thickness of the two seams is 17 m, with seam I 10 m thick and seam II 7 m thick. The intermediate medium has a thickness of approx. 6 m.

### 2.5.5 Erdenes Oyu Tolgoi LLC

An Investment Agreement and a Shareholders Agreement were signed in 2009 and 2010, respectively, by and between the Government of Mongolia, Rio Tinto International Holdings Ltd and Ivanhoe Mines LLC for the development of the Oyu Tolgoi copper-gold deposit.

The Mongolian government assigned the state-owned mining holding company "Erdenes MGL LLC" to manage the 34% state-owned shares of Oyu Tolgoi LLC. Based on the 348th Government Decision from 2011 and the 28th Board Decision of Erdenes MGL LLC from 2011, Erdenes Oyu Tolgoi LLC was established as a subsidiary of Erdenes MGL LLC. The mission of Erdenes Oyu Tolgoi LLC in terms of the partnership agreement is to hold the 34% interest in Oyu Tolgoi LLC and to take care of the operation of Oyu Tolgoi LLC in order to profitably use the interest during the operation.

The governing body of Erdenes Oyu Tolgoi LLC consists of the general meeting of shareholders, the board of directors and the managing directors. The decision-making authority is vested in the General Meeting of Shareholders, which is fully chaired by the Board of Directors of Erdenes Mongol LLC. The Board of Directors of "Erdenes Oyu Tolgoi" LLC consists of seven members in accordance with the 28th Board Resolution from 2011.

### 2.5.6 Erdenes Silver Resources LLC

The Salkhit silver-gold deposit is located in the Gurvansaikhan Soum of Dundgovi Province, 280 km south of Ulaanbaatar and 70 km northeast of Mandalgovi city center. Based on the Decision of the Government of Mongolia No. 133 of 2014 and the Decision of the Board of Directors of Erdenes Mongol LLC No. 15 in 2019,

Erdenes Silver Resource LLC was established in May 2019. The business purpose includes the following areas:

- Consulting
- Prospecting and exploration
- Extraction and processing
- Execution of contractual drilling works

Erdenes Silver Resource LLC was initially granted an exploration license on October 17, 2019 over an area of the size of 2887.85 ha named "Salkhit" in Gurvansaikhan Soumum, Dundgovi Province and subsequently mining license MV-021483. In the first quarter of 2020, Erdenes Silver Resource LLC internally developed an organizational and operational work plan to implement a successful private sector benchmark project and economically exploit the deposit in mid-2020.

### 2.5.7 Erdenes Rare Earth Elements R&D Unit

Based on the Decision of the Mongolian Government No. 176 dated 01.05.2019 and the Board Decision of Erdenes Mongol LLC No. 16 dated 10.05.2019, the Rare Earth Research Project Unit was established with the following objective:

- Improve the legal environment for geology and exploration of rare earth elements in Mongolia;
- Identify state interests in rare earth deposits and provide support and participation to private stakeholders.
- Conduct detailed studies regarding geological exploration and processing technologies
- Conduct investment and economic benefit studies.

### 2.5.8 Erdenes Commodity Exchange LLC

In accordance with Article 3.7.5 of the "State Policy Paper on the Mineral Resources Sector" approved by the Parliament of Mongolia for the establishment of a mineral exchange to regulate exports through a unified policy", Erdenes Mongol LLC, together with the private company MSH Clearing LLC, established Mineral Bayalgiin Birj LLC on June 12, 2019.

The main activity of the company is to create a market that has international standards for open and transparent trading of commodities through regulated and organized markets and complies with government policies in the field of mineral resources. The Board of Directors elected the members of the Company's management on June 27, 2019 and appointed the management on June 28, 2019.

### 2.5.9 Erdenet Mining Corporation (ERDENET) JSC

Erdenet Mining Corporation is a mining company located in Erdenet, Mongolia. It was built in 1974 to exploit Asia's largest copper ore deposit. At the same time, this is the fourth largest copper mine in the world. Erdenet Mining Corporation was a Mongolian-Russian joint venture until about 2016 and has a large share of Mongolia's revenues. Erdenet mines 22.23 million tons of ore annually and produces 126,700 tons of copper and 1954 tons of molybdenum. Mining operations account for 13.5% of Mongolia's GDP and 7% of tax revenues. Approximately 8000 people are employed in the associated operations.

Geological surveys at Erdenetiyn-Ovoo first began in the late 1950s. The copper-molybdenum ore body consists of the minerals chalcosine (Cu2S), chalcopyrite (CuFeS2), turquoise, bornite (Cu5FeS4), brochantite (Cu4SO4(OH)6), azurite (Cu3(CO3)2(OH)2), molybdenite (MoS2), delafossite (CuFeO2), tenorite (CuO), sericite and other minerals.

For a long time only copper concentrate was produced and exported in Erdenet. Since 2005, pure copper has also been produced. Today, the open pit mine has reached a length of 2500 m, a width of about 1500 m and a depth of about 120 m.

### 2.5.10 Bor Undur JSC

The largest fluorite deposit in the country covers an area of 70 km$^2$. The ore bodies are grouped in the four deposits Bor-Undur, Adag, Bayan-Ulaan, Undur-Ovoo at intervals of 2 to 10 km. Of these, Bor-Undur is the largest deposit. The Bor-Undur basement and the Alagsav volcanic depression are the main geological-structural features of the ore field and connect the individual deposits.

The ore bodies are represented by high-angle, vein-like zones of mineralization consisting of breccias as host rocks with fluorite and quartz-fluorite cement. Ore body thickness varies from 0.5 to 3 m; they extend from 100 to 3400 m along strike and 20 to 350 m down dip.

The average grade of the main constituents varies within the following limits:

- $CaF_2$ 28–44%
- $SiO_2$ 40–55%
- $CaCo_3$ 0.8–2.2%
- $BaSo_4$ up to 0.05%
- $Al_2O_3$ 1.5–2.7%
- $Fe_2O_3$ 0.04–2.7%
- S up to 0.04%
- P 0.01–0.04%

The other two deposits, Bayan-Ulaan and Undur-Ovoo, are characterized by modest parameters. They can be considered as subsidiary deposits of the Bor-Undur field.

### 2.5.11 Mongolyn Alt Corporation (MAK)

Mongolyn Alt (MAK) LLC is one of the largest private business groups in Mongolia with its origins in gold mining and later expansion into coal and copper mining. MAK was founded in 1993 by Nyamtaishir Byambaa.

MAK operates the mining operations listed below:

- Eldev coal mine
- Erdenet coal mine
- Naiirin (or Naryn) Sukhait open pit coal mine
- Tsagaan Suvarga copper and molybdenum project
- Aduun Chuluun thermal coal mine

MAK also has a stake in the airline Hunnu Air.

### 2.5.12 Mongolian Mining Corporation (MMC) "Energy Resource" LLC

Mongolian Mining Corporation (MMC) is a Mongolian coking coal producer listed on the Hong Kong Stock Exchange. It is the largest coal mining company in Mongolia and owns two mining operations in the Gobi Desert, namely the Ukhaa Khudag and Baruu Naran open pits.

MMC was incorporated in Hong Kong on May 18, 2010, as a limited liability company with its registered office in the Cayman Islands. Prior to incorporation, the business was operated by Energy Resources LLC, a limited liability company incorporated under Mongolian law on April 22, 2005. Energy Resources LLC is now a wholly owned subsidiary of MMC.

MMC is engaged in open pit mining with Energy Resources LLC at the Ukhaa Khudag (UHG) deposit in the Tavan Tolgoi coal formation in South Gobi and operates the Baruun Naran open pit mine through its wholly owned subsidiary Khangad Exploration LLC. According to MMC's 2014 annual report, Odjargal Jambaljamts, the company's chairman, is the company's largest shareholder. Kerry Group Limited, a Hong Kong group of companies, also owns a substantial stake in MMC (Fig. 13).

## 2.6 *Project Development*

The rise in the share prices of mining companies operating in Mongolia and registered abroad—such as Xanadu Mines, Mongolia Energy Corporation (MoEnCo), Mongolian Mining Corporation (Energy Resources) and others—is a sign that foreign investors here assume stable market conditions and legal certainty.

The highest growth on the Hong Kong Stock Exchange was recorded by MoEnCo, and although the other companies did not reach the 400 percent increase

| Deposit name | Type of mineral | Companies |
|---|---|---|
| Tavan Tolgoi | Bituminous coal | ERDENES MONGOL; MONGOLIAN MINING CORPORATION |
| Nariin Sukhait | Bituminous coal | MAK |
| Baganuur | Brown coal | БАГАНУУР ХК |
| Shivee Ovoo | Brown coal | ШИВЭЭ ОВОО |
| Oyu Tolgoi | Copper, gold | RioTinto |
| Tsagaan Suvarga | Copper, molybdenum | MAK |
| Erdenet | Copper, molybdenum | Erdenet Mining Corporation |
| Asgat | polymetallic deposit (silver) | ERDENES SILVER RESOURCE |

**Fig. 13** Mongolia's main mining companies

in a single day, the shares of Energy Resources and SouthGobi Resources (SouthGobi Sands) also rose between 32% and 45%. It is now up to the new government to ensure that investors do not lose their confidence again.

Yet this new government faces an enormous challenge. The global economy has stalled due to the pandemic. All the more reason for Mongolia to attract foreign investment to repay substantial debts at a time of declining revenues and to make progress on megaprojects. A major advantage for the new government is that the ruling party's large majority in parliament makes it easier to implement policy decisions. The previous government initiated urgently needed projects and created framework conditions for the economy.

Over the next four years, the megaprojects must now be developed to create jobs and generate revenues. Substantial investments must be acquired to develop new raw material deposits and strengthen the mining sector.

### 2.6.1 Tavan Tolgoi

D. Sumiyabazar, Minister of Mines and Heavy Industry in the previous government, estimates that Mongolia's economy will grow by 10 percent when mining of the deposit reaches full production. Export revenues could increase by $2.8 billion and between $0.4 billion and $1.1 billion would benefit the state budget. About 5800 new jobs could be created.

Successful development of the deposit would also mean that the citizens of Mongolia, who each received 1072 shares in the deposit from the state in 2008, would receive dividends from it. However, this would require a processing plant, a power plant, a railroad and paved roads, and an initial public offering on an international stock exchange. A working group headed by the Minister of Mines and Heavy Industry determined that all of these and other measures to further develop the deposit would require a total investment of $7.11 billion.

The following table provides details of the proposals at that time, although they were subject to some changes over time (Table 7).

Considerable advance work had been done for the IPO planned for this year before the whole thing was put on hold before the election. No explicit reasons were given, but both domestic political considerations and global uncertainty due to the pandemic must have played a role in the government's decision.

There was no word from the new government on what it will do about the IPO or how it will finance the development of Tavan Tolgoi. The IMF believes that the global economy is facing a deep crisis and that a reasonable recovery will take a long time. An updated Asia Development Bank report forecasts that Mongolia's economy will contract by 1.9 percent in 2020 but grow by 4.7 percent in 2021. If the

**Table 7** Future projects at Tavan Tolgoi

| Project | | Term of implementation | Estimated cost |
|---|---|---|---|
| 1. | Conveyor system | 2018–2020 | $75 million |
| 2. | Mining equipment | In 2021, 35 million tonnes to be mined | $1.3 billion |
| 3. | Exploration work | 2019–2030 | $150 million |
| 4. | Coal processing plant | 2019–2021 | $50 million |
| 5. | Water supply | | $150 million |
| 6. | 240-km railway on the Tavantolgoi-Gashuunsukhait route | | $824 million |
| 7. | Basic structure of a railway on the Tavantolgoi-Zuunbayan route | | $1.24 billion |
| 8. | 250-km asphalt road on the Tavantolgoi-Gashuunsukhait route | | |
| 9. | Special purpose road for heavy trucks on the Tavantolgoi-Oyutolgoi-Khangi route | | |
| 10. | 270-km road to the Tsagaan Del border check point | | |
| 11. | Power plant with 45 MWt capacity | 2018–2021 | $1 billion |

Tavan Tolgoi IPO takes place, it would be under very different conditions than before COVID-19.

One way to inject more money into Mongolia's economic cycle is to pay dividends on the 1072 shares and allow the shares to be traded. Erdenes Tavan Tolgoi no longer has any debt and has made a profit in the last two years with sales revenues of $1 billion. Part of this profit was spent on the construction of the Tavantolgoi-Zuunbayan rail line, and it remains a priority project as coal exports need to be increased to compensate for the loss of revenue.

### 2.6.2 Oyu Tolgoi

The underground mine currently being developed contains 80 percent of Oyu Tolgoi's total resources. If full production starts there, about 1.8 million tons of Cu concentrate would be mined, allowing Mongolia to generate an additional $1 billion or so in tax revenues annually.

Equally important for Mongolia, successful development of the underground project by the operator would restore foreign investor confidence in Mongolia. The government's statement that the mine will be operational in the next four years is already a very positive message, as it makes it clear that the government wants the project to proceed without major political interference or obstruction. This is in line with Parliament's decision in November 2019 to proceed with the development of the operation as part of the overall project.

How this will be done is the basis of negotiations between a Mongolian government working group and Rio Tinto representatives. The new government must continue the talks and bring them to a satisfactory conclusion. Rio Tinto has indicated that the design of the underground mine will have to be changed because the geotechnical conditions would be different than previously assumed. In addition to an extension in the execution, this will result in a significant increase in costs.

Detailed estimates of time and cost are expected in the second half of the year. The work is further delayed because Rio Tinto cannot bring its personnel into the country due to the quarantine required by the pandemic. However, it is still currently expected that the underground operation will start production in 2023.

### 2.6.3 Tsagaan Suvarga

When production begins at this copper-molybdenum deposit, between 200 and 250 billion MNT in taxes and royalties will flow into the state budget annually, and about 1100 additional jobs would be created. This makes it a significant mining project.

A few years ago, MAK acquired a $450 million loan from the European Bank for Reconstruction and Development (EBRD). Since Tsagaan Suvarga was a deposit of strategic importance, the state would own 34% of the shares, but would also bear a corresponding share of the development costs, estimated at $1.1 billion.

Parliament discussed the issue and decided not to take this risk. In accordance with Parliamentary Resolution No. 54, the government signed an investment agreement with MAK guaranteeing that tax rates would remain stable for the duration of the project and that the company would pay all applicable taxes for 24 years. When the EBRD withdrew its offer of financing, citing domestic political disagreements as the reason, the project remained frozen for some time until MAK resumed work with $400 million in loans from international banks and financial organizations. To date, 50 percent of the work has been completed, including the purchase of all equipment. Tsagaan Suvarga is a good example of how the private sector can develop a mega-project without government support.

### 2.6.4 Kharmagtai

The share price of Australian-listed Xanadu Mines has risen since the election, as the foreign investor in this copper-gold project has been promised government support. Xanadu's ongoing drilling program has delivered positive results.

The Kharmagtai deposit, including Oyut Ulaan and Shar Chuluut, is located in the South Gobi porphyry copper province, which also hosts Oyu Tolgoi and Tsagaan Suvarga. Resources will not be announced until exploration is completed and approved by the Mineral Resources Professional Council.

### 2.6.5 Asgat

Asgat was among the deposits whose mining license was nationalized by the previous government. When the license for this deposit of strategic importance was transferred from Mongol RosTsvetmet to Erdenes Mongol in 2019, the government declared that it would be put into economic circulation without much delay. Accordingly, the Emergency Management Authority and Asgat Kent Company opened the mine's shaft and took 20 tons of samples from it for laboratory analysis late last year.

The results of the analysis are important because Asgat has reserves of 2359.3 tons of silver and 356 kg of gold, as well as lead, zinc, cadmium, arsenic and antimony. The presence of the last two toxic minerals in the mixed metals ores resulted in the deposit never being mined, as the level of toxicity is more than tolerable and there is no reliable technology to separate the toxic minerals from the concentrate.

Even if a solution to this problem is found, the problem of processing the toxic materials in the processing plant and removing the toxic waste materials still remains.

Another difficulty is that the deposit is located partly in Mongolia and partly in Russia, so an international arrangement must first be made here.

So, despite the government's stated intention, which even then was seen as populist rather than pragmatic, it is unlikely that work at Asgat can begin anytime soon, or even during the new government's term. Speaking at the Discover Mongolia

Forum 2019, P. Gankhuu, CEO of Erdenes Mongol, said Asgat would require $127 million for development and a feasibility study was being prepared.

### 2.6.6 Salkhit

This deposit is an example of how MNT 800 billion in proceeds for the sale of the mining license in 2018 had already been distributed to the population before production began. Since Salkhit is one of the strategic deposits, the government bought back the license without having the financial resources to do so. Therefore, it has issued bonds to be repaid with the sales proceeds from the production of the Salkhit deposit. The largest lenders, Khan Bank and State Bank, bought 88.6 percent of the bonds and seven other commercial banks bought the rest.

An updated feasibility study indicates that even if Salkhit begins production and sales, the likely cumulative profits at the time the bonds mature would be just enough to cover about 1/3 of the amount needed.

The mining license is now held by Erdenes Silver Resource, which sold a 47.2 kg silver ingot from Salkhit some time ago, but it is not clear when and where the ore was mined and processed, or where the ingot was cast.

### 2.6.7 Heavy Industries

Work on the oil refinery in Dornogovi Aimag is progressing well so that oil production can begin in 2023. The investment volume is 1.24 billion US dollars. Much of the infrastructure has already been completed and 550 apartments are currently being built. Construction of the refinery itself has been delayed by the fact that the foreign experts who are to supervise the work cannot come to Mongolia because of the pandemic.

The refinery is expected to form the basis for a diverse petrochemical industry in Mongolia. As the prime minister said in his speech at the groundbreaking ceremony for the refinery, "This opens up the possibility of producing everything from lipstick to machinery in Mongolia."

During the socialist era, Mongolia produced ten industrial items and now, some three decades later, the industrial park is expected to house production facilities for some 30 products. This requires significant investment.

The government plans to build an iron ore processing plant and a coke plant in the next four years. A beneficiation plant for processing iron ore using the wet magnetic process is being built in Altanshiree Soum and is scheduled to start production next year. It will have a capacity of 2 million tons of iron ore per year and recycle 90 percent of the water used in production. The construction of a smelter and a precious metals and gold refinery also continues to be protected. However, no investor has yet been found.

### 2.6.8 TT-ZB Railway

Die Regierungspartei teilte den Wählern mit, dass in den kommenden vier Jahren 730 km neue Eisenbahnlinie gebaut werden würden. Das klingt ein wenig zu optimistisch, da seit ihrer Verabschiedung vor zehn Jahren auf keiner der Strecken der staatlichen Eisenbahnverkehrspolitik tatsächlich ein Kilometer Eisenbahn gebaut wurde.

Dennoch wurde an der 414 km langen Eisenbahnstrecke Tavan Tolgoi-Zuunbayan, die zwar nicht im Aktionsplan der Regierung enthalten war, aber ein Lieblingsprojekt des Präsidenten ist, umfangreiche Arbeiten ausgeführt.

Die Bahnlinie soll 2021 für den Verkehr freigegeben werden. Die Kosten belaufen sich auf 1,2 Milliarden US-Dollar, von denen 750 Milliarden MNT aus den Gewinnen von Erdenes Tavan Tolgoi im Jahr 2018 stammen. Ein Betrag, der ansonsten als Dividende an alle Bürger ausgeschüttet worden wäre.

Die Bahn ist an die Transmongolische Eisenbahn angeschlossen und kann jährlich bis zu 30 Millionen Tonnen Fracht sowohl in nördlicher als auch in südlicher Richtung befördern. Insgesamt 37 Lagerstätten, darunter Tavan Tolgoi, Tsagaan Suvarga, Kharmagtai, Manlai und Kharaat Mountain, liegen an dieser Achse.

### 2.6.9 The TT-GS Railway

Erdenes Tavan Tolgoi estimates that about 30 million tons of coal per year can be exported across the Gashuunsukhait border once this railway is built. This would require a doubling of current production levels.

Coal transportation costs would drop to 25% of current levels, saving about $150 million. But rail transportation would also minimize the difficulties created by the current pandemic, for example.

With the benefits at hand, one would expect Mongolia to give this project top priority. In fact, however, political differences and repeated rethinking have meant that work has not yet begun. For example, the choice of gauge is a recurring point of contention.

Also, the commissioning of Bodi International in 2019 to implement the project seems to be going nowhere, as the company has no experience with such projects. In addition, the feasibility study prepared eight years ago has never been updated, so the cost estimate may no longer be valid.

### 2.6.10 The Nariinsukhait-Shiveekhuren Railway

On this railroad, too, execution of the work was postponed from 2008 to 2018 because the gauge issue could not be resolved. Subsequently, the project was awarded to Shiveekhuren Railway Company, a specially formed joint venture, with 51% owned by Mongolian Railway and 49% owned by a consortium consisting of Osokh Zoos, SouthGobi Sands, and Qinhua MAK, all companies that are also included in the Nariinsukhait Deposit Group. The Shiveekhuren-Ceke section of the line was scheduled for completion in 2019, but nothing has happened on the ground so far.

### 2.6.11 The Zuunbayan-Khangi Railway

This railroad line was included in the state policy two years ago, as both Russia and China showed a corresponding interest. Khangi is the gateway to Bugat in China's Inner Mongolia Autonomous Region. Bugat is home to the country's largest steel mill, which buys coking coal from Energy Resources. This center of the steel industry offers great opportunities for exporting Mongolian coal and iron ore.

President Putin referred to the route at the presidential summit in Bishkek. During his visit to Mongolia in 2019, he was told that part of a promised Russian loan would be used to finance railroad construction. With an investment of about $1 billion, the broad-gauge railroad would be able to carry up to 21 million ton of freight per year.

### 2.6.12 The Erdenet-Ovoot-Artssuuri Railway

This project was also included in the state policy in 2018. With an investment volume of around 2 billion US dollars, construction is expected to take 4–5 years.

Aspire Mining, a company listed in Australia, plans to build a 547 km section of the line. At the end of 2019, Ts. Tserenpuntsag made an additional investment of AUSD 33.5 million in Aspire Mining. He now holds 51 percent of the shares and is a member of the new parliament.

In 2016, the railroad was included in the Northern Railway Corridor of the Economic Corridor Establishment Program for Mongolia, the Russian Federation and the People's Republic of China. Russia is interested in connecting the railroad line to the planned Kysyl-Kuragino Railway, as this would allow it to export coal from the Ulug-Khemsk Basin in Tuva to China.

### 2.6.13 Tavan Tolgoi Power Plant

The government's decision to build the Tavan Tolgoi power plant alone promises several benefits for Mongolia.

Construction costs are said to be 30% lower than those offered by Rio Tinto. This means that the generated 300 MW electricity can be sold 50% cheaper. This, in turn, could significantly reduce operating costs at Oyu Tolgoi and could result in termination of the contract with the Chinese power company. In addition, dividends from OT would be due earlier.

Power generation is scheduled to begin in 2025 if construction is completed as planned.

### 2.6.14 Shivee-Ovoo Energy Export Project

This is a very ambitious project as the planned plant will generate electricity from both renewable and non-renewable sources and supply it to southern Mongolia, the Hebei region of northern China and the power grid of Northeast Asia. The project group was founded in 2015 at Erdenes Mongol. In the same year, a contractual agreement was signed with the State Grid Corporation of China, a state-owned company that supplies electricity to 1.1 billion consumers spread over 88 percent of China's territory.

The study was completed at the end of 2017 and presented to the Mongolian side. It recommended a 5280 MW plant with transmission lines, water supply and an open pit mine. The study estimates that $3.83 billion will be required to construct the facility, $1.59 billion to install the transmission lines, $1.4 billion to develop the coal mine and $237.4 million to dollars are needed to build the water supply project. This results in a total investment volume of 7.1 billion US dollars and would be the largest investment in Mongolia after Oyu Tolgoi.

The Chinese side bears overall responsibility for the financing and implementation of the project.

## 2.7 *Investment Climate in the Mining Sector*

The Ministry of Finance of Mongolia collects data on investments from the state and local budgets as well as investments financed by foreign aid and loans. These are transmitted to the National Statistics Office, while Mongolbank compiles data on foreign direct investment. The calculation is based on balance of payments statistics.

Data on local budget investment and investment by institutions and individuals from their own resources are collected by the statistical departments of the aimags and the capital. Investment data are compiled by classification into construction, machinery and equipment, deposit exploration, mining, and other (design and other intangible assets).

Construction activity includes construction of houses, bridges, banks, as well as creation of fixed capital through expansion, repair, restoration. Machinery and equipment includes all types of vehicles, mechanical equipment, electrical equipment and their spare parts. Other includes all types of intangible assets, design and computer software and databases (Fig. 14).

### 2.7.1 Industry

This section contains statistics on the state of the industrial sector by number of establishments, their size, expenses, production volume (in volume and value), turnover and main indicators of fixed assets and financial statements. These statistics are

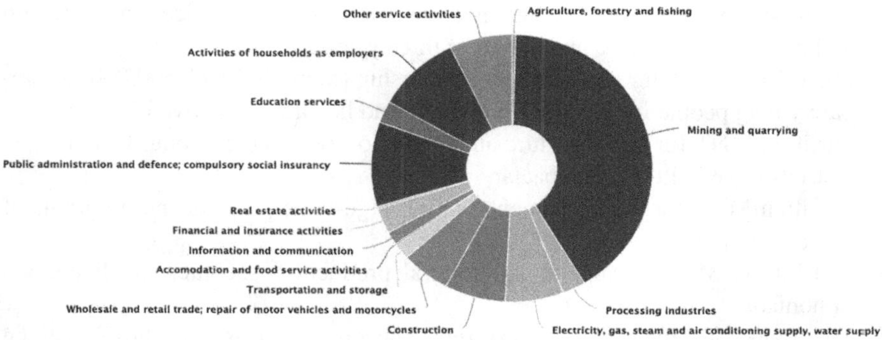


**Fig. 14** Investments by sectors 2020

compiled using monthly and annual establishment reports and the results of censuses and surveys. A monthly index of industrial production is calculated and published in the Statistical Bulletin.

Total industrial production was estimated at constant 1967, 1986, 1995, and 2000 prices. From January 2008, it was estimated using 2005 prices as the base year. For the monthly publication, about 300 industrial products are recorded and the production volume index is calculated. The production output refers to the products and services for the respective establishment market and the products and services for their end use.

The sale of products is defined as the amount, of production and services exported sold in the current year. Companies producing different types of products and services are classified according to the products and services they mainly produce.

Once a year, the Federal Statistical Office compiles the balance figures of electricity, heat and coal based on information from power plants and mines. The amount of electrical energy produced and imported or exported, coal quantity, own consumption of electricity plants and energy losses in the transmission and distribution of electricity are classified according to the main sectors of the economy and included in the balance.

## 2.8 Mongolian Funding Programs

The economy, which had grown by 5.2 percent in 2019 due to the pandemic, contracted by about 6 percent in 2020, as expected. As a result, the number of employees in more than 46,000 companies decreased by 70,000 in the third quarter of 2020. Against this backdrop, the government has decided to implement policies and targets to protect the health of citizens and revitalize the economy by 2023.

The following measures will be implemented in the MNT 10 trillion plan to protect citizens' health and revitalize the economy:

- 2 trillion MNT loan for small and medium enterprises and service providers with an interest rate of 3% pa and a term of three years.
- 500 trillion MNT for a two-month scholarship (monthly MNT 500,000) to prepare young people for jobs and help them lead healthy and active lifestyles.
- 3 trillion MNT for infrastructure and land allocation in the "Youth I, II and III" districts to be built in Ulaanbaatar.
- 2 trillion MNT for the implementation of large-scale projects and programs of strategic importance
- 500 billion MNT to support agricultural production, income, and livelihood of nomads
- The Bank of Mongolia's annual repo-financing of 1 trillion MNT will be increased to 2 trillion MNT (from 300 million MNT for small and medium enterprises to 500 million MNT and from 1 billion MNT for non-mining exports to 3 billion MNT).

## 2.9 Bilateral Cooperation and Projects

The Ministry of Mines and Heavy Industry of Mongolia is expanding its collaborations by organizing major forums for foreign and domestic investors. Currently, the Ministry of Mines and Heavy Industry (MMHI) participates with partner organizations in forums such as Discover Mongolia and Invest Mongolia.

In the area of development cooperation, the MMHI conducts joint projects with Germany, Japan, Canada and Australia. For example, SESMIM projects have been implemented by the Canadian government since 2016 until 2022 to increase the capacity of Mongolia's social sector.

It also aims to improve transparency and public sector policies in extractive industries and mining with support from the Ministry of Mines and Heavy Industry.

The Gesellschaft für Internationale Zusammenarbeit (GIZ) of BRD, in cooperation with the Ministry of Mines and Heavy Industry, is implementing the "Minerals and Raw Materials Initiative," "G7 Fast Track Partnership," and "CONNEX Initiatives" programs that support mining agreement procedures.

Joint projects with major foreign mining companies, such as Oyu Tolgoi LLC's underground mine, Gatsuurt gold deposit, Areva Mines LLC's uranium mine, the opening of Tsagaan Suvarga copper concentrator, work to complete the Tavan Tolgoi agreement and start operations are underway. The Ministry of Mines and Heavy Industry is actively pursuing a policy to stabilize the fiscal and legal environment in order to attract foreign investment.

In terms of foreign cooperation, the following activities are underway. A memorandum of understanding between the Ministry of Mines and Heavy Industry and the Ministry of Investment and Development of the Republic of Kazakhstan is ready for signing. In addition, the Ministry of Mines and Heavy Industry is discussing a memorandum of understanding with the foreign ministries of the United Kingdom of Great Britain and Northern Ireland on cooperation in the development of Mongolia's extractive sector.

In the field of development cooperation, joint projects are already being implemented with Germany, Canada, and Australia.

### 2.9.1 MERIT Project

The US$11 million MERIT project is supported by the Government of Canada and will run over six years from 2016 to 2022 to improve raw materials sector management. The project aims to strengthen three key capacities:

- Ensure effective policy development and establishment of key government agencies,
- Improving relationships with key stakeholders in at least two Aimags at the local and regional levels;
- Support the mining industry by qualifying two training institutions.

The project aims to provide technical and methodological support to strengthen the capacity of government agencies and to develop and implement policies and procedures. It also aims to increase the involvement of stakeholders and local communities in decision-making and to improve the coordination of human resource needs of government agencies and training of educational institutions.

### 2.9.2 SESMIM Project

The US$8.4 million SESMIM project, supported by the Government of Canada in partnership with the Ministry of Mines and Heavy Industry, aims to increase the capacity of Mongolia's social sector and improve public sector policies over the five-year period from 2015 to 2020.

Through the training and advisory program, the SESMIM project will help strengthen the capacity of key ministries and agencies regarding their cooperation and coordinate the implementation of policies and regulations regarding extractive sector management within and among cooperating ministries.

### 2.9.3 Program: Minerals and Raw Materials Initiative

The program, implemented by the Deutsche Gesellschaft für Internationale Zusammenarbeit in cooperation with the Ministry of Mines and Heavy Industry, supports Mongolia's national, resource-based development. The program aims to build on the capacity of national and local partners to create broad and sustainable growth using the potential of the mineral and raw materials sector. The program contributes to the implementation of the Mongolian-German agreement on cooperation in the field of raw materials and technology, which was signed during the visit of German Chancellor Angela Merkel to Mongolia in 2011.

Within the framework of this program, the foundation and establishment of the "Deutsch-Mongolischen Hochschule für Rohstoffe und Technologie" (DMHT) or "German-Mongolian Institute for Resources and Technology" (GMIT) took place. Lectures started in 2013 and the first graduation of Bachelor students took place on June 20, 2018.

As a state university, GMIT aims to combine a new quality of education with the aspiration of leading, international standards and establish it in Mongolia. The country's development is to be advanced in the field of commodity engineering and teaching and research are to be combined.

### 2.9.4 Small-Scale Mining

Within the framework of Mongolian-Swiss cooperation, a "Sustainable Small-Scale-Mining" project was funded with MNT 2.8 billion between 2015 and 2018 in cooperation with the former Ministry of Mines with a grant from the Swiss Agency for Development and Cooperation.

Since the start of the project, much progress has been made in small-scale-mining. The legal framework has been established and the issue of micro-mining is reflected in the state's natural resources policy. With the formalization of small-scale-mining, state services to miners have improved. Work safety conditions have improved, and the number of accidents has decreased. The amount of gold deposited in the state treasury and royalty revenues have increased sharply. The organization of small-scale-mining workers has improved cooperation between local authorities and miners.

## *2.10 Tendering Process*

Exploration licenses are awarded through a tender process organized by MRPAM (Mineral Resources and Petroleum Authority). When a tender is announced, eligible interested parties submit their bids with the required documentation and pay a bid fee to MRPAM. MRPAM then evaluates the technical and financial proposals of the proposers and selects the best. If two or more applications have equal scores, the MRPAM selects the applicant who submitted its proposal first.

MRPAM notifies the selected applicant, who can either reject or accept the license. If the applicant accepts, it must make certain payments before the license is issued.

Mining licenses may be issued through a competitive bidding process or at the request of an exploration license holder. In principle, if an exploration license holder meets the relevant requirements and applies for a mining license for the relevant area within the exploration license, MRPAM should grant a mining license on a priority basis. If the requested license is not covered by an existing valid exploration license, MRPAM may issue a mining license through an open tender process.

Mongolia continues to have a Tender Law of Mongolia for the purchase of goods and trades and services with state and local ownership. In general, the private sector also follows these tendering procedures. The most important points are listed below:

Article 1. Purpose of the Act

1.1 The purpose of this Act is to plan and organize the procurement of goods, trades, and services with state and local funds, monitor the process, resolve complaints, and clarify relationships that arise in connection with the violation of this Act.

Article 2. Legislation on procurement of goods, trades and services with state and local funds.

The legislation on procurement of goods, trades and services with state and local funds shall be in accordance with the Constitution of Mongolia, the Civil Code, the Law on Transparency of Information and Right to Information, the Law on Transparent Accounts, this Law and in accordance with the laws issued therewith.

2.2. If an international treaty to which Mongolia is a party provides otherwise than this Law, the provisions of the international treaty shall prevail.

Artikel 9. Beteiligung ausländischer Personen an Ausschreibungen

9.1. A foreign entity shall have the right to participate in the tender in accordance with the procedures established by this Act.

9.2 The contracting entity may not prohibit foreign entities from participating in tenders for construction works with a cost budget exceeding MNT 10,000,000 and for goods and services with a cost budget exceeding MNT 100,000,001.

9.3 A foreign legal entity may participate in the tender only in accordance with 9.2 of this Law.

## 3 Final Thoughts

Mongolia is a developing country with a small population and huge raw material reserves of coal, copper, gold, silver, iron, fluorite, uranium, zinc, and rare earths. This opens enormous potential for German companies in the mining and mining supply industry.

- Mongolia has the raw materials and Germany has the technology.

German engineers and German industry have an excellent reputation in Mongolia, especially in the mining sector. For example, German mining engineers, geologists and hydrogeologists were consulted by the Mongolian government during negotiations between the Mongolian government and Rio Tinto over cost increases and delays in commissioning the Oyu Tolgoi underground mine in early 2021.

- More than 1% of the population speaks German.

There are several projects that already have a commitment for financial support from the Mongolian government based on detailed studies. The country is on the move, and it has been recognized that the future lies in the raw materials industry. However, there are too few trained engineers and skilled workers in the country to be able to use the raw material reserves for value creation in the country itself and without foreign assistance.

- Mongolian engineers have been trained to German standards at the German-Mongolian Institute for Resources and Technology (GMIT) in Nalaikh since 2013.

## References

Монгол Улсын Их Хурлын тогтоол, Дугаар 18, Төрөөс эрдэс баялгийн салбарт баримтлах бодлого батлах тухай, 2014 оны 01-р сарын, p 16

https://www.legalinfo.mn/law/details/63. Accessed 22 May 2023

https://mrpam.gov.mn/home/. The implementing agency of Mongolia Government, Mineral Resources and Petroleum Authority (MRPAM). Accessed 22 May 2023

Монголын Үйлдвэрлэлийн Геологичдын Холбоо, Хайгуулчин сэтгүүл, 2015 он, дугаар 53, хуудас, pp 8–53

Монголын Үйлдвэрлэлийн Геологичдын Холбоо Хайгуулчин сэтгүүл, 2019 он, дугаар 61, хуудас, pp 81–83

С.Дьяков, С.Санждорж, Г.Жамсрандорж, 2016 он, Оюу толгойн ордыг нээсэн тухай, pp 216–232

Монгол Улсын Шинжлэх Ухааны Их Сургууль Геологи, Уул Уурхайн Сургууль, С.Жаргалан, 2021 он, Газрын Ховор Металл (Монгол орны тархалт, хэтийн төлөв)

Technical Report Oyu Tolgoi 2020 Technical Report Turquoise Hill Resources Ltd. Umnugovi Aimag, Mongolia. https://turquoisehill.com/site/assets/files/5193/oyu_tolgoi_2020_technical_report_ni43-101_finalv2_unsecured.pdf. Accessed 22 May 2023

Уул уурхай, хүнд үйлвэрийн сайдын 2018 оны А/28 дугаар тушаалын хавсралт, *УУХҮС-ын 2020 оны А/16 дугаар тушаалаар баталсан нэмэлт, өөрчлөлт,* ТУСГАЙ ЗӨВШӨӨРӨЛ ОЛГОХ СОНГОН ШАЛГАРУУЛАЛТЫН ЖУРАМ. https://www.legalinfo.mn/annex/details/8297?lawid=13224. Accessed 22 May 2023

Government of Mongolia, Ministry of Mining and Heavy Industry. http://mmhi.gov.mn/. Accessed 22 May 2023

Mongolian Statistical Information Service. https://www.1212.mn/default.aspx. Accessed 22 May 2023

Mongolian Law on Investment, October 3, 2013. https://mrpam.gov.mn/public/pages/16/investmentlaw_ENG.pdf. Accessed 22 May 2023

Your Guide to Invest in Mongolia, Investment Guidebook 2020

Exportbericht Mongolei April 2019

Эндрю Бауэр, Намхайжанцангийн Дорждарь, “БАРЬЦГҮЙ ӨСӨЛТ” “ЭРДЭНЭС МОНГОЛ”-ЫН ҮНЭЛГЭЭ

Mongolia Fourteenth EITI Reconciliation Report 2019, Mongolia Extractive Industries Transparency Initiative (Mongolia EITI) Ulaanbaatar city 2020

Erdenechimeg E, Asralt B, Khurelbaatar G, Distribution of Mongolian Mineral Resources, Transportation and Logistics Analysis. Asian Infrastructure Research Institute, Transportation Department, MTS of Mongolian University of Science and Technology, Ulaanbaatar, Mongolia

www.mongolianminingjournal.com. "Mongolian Mining Journa" 2020/1-12/ 2021/1-8/. Accessed 22 May 2023

Bundesanstalt fur Geowissenschaften und Rohstoffe (BGR) (Federal Institute for Geosciences and Natural Resources) and Mineral Resources Authority of Mongolia (MRAM) (2013) Rare Earths of Mongolia: Evaluation of Market Opportunities for the Principal Deposits of Mongolia, Technical Report

Turquoise Hill Resources Ltd. (2020) Oyu Tolgoi 2020 Technical Report

# The Importance of Sustainability Criteria for the Professional Education of Geo-engineers and the Role of Academic Institutions

**Christoph Dauber, Peter Goerke-Mallet, Benedikt Gräfingholt, and Christian Warnecke**

## Contents



**Abstract** The sustainable use of geo-resources is an important prerequisite for the acceptance of mineral extraction in our society, both nationally and globally. A growing world population—9.8 billion people are forecast for the year 2050—requires the extraction of additional mineral resources. In contrast, the acceptance of mineral extraction in our society has decreased significantly. That development can be counteracted by raising the awareness for sustainability in mining. For this purpose, academic programs in mining engineering teach their students to manage future extraction of minerals in an environmentally friendly manner. Global annual surveys in the mining industry show that sustainable mining and processing has become a focus of industry executives. Environmental, Social and Governance (ESG) criteria and license to operate are perceived as significant risk factors.

The Georg Agricola University (THGA) in Bochum, Germany, takes a holistic approach to sustainability. THGA's motto "sustainability: engineering for a better world" reflects the training of its students as well as the focus of the university.

C. Dauber · P. Goerke-Mallet (✉) · B. Gräfingholt · C. Warnecke
Technische Hochschule Georg Agricola University, Bochum, Germany
e-mail: christoph.dauber@thga.de; peter.goerke-mallet@thga.de; benedikt.graefingholt@thga.de; christian.warnecke@thga.de



W. Frenz, A. Preuße (eds.), *Yearbook of Sustainable Smart Mining and Energy 2022*, Yearbook of Sustainable Smart Mining and Energy - Technical, Economic and Legal Framework 2, https://doi.org/10.1007/978-3-031-41873-0_12

The paper illustrates how universities can implement the **S**ustainable **D**evelopment **G**oals (SDG) in the project work, in the professional training and education of geo-engineers to maintain "social license to operate."

The following presents three examples of SDG implementation at THGA.

The first example covers the post-industrial phase of coal mining in the Ruhr region. About 10 billion t of hard coal were produced in the past 150 years. Mining ended in 2018 but the industry continues to manage the risks of rising mine water levels and methane releases into the environment. THGA established the Research Center of Post-Mining (FZN), strives to objectify the public discussion of post-mining environmental impacts through research activities, extensive monitoring and reporting. The second example is storage of natural gas, crude oil and helium in salt caverns near the town of Gronau-Epe close to the German-Netherlands border. There, salt extraction has created large caverns in which is stored. Disputes between cavern operators and local residents are generated by ground movements, which in the long term may cause subsidence, damage to buildings and altered ground water levels. THGA has significant expertise in raw material extraction and process engineering and supports these research activities.

The third example outlines the project "green campus" at THGA. With its envisioned Green Campus, the THGA can become a sustainable oasis of knowledge. In the planning process the 17 Sustainable Development Goals of the United Nations play a decisive role.

## 1 Introduction

The extraction of mineral resources is one of the oldest activities with which humankind has sought to secure its existence, life and survival. Prehistoric humans mined flint, obsidian and hematite to produce weapons, tools and color pigments. Quantity and variety of raw materials have increased enormously since. This development will continue; by the year 2050, the world's population will rise by 2.2 billion to a total of 9.8 billion[1] and the demand for raw materials is generally proportional to population growth and standard of living. The past two centuries were characterized by fossil raw materials that fueled worldwide and profound industrialization. Now, fossil fuels must be substituted by renewable energies in order to limit global warming from carbon dioxide emissions. Renewable energy requires a different palette of raw materials that must now be mined in larger quantities. Examples include battery metals such as lithium, cobalt, nickel and rare earths, which many countries consider to be "critical minerals". The importance of primary raw material production for the well-being of humanity is indisputable and, in view of the energy transition, will grow in the coming decades. This is strongly contrasted by the significant decline in social acceptance for all activities associated with the extraction of

[1] United Nations (2022).

mineral resources. This perception applies not only to industrialized countries, but also to emerging economies in South America and Asia.

In earlier centuries, mining including extracting oil and gas, was a welcome driver of social advancement and prosperity. Its impact on landscape, nature and the environment was more or less accepted. Today, almost every mining project is regarded as undesirable, polluting the environment, and must be combated. Since public participation has become increasingly important in the mining approval process—a development that makes sense and is to be welcomed—this often leads to lengthy reviews and delays in permitting or failure of approval altogether. In self-criticism, it must be admitted that the poor reputation of the mining and extraction industries in public opinion is partly self-inflicted. There are too many examples of mining- induced environmental catastrophes such as the fire and oil leak on the Deep Water Horizon oil platform in the Gulf of Mexico[2] or the Brumadinho tailings dam failure in Brazil, in which 270 people lost their lives.[3]

In addition, interest in and appreciation of ecological issues has increased significantly in our society. Terms such as sustainability, circular economy and ecological footprint have found their place in society. Particularly the younger generation regards them as essential components of our way of life. Considering our ethical responsibility towards future generations, this development is to be welcomed without reservation, and to be further promoted. In the context of this discussion, an impeding factor is that mining activities are never "sustainable" in the classic sense, as mineral deposits are not renewable. Extraction of mineral resources alters the natural environment and can never be fully restored. Of course, there are many examples worldwide that prove the damaging influence of mining activities, impacting landscape, flora, fauna and water bodies. On the other hand, there are numerous mining operations—ongoing and closed—that demonstrate successful, environmentally-friendly and ecologically acceptable restoration of mining areas. Concerning specific mining projects, mining companies and affected citizens, NGOs and political groups are often irreconcilably opposed to each other. The mistrust on both sides is often deep and the dispute so emotional that a fact-based dialogue seems impossible. For the social and sustainable shaping of our future, it is essential that the deep rift between the mining industry and the public will be at least reduce, if not closed. At this point, universities and mining schools are partners that can help ensure that such disputes continue in a less emotionally charged atmosphere by providing fact-based information and subject matter expertise. Universities and colleges, including those offering mining engineering programs, enjoy a rather good reputation in society. In the public perception, science has a high status; ideally, it is not subject to directives and is not guided by economic interests. Students and researchers who are preparing for their professional life and critically reflecting on tradition, ensure that universities are perceived positively. For this reason, universities enjoy a sense of trust, in contrast to representatives of the mining industry.

[2] National Commission (2011).

[3] BBC News (2021).

Such trust is an indispensable prerequisite for gaining public acceptance for mining projects.

In this context it is important to make clear that humans are part of the “game”. The raw materials demand per person and annum is in Germany 16 t and in the United States 20 t.[4] Humans are consumers of all mined products, and consumption increases with advancing standards of living that we all desire. In this respect we humans are part of the problems. But human also have the skills to develop solutions. The provision of geo-resources requires experts with skills in process monitoring, risk and opportunity management, creating transparency and communication. The first step in this direction is to create awareness of sustainability among future leaders, starting with our students. Miners and mining companies are beginning to integrate environmental, social and governance criteria (ESG) into corporate strategies, decision-making and stakeholder reporting as the issue becomes a priority for shareholders and investors:

1. Mining and geo-resource engineers who manage their impact on the environment and communities can build a positive legacy.
2. Managing social impacts and ensuring a “just” transition matters. This is an issue that will increasingly come to the fore as the energy transition accelerates.
3. Transparent supply chains can demonstrate human rights commitments. Progressive planning for closure including the circular economy should be adopted across life of mine.

Against this backdrop the professional education of geo-engineers is a crucial task for the mining industry and the universities as well. The efforts made at the Research Center of Post-Mining at THGA to enhance public acceptance and awareness regarding mineral resources are described in two industry related examples. A third example reflects the own attempts for sustainability on THGA’s campus. The three examples are:

1. Hard coal: Mine and ground water in the Ruhr area
2. Natural gas: Gas storage cavern Gronau-Epe
3. THGA Green Campus: Engineering for a better world

## 2 Mine Water in the Ruhr Area

The long lasting hard coal mining history in the Ruhr area left behind two major perpetual tasks related to two different water bodies. The first is the mine water body in the former underground coal mines in the Ruhr district. Due to the deep underground coal mines the water level is at a depth of more than 500 m. When the former coal producer RAG stops pumping activities at the Zollverein colliery (see Fig. 1, “ZO”) in a depth of nearly 1000 m the mine water level will rise. It is

[4] Umwelt Bundesamt (2022) and USGS (2022).

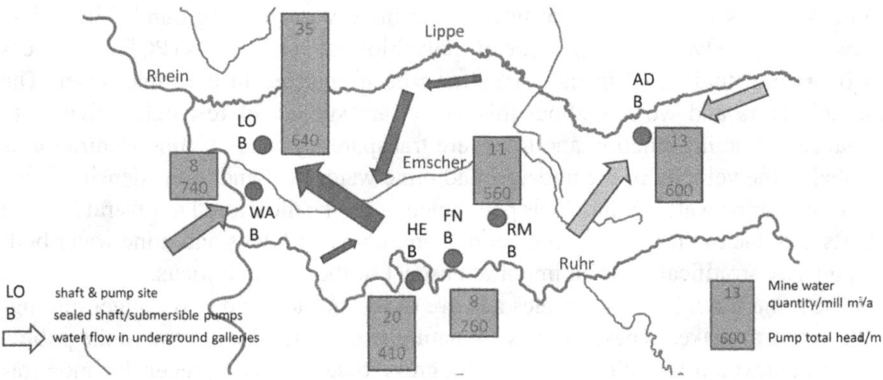


**Fig. 1** RAG-concept of long term mine water management in the Ruhr area (© THGA)

predicted that the rebound to 600 m below surface will take some 15 years. This concerns vast areas underground in the western and northern part of the Ruhr deposit.

The final concept of RAG aims to focus the pumping activities at six shaft sites (Fig. 1). Each column in Fig. 1 represents a shaft where submersible pumps have been or will be installed. The first number, for example at shaft LO, names the amount of mine water to be pumped at this shaft per year (35 million $m^3/a$). The second number indicates the level from which the mine water must be lifted (640 m). The arrows in different colors illustrate the flow of mine water in underground galleries. The letter B indicates an active pumping site, the letter S shows shaft sites where additional mine water pumping is possible if the concept fails. The total amount of mine water which must be pumped at six shafts is about 80 to 90 million $m^3$ per year.

The second water body is the ground water horizon directly at or below the surface. Due to the subsidence evoked by the mining process, vast ponding polder areas have formed that can no longer drain naturally. Nearly 1000 $km^2$ in the Ruhr area have no direct drainage to creeks or rivers. Thus, pumping is necessary to keep the polder areas dry and to protect the population and the infrastructure. In total the amount of pumped polder water is around 800 million $m^3$ per year.[5] Water management in both water bodies has been financed by the RAG Foundation since 2019. To support these ongoing tasks with scientific research the Research Center of Post-Mining (FZN) at THGA University was founded in 2015. Its aim is to provide scientific support to the cessation process of coal mining in the Ruhr, Saar and Ibbenbueren area.[6] Additionally, the Research Center provides comprehensive information about mine water issues to the public via its homepage.[7] There are numerous technical challenges that include maintaining the functionality of the

[5] RAG (2016).

[6] Melchers et al. (2019).

[7] Forschungszentrum Nachbergbau.

mine openings that conduct the flow of the mine water underground. The public view links mine water with the issue of 'polychlorinated biphenyls (PCB)', as traces of hydraulic fluids used in the mines have been detected in the mine water. The public's fears and worries about this aspect are subject of research activities to enhance public information and to ensure transparency. The PCB-concentration is related to the velocity of the underground mine water flow and to the density stratification in mine water bodies. This phenomenon can be monitored in several flooded shafts and due to the possible contact of drinking water layers and mine water bodies, density stratification is an important aspect of the research focus.

The mine water rebound creates a heave of the surface, and may trigger mining-induced earthquakes. These aspects are naturally of particular interest to the public. In this context a proof of concept is in progress to develop comprehensive monitoring measures for the critical infrastructure of communities in the Ruhr area. Another extensive research project deals with the evaluation of mine water rebound processes in Germany and other countries in Europe. A sensitivity analysis studies the parameters that influence the rebound process as well as the behavior of released methane. All research projects are linked to geo-monitoring measures and interpretation of spatio-temporal data. The scientific objective is to enhance the understanding of the ongoing processes; the second equally important objective is ensuring reliable and trustworthy communication with the public. In this respect the 'Forum Mining and Water'[8] should be mentioned. This forum is a non-profit foundation and promotes science and research in the field of hydrogeology with the focus on mine water. Particular attention is paid to the opportunities and risks of a rise in mine water after the cessation of coal mining activities and its ecological and economic consequences. The foundation is both operational—i.e., carries out research projects—and promotional—i.e., awards grants to third parties for specific projects. The Forum operates independently and determines its own activities within the framework of the foundation's mission. All the results of its activities are presented and published in a transparent and accessible manner. The director of the FZN and vice president of THGA is also the vice-chairman of the Forum.

Ten years ago, another platform for communication between experts and the public was established. The conference is called "NACHBergbauzeit in NRW" (Post Mining Period in NRW) and takes place every two years at the THGA. While the organizers are the mining authority of North Rhine-Westphalia (NRW), referred to as the Ruhr area, and the THGA, the conference program deals with different topics about mine water, polder water and structural changes. The conference enhances permanent networking, communication, a continuous exchange of ideas and problems and the creation of transparency. As the access is not restricted the audience covers experts, politicians, representatives of the communities and citizens' initiatives. In addition, another small detail of the communication regime should be mentioned. At the local adult education center in Bochum a seminar called Post-Mining has been organized and the speaker is a professor of the THGA.

[8] Forum Bergbau und Wasser (2022).

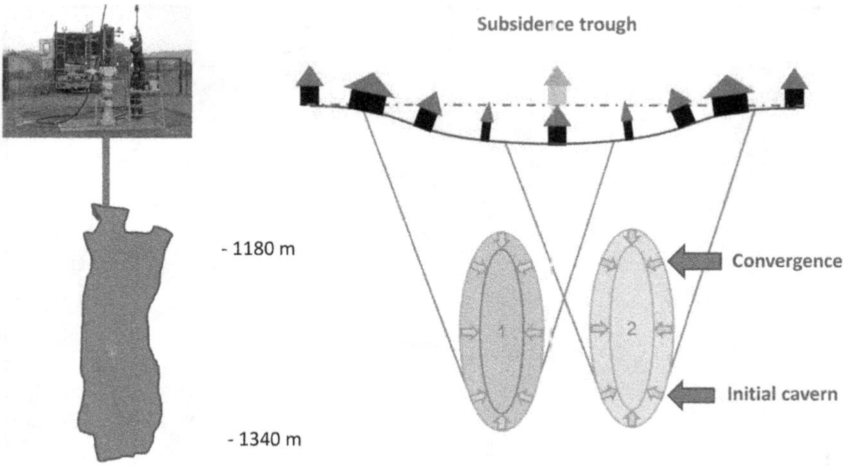


**Fig. 2** Salt cavern and convergence (© THGA)

## 3 Cavern Storage Gronau-Epe

In Germany, a total of 272 storage caverns with a natural gas volume of about 20 billion m$^3$ are currently in operation at numerous locations. This form of storage has considerable safety advantages and requires a low surface area. One of these storage facilities, the Gronau-Epe cavern complex, is located in the north of the Ruhr area near the border with the Netherlands. In the 1960s, drilling operations in search of natural gas explored a shallow rock salt formation up to 400 m thick at a depth of 1000 m. Since 1972, several million tons of sodium chloride in the form of halite brine have been extracted from this deposit. The caverns created by brine extraction are used as storage facilities for gases and crude oil.

The Gronau-Epe cavern facility comprises about 100 caverns spread over an area of 22 km$^2$. The storage volume is about 4 billion m$^3$ of natural gas and 1 million m$^3$ of crude oil. Helium is stored in one cavern, and the storage of hydrogen is currently in the permitting process.[9] The salt caverns are up to 250 m high, have a diameter of about 60 m and a volume of more than 600,000 m$^3$ (Fig. 2). About 60% of the cavern cavity is used as storage space; the remaining volume is filled with cushion gas under a minimum pressure of 70 bar. The caverns are operated at a maximum pressure of 200 bar; accordingly, one cavern can store about 50 million m$^3$ of natural gas in regular operation.

Both the production of the cavern reservoir by brine extraction and its further use as a storage site lead to stress changes in the reservoir walls and in the overburden. Due to the associated convergence, i.e., the movement of the salt rock into the cavity, a cavern loses on average about 1% of its volume per year. This annual volume

[9] SGW (2022).

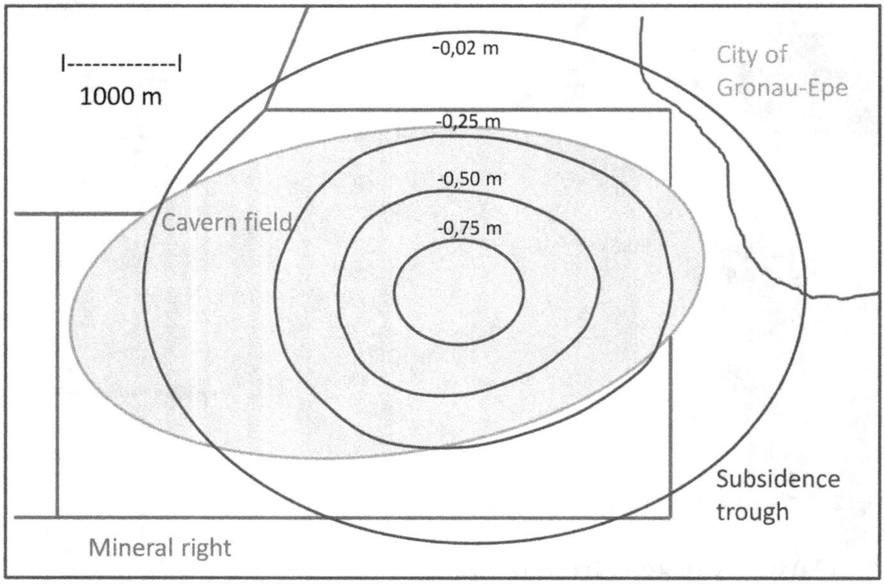


**Fig. 3** Subsidence around the cavern storage at Gronau-Epe (© THGA)

loss of about 6000 m$^3$ is transmitted through the rock mass to the surface. The resulting subsidence affects an area of about 5 km$^2$ at a limiting angle of about 35°. In the Gronau-Epe field, the maximum surface subsidence has currently reached 0.85 m (Fig. 3). Prediction methods have been developed for ground movements above caverns. However, it has been demonstrated that the actual convergence behavior of a cavern depends to a large extent on pressures which are a function of storage and withdrawal of the storage medium. In the salt caverns, it was customary to fill them to a pressure of 200 bar in summer. In winter, gases were withdrawn down to a minimum pressure level of 70 bar. Due to changes in the international gas markets, storage and withdrawal cycles have become more dependent on market supply and demand. As measurements indicate, this is associated with higher convergence rates in individual cases and thus greater ground movements.

The responsible authorities, the public and, in particular, home- and landowners in the area affected by the cavern facility expect a reliable forecast of future ground movements. Additionally, there is the question of the final safekeeping of caverns. Scientists must assess how the convergence process can be stopped or kept to a minimum when the cavern operation is terminated. Mode of operation and final safekeeping are of considerable relevance for the subsidence prognosis, which is not only associated with potential subsidence damages, but may also have an impact on flood protection at local water bodies. The special features associated with the operation of storage caverns have triggered a climate of mistrust, concern and even fear among the affected population. The resulting protest was led by a citizens' initiative formed after a spillage at an oil cavern in Gronau-Epe.[10] Both the citizens' initiative and the city council of Gronau showed great interest in involving a scientific institution in the communication process. Employees of the FZN were invited to meet with members of a citizens' initiative, where they had the opportunity to inform and answer numerous questions. The establishment of a research cooperation was discussed with all parties involved and agreed by contract in the summer of 2021, in the presence of the mayor of the city of Gronau. The city council, the citizens' initiative and the THGA are members of this research cooperation together with the company EFTAS from Münster. Among other things, EFTAS specializes in satellite-based remote sensing and is thus an important partner for monitoring ground movements at the surface of the cavern field. Topographic changes above the cavern site are caused by mining activities, but are also influenced by other factors, such as the nature of the soil layers, the kind of land use and extreme weather events. For the most part, subsidence movements occur slowly, but they are complex and require precise, sophisticated monitoring. Appropriate data are provided by orbiting satellite sensors (EU Copernicus Program), as well as by aircraft and drones, and enhances with terrestrial surveys. Data interpretation also requires a high level of expertise and is often difficult to communicate. As the first citizens' meetings in Gronau have shown, the engagement of academic representatives is met with a high level of acceptance. The technical expertise of the FZN working group is crucial in this context. For example, the members of the team have many years of practical experience in the storage industry, hydrogeology, rock mechanics, mine surveying and mining damage science, as well as communication in a tense environment.[11]

The research cooperation creates a platform for dialogue and discussion of issues at eye level. The FZN team organizes regular meetings with representatives of the research cooperation and with the steering committee, which consists of representatives of the district, the municipality, local political bodies and the citizens' initiative. The public homepage of the research cooperation,[12] which offers detailed information to interested parties and is maintained by the FZN, also serves for transparency.

## 4 Green Campus: Engineering for a Better World

In line with the THGA's mission statement "Engineering for a better world", the next generation should find a world worth living in and have better opportunities than we have today. Therefore, it is a central concern of the university to advance the topic of sustainability and to make a significant contribution to a sustainable future.

[10] Ministerium für Wirtschaft, Innovation, Digitalisierung und Energie des Landes Nordrhein-Westfalen (2019), pp. 66–69.

[11] Rudolph et al. (2022).

[12] Monitoring Epe (2022).

Sustainable management of the earth's resources has developed from a niche topic into a megatrend that is supported by a high level of social interest and, in its thematic breadth (resource efficiency, circular economy, sustainable supply chains, climate change, avoidance and disposal of waste, etc.), offers a diverse field of activity for research and teaching at the THGA.[13]

As the oldest university in Bochum, the THGA faces challenges in creating a sustainable campus—the historic building and long-established structures are just a few of them. The university's central location, which is located on the northern edge of Bochum city center, offers great opportunities to actively engage society in the transformation process. After all, a sustainable campus should be more than just $CO_2$ neutral. It is important to direct a holistic view on all resources, including the personal commitment of everyone. In this way, a campus, originally established in 1899, can become a place of encounter where students find a space to engage with the topic of sustainability; where scientists and scholars realize forward-looking projects; where employees experience sustainability in their daily work and where the broad urban society can interact with all university members to develop new ideas for the future of Bochum and the Ruhr area. With its envisioned Green Campus, the THGA can thus become a sustainable oasis of knowledge in the northern city center. In the planning process the 17 Sustainable Development Goals of the United Nations play a decisive role. Construction measures are implemented in such a way that they are $CO_2$ neutral and improve the microclimate of the neighborhood. Green roofs, unsealed surfaces and newly planted trees create sustainable study and working conditions.[14] (Fig. 4).

Due to its long history and close links with the industrial heritage in the Ruhr area, the university has a special connection to the sustainability goal. Whereas primary raw materials and the coal and steel industry used to be the focus of the university, now it is the challenges of post-mining, resource efficiency, sustainable solutions and technologies for circular value creation processes and the transformation of energy supply. The structural change process in the Ruhr area is not complete; the transformation of the economy and society towards sustainability that meets the needs of present and future generations alike must continue to be driven forward with great commitment.

Therefore THGA pursues a holistic sustainability approach that focuses on all three perspectives of sustainability in the context of the 17 United Nations Sustainable Development Goals (SDGs): The University's goal is to promote ecological sustainability, social justice, and economic performance in equal measure in the long term. Understanding of sustainability must be distinguished at different levels of the university: At the individual or micro level, the aim is to promote sustainability awareness and the ability to act in a sustainability-oriented manner among all university members and individuals of the THGA-family. Acting responsibly as a sustainable university requires an awareness of sustainability among all

[13] Technische Hochschule Georg Agricola (2022), pp. 63–68.

[14] Jahresbericht Technische Hochschule Georg Agricola (2021), p. 48.

**Fig. 4** Illustration of the envisioned THGA Green Campus in Bochum (© THGA)

university members and must also be reflected in the infrastructure, especially in the building, the use of energy and the handling of resources and waste avoidance (sustainability within the university). At the university level, it is about the institution itself as an organization, including its committees and academic departments. The further development of the university as an organization is oriented towards the SDGs in teaching, research, transfer and third mission and thus contributes to the sustainability of the economy and society. Responsible use of resources, efficiency of production facilities and processes, efficient recycling processes, avoidance of material input into the environment are topics that are of central importance in research and teaching at the THGA. This ensures the THGA's sustainable existence and relevance (and the sustainability of higher education development).

The recognition of trends and dealing with disruptive developments play just as an important role as the conscious strengthening of unique selling points, concepts for human resources development and the thematic orientation and further development of study programs. The sustainability approach of the THGA goes beyond these inward-looking perspectives, however, in that the THGA takes sustainability issues proactively into dialogue with society to the outside world. In the sense of a transformative university, it thus initiates processes of change in society, politics, and business (trough sustainability in dialogue).

The various perspectives of sustainability are intensively discussed and further developed at the THGA. An important role is played by the "360°" working group. Sustainability has also found an organizational home through the establishment of a sustainability office. The broad commitment and openness of the staff regarding the topic are proof of the high value placed on sustainability at the THGA. It is still a question of further deepening the general awareness and the concrete individual reference to the UN's 17 sustainability goals among the university staff. The university's mission statement emphasizes the great importance of these issues.

As a scientific institution, the THGA contributes to tackling social, ecological, and economic challenges through teaching, research, and knowledge transfer. It is a driver for practical innovations and supports start-ups working for a sustainable future based on the 17 SDGs. Aspects of sustainability are already an integral part of the curriculum of many courses. Some degree programs—for example Material Engineering and Industrial Heritage Conservation and Climate Change Management (accreditation planned)—deliberately focus on the core areas of sustainability. Topics such as energy, raw material extraction, environment, transformation, and post-mining, which are given high priority at the THGA, position the university as a competent contact and change agent in matters of sustainability.[15]

To accomplish its strategic goals concerning sustainability THGA has defined a catalogue of measures such as: raising awareness and activating of all university members; definition of a roadmap to sustainable and climate-neutral THGA; implementing and making sustainability visible as a cross-cutting issue in all courses of study; and creation and implementation of the "Sustainability in Dialog" concept with all stakeholders at all levels (local, regional, national, international). These measures will ensure that the holistic approach chosen by THGA will contribute to achieving the mission statement of the University "Engineering for a better world".[16]

## 5 General Conclusion

The participation of universities in dialogue processes for mining projects is complex and often a tightrope walks between the interests of the parties involved. In general, universities with a mining focus are industry-friendly since their graduates usually start their careers there and industry certainly awards research and service contracts to universities. On the other hand, public universities are committed to the society as a whole and must place the interests of the public above individual interests, such as those of companies or academic institutions. How can universities contribute to improve the social acceptance of mining projects? The suggestions outlined below are not fundamentally new; they essentially relate to the core task of universities, i.e., teaching (first mission), research (second mission) and public outreach (third mission).

This is supplemented by the topic of public relations, which should address not only university-specific but also industry-specific issues. In the past, research at

[15] Technische Hochschule Georg Agricola (2022), pp. 63–68.

[16] Technische Hochschule Georg Agricola (2022), pp. 63–68.

mining universities mostly focused on deposit and production related topics—e.g., drilling and blasting, haulage, ventilation, automation, safety, process control etc. These focal points still have their justification but are usually focused on tasks within the industry and are of less importance to the public. In view of the legitimate interest of the public, the research spectrum at mining universities must be expanded to include topics that are relevant to the concerns of all citizens. This has already been implemented at many universities; three main research areas relevant to this causality are mentioned below as examples:

4. Monitoring of mining operations and mining regions
5. Decarbonization of extraction and processing operations
6. Transformation of society and infrastructure in mining regions.

The monitoring of environmentally relevant parameters—e.g., water levels and water quality, emissions, vegetation, ground movements etc.—is a complex research area that has received significant impetus from new technologies. For instance, the use of geo-data, which are constantly and often cheaply obtained by satellites, or aerial surveys with well-equipped drones. However, it is important that monitoring activities not only focus on the active operating phase, but also continue for many years after the extraction of raw materials has been completed. University should also ensure that monitoring data and results are made accessible to the public, in joint agreement with industry, the supervisory authority and those affected. It will only be possible to gain the trust of the affected citizens and create the conditions for an objective dialogue by transparency.

Today decarbonizing is a high-priority topic due to current efforts to combat climate change. With their expertise, mining universities can contribute to reduce $CO_2$ emissions sustainably while extraction and processing mineral resources (Scope 1 and 2). In the meantime, numerous mining companies have organized specific actions for $CO_2$ reduction and are offering to partner with universities for climate projects—a topic very much in line with the traditional, technical related focus of mining universities.

The social and infrastructural changes that mining causes in the affected regions have not been the focus of mining universities so far. The topic of structural transformation is less technology-driven, but encompasses sociological, industrial policy and regional planning aspects. When mining withdraws from regions on a large scale, this has sometimes led to massive social upheaval. The mining industry, but also the political decision makers concerned, are responsible for successfully shaping the transformation process following the closure of mining operations. If this is unsuccessful, does not succeed, the public associates the negative effects of structural change with the extraction of mineral resources.

Another area requiring more attention from universities is public advocacy for mineral resources issues. Public relations work is often limited to issues of the university and research topics which are investigated at the mining institutes. Contributions describing mineral resources as necessary for the further development of society are rare. Equally rare is the information on current mining projects, in a holistic view with both positive aspects and possible impacts. Digital media such as the internet, Facebook, Twitter, Instagram, etc. offer the opportunity to reach out to a large group of interested parties promptly, with manageable effort and in an understandable way (see examples 1 and 2). For example, it is easy enough to set up an info blog that regularly reports on mining activities in the region. In addition, universities should organize public lectures, discussions, and statements about mineral resources extraction on a regular basis. Disputes with mining opponents is seldom easy, often emotional and frustrating. Nevertheless, for their own self-image and for their public reputation, it is important that universities do not shy away from confronting critics. At this point, one can gain the trust of many stakeholders if you take a stand authentically and with understandable information.

Teaching is another area that should be mentioned. Mining universities focus on training young engineers to optimize relevant production procedures in mining operations, according to scientific standards. The fact that these actions are often associated with social or ecological changes has not been a focus of the curriculum so far. There is no doubt that universities provide knowledge about the effects on nature, water bodies and the topography that are associated with the extraction of mineral resources. However, the evaluation of operational procedures is mostly based on technical or economic criteria, rather than on sustainability factors. A change of perspective, the struggle for a compromise that also accommodates the interests of the public, or the search for less environmentally affecting solutions are not a regular part of the curriculum. It is certainly challenging to convey this message to students in the context of a lecture, e.g., Sustainable Development of Mining Projects. Here, seminars or exercises with case studies and role reversal are certainly better suited to impart technical, but also social competence for mining-related conflicts. Universities should also illustrate clearly how important transparency and timely information are for public acceptance of mining projects. Universities should aim not only to prepare students technically for their professional life, but also enable them to include the interests of the public in planning and operational procedures. Above all, it should convey to the students the sensitivity for such conflicting goals and the search for compromises.

In addition to the above-mentioned core tasks of teaching ("First Mission") and research ("Second Mission"), the THGA sees its mission within the framework of the so-called "Third Mission" as being socially engaged and in a lively exchange with civil society, companies, and politics. Beyond its regular educational mission, the university is committed, for example, to special target groups such as those who have moved up the educational ladder or people with refugee experience and promotes STEM education for children and young people.

The focus is also on increasingly transferring research results to society. Current findings and scientific expertise are made available for solving societal issues. Finally, the THGA also wants to be a platform for the general public that is open to all those interested in dialogue. To this end, THGA and the worldwide network of mining universities must create spaces and suitable formats to discuss mining operations.

## 6 Summary

Using three examples associated with major extractive industries and universities, the paper describes opportunities for mining universities to improve the sustainability and the public acceptance of mining and mining related projects. In this context, research priorities should focus on the monitoring of mining activities, decarbonization and structural transformation. Independent of all university activities, however, corporate responsibility is applicable at first, to ensure a sustainable consensus between privately driven mining projects and the public interest.

The public relations work of universities should not shy away from reporting on mining-related topics and documenting the value of mineral resources extraction for society. And, of course, the university's primary goal is to educate responsible graduates who have internalized the principle of sustainability in their thinking and actions. Special bachelor and master course can offer new insights on how to manage mining in a sustainably manner.

## References

BBC News (2021) Vale dam disaster: $7bn compensation for disaster victims. https://www.bbc.com/news/business-55924743. Accessed 22 May 2022

Forschungszentrum Nachbergbau. https://fzn.thga.de/. Accessed 22 May 2022

Forum Bergbau und Wasser (2022). https://forum-bergbau-wasser.de. Accessed 22 May 2022

Melchers C, Westermann S, Reker B (2019) Evaluierung von Grubenwasseranstiegsprozessen – im Ruhrgebiet, Saarland, in Ibbenbüren sowie weiteren deutschen Steinkohlenrevieren und dem angrenzenden europäischen Ausland. Projektbericht. (Berichte zum Nachbergbau 1). Selbstverlag des Deutschen Bergbau-Museums Bochum, Bochum. https://doi.org/10.48771/43qw-vp33. Accessed 22 May 2022

Ministerium für Wirtschaft, Innovation, Digitalisierung und Energie des Landes Nordrhein-Westfalen (2019) Jahresbericht 2019 der Bergbehörden des Landes Nordrhein-Westfalen, pp 66–69. https://www.bra.nrw.de/system/files/media/document/file/jahresbericht_2019_berg.pdf. Accessed 22 May 2022

Monitoring Epe (2022) Forschung und Transfer für Epe 2022. https://monitoring-epe.de/. Accessed 22 May 2022

National Commission on the BP Deepwater Horizon Oil Spill and Offshore Drilling (2011) Report to the President. https://www.govinfo.gov/content/pkg/GPO-OILCOMMISSION/pdf/GPO-OILCOMMISSION.pdf. Accessed 22 May 2022

RAG (2016) Deutsche Steinkohle 2016: Grubenwasserhaltungskonzept NRW [unpublished]

Rudolph T, Goerke-Mallet P, Poplawski M, Müterthies A, Teuwsen S, Yang C (2022) Monitoring Ground Movement – the big jigsaw of the cavern gas-storage Epe, Germany. EAGE GET2022 3rd EAGE Global Energy Transition. Conference & Exhibition

Salzgewinnungsgesellschaft Westfalen (SGW) (2022). https://www.solvay.de/standorte/ahaus-epe. Accessed 22 May 2022

Technische Hochschule Georg Agricola (2021) THGA mach Sinn. Studieren. Forschen. Verbessern. Jahresbericht 2021, p 48. https://www.thga.de/fileadmin/website/Formulare/Publikationen_divers/Jahresberichte/THGA_Jahresbericht_2021_mit_Plakat.pdf. Accessed 22 May 2022

Technische Hochschule Georg Agricola (2022) Hochschulentwicklungsplan (HEP) 2022-2028. Zukunft seit 1816, pp 63–68. https://www.thga.de/fileadmin/website/Formulare/Publikationen_divers/THGA_Hochschulentwicklungsplan.pdf. Accessed 22 May 2022

Umwelt Bundesamt (2022) Rohstoffkonsum steigt wieder an. https://www.umweltbundesamt.de/presse/pressemitteilungen/rohstoffkonsum-steigt-wieder-an-auf-161-tonnen-pro. Accessed 22 May 2022

United Nations (2022) Department of Ecological and Social Affairs. https://www.un.org/en/desa/world-population-projected-reach-98-billion-2050-and-112-billion2100#:~:text=COVID%2D19,World%20population%20projected%20to%20reach%209.8%20billion%20in%202050%2C%20and,Nations%20report%20being%20launched%20today. Accessed 22 May 2022

United States Geological Survey (2022) How many pounds of minerals are required by the average person in a year. https://www.usgs.gov/faqs/how-many-pounds-minerals-are-required-average-person-year. Accessed 22 May 2022

Printed in the USA
CPSIA information can be obtained
at www.ICGtesting.com
LVHW020051091223
765924LV00005B/194

9 783031 418723